Generation and Application of Ultrahigh Laser Fields

Generation and Application of Ultrahigh Laser Fields

Alexander A. Andreev

Nova Science Publishers, Inc.
New York

Senior Editors: Susan Boriotti and Donna Dennis
Coordinating Editor: Tatiana Shohov
Office Manager: Annette Hellinger
Graphics: Wanda Serrano
Book Production: Matthew Kozlowski, Jonathan Rose and Jennifer Vogt
Circulation: Cathy DeGregory, Ave Maria Gonzalez and Raheem Miller
Communications and Acquisitions: Serge P. Shohov

Library of Congress Cataloging-in-Publication Data

Andreev, Alexander A.
Generation and application of ultrahigh laser fields / Alexander Andreev.
p. cm.
Includes index.
ISBN 1-59033-142-7
1. Laser beams. 2. Laser beams – Industrial applications. I. Title.
QC689.5.L37 A54 2002
621.36'6—dc21 2001058726

227 Main Street, Suite 100
Huntington, New York 11743
Tele. 631-424-6682 Fax 631-425-5933
e-mail: Novascience@earthlink.net
Web Site: http://www.nexusworld.com/nova

Printed in the United States of America

CONTENTS

ACKNOWLEDGMENTS

The author acknowledges the help of many colleagues from the Institute for Laser Physics over the many years during which most of these notions were formulated. These include Drs. K.Yu Platonov, Yu.V. Rozhdestvenskii, V.E. Yashin and A.N.Sutyagin. I want to thank Profs. A.A.Mak and N.N.Rozanov for their support and assistance.

The author is greatly indebted to many people. I want to express my appreciation to my coauthors Drs. A.R Bell, G.Bonnaude, J.C.Gauthier, P.Gibbon, E.Forster, J.Limpouch, E.Levfebre, A.A.Levkovskii, I.A.Litvinenko, H.Nakano, V.N.Novikov, Yu.M.Pis'mak, H.Ruhl, R.Salomaa, K.Tanaka, V.T.Tikhonchuk, U.Teubner, V.E.Sherman and A.G.Zhidkov. I want to acknowledge my indebtedness to my colleagues at Vavilov State Optical Institute whose continual assistance over the years has been invaluable. Particularly, I'd like to express my thanks to Drs. A.V.Charukchev, V.M.Komarov, N.A.Solov'ev, V.E.Semenov, A.N.Shatcev, V.D.Vinokurova and I.V.Kurnin, A.G.Samsonov, A.N.Semakhin.

Finally, my thanks go to my wife who typed some parts of the manuscript and its revisions with great skill and accuracy and also K.Wilkinson edited some parts of the manuscript.

PREFACE

This book is devoted to my wife and daughter

The world is in a constant state of flux. Nothing remains static or stable for very long. One of the key examples of recent fast changes in the world is a laser. This book attempts to put into perspective some fundamental changes that have taken place in laser physics during the past ten years. The main laser parameter - laser intensity – dramatically increased by five orders of magnitude over the past decade. It is difficult to find another area in modern science where progress is so evident. Current laser intensities still far from the theoretical limit, and this limit will be discussed in this book. It is very significant that for providing such high laser intensities new technologies such pulse compression, are not so expensive - and permit to be used in many laboratories around the world. The spreading of such technique in science and industry is continuing now because of the wide range of potential applications.

This book is concerned with this rapidly developing area of physics: the generation of super-strong laser fields and their applications. The book is intended for graduate university students, as well as for those professionally working in this research area. The aim was to describe the new lasers and the phenomena involved in the interaction between ultra-high-power laser radiation and over-dense plasma. Most of this book is based on recent papers by the author describing the physics of high-temperature, over-dense laser plasma and some of its applications, in particular its use for laser-induced nuclear reactions. Work in this area has lately been very active, so it is premature to present this material as a well-established knowledge. I have just tried to introduce the reader to the state of the art and essential aspects of the problem, thus facilitating a more detailed study of particular issues concerning the generation of super-strong laser fields and their applications.

The topics contained in this book are meant to stimulate your thinking. I hope the different issues raised here serve to pique your interest, expand your horizons, and in general make your think about issues, which normally you would not. And most importantly, I hope this book is useful to you, and meets your needs.

Chapter 1

General Introduction

One of the major problems in laser development is the concentration of radiation energy in both time and space, because this is hampered by some linear and nonlinear effects — aberrations of optical elements, effects of high power pulses, optical breakdown, and stimulated scattering [1-2]. A radical suppression of nonlinear effects, which are particularly hazardous to ultra-short laser pulses, has become possible only with a novel laser architecture based on pulse compression. In this approach, relatively long pulses are enhanced and their nonlinear effects become considerably weaker, with the pulse being compressed in time at the laser output. An ultra-short pulse of high intensity interacts with a limited number of optical elements, such as the focusing lens or mirror, the input window of the target chamber, and some diagnostic units. This preserves the short duration of the pulse and makes its precise focusing possible.

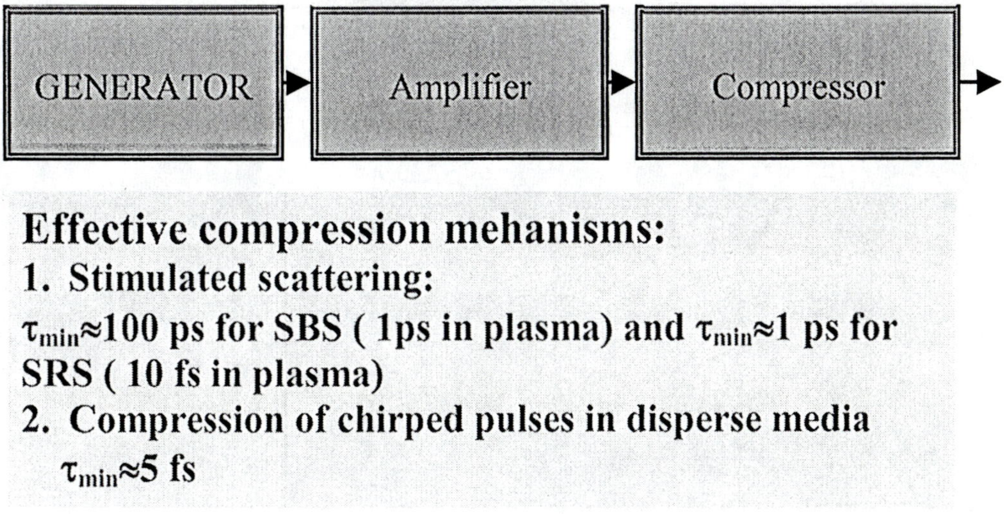
GENERATOR
Amplifier
Compressor
Effective compression mehanisms:
1. Stimulated scattering:
τmin≈100 ps for SBS (1ps in plasma) and τmin≈1 ps for SRS (10 fs in plasma)
2. Compression of chirped pulses in disperse media
τmin≈5 fs

Fig. 1.1. Laser scheme with pulse compression.

The obvious advantages of this laser architecture were recognized long ago [3,4], but its implementation came only with the advent of effective ways of pulse compression. At present, there are two elaborated compression techniques (**Figure 1.1**). One is to use stimulated light scattering by sound waves (Mandelshtam-Brillouien stimulated scattering) or molecular vibrations (stimulated Raman scattering) [1]. The minimum duration of the pulse compressed in this way is eventually limited by the relaxation time of the vibration medium and is about 1 picosecond for SRS (it is 10 fs for plasma). In the other approach, free from this limitation [4], the compression of a laser pulse with a linear frequency modulation (so called chirped pulse) occurs in a dispersive delay line often consisting of a pair of diffraction gratings (**Figure 1.2**) [5,6]. This method has provided record high radiation powers during the last ten years.

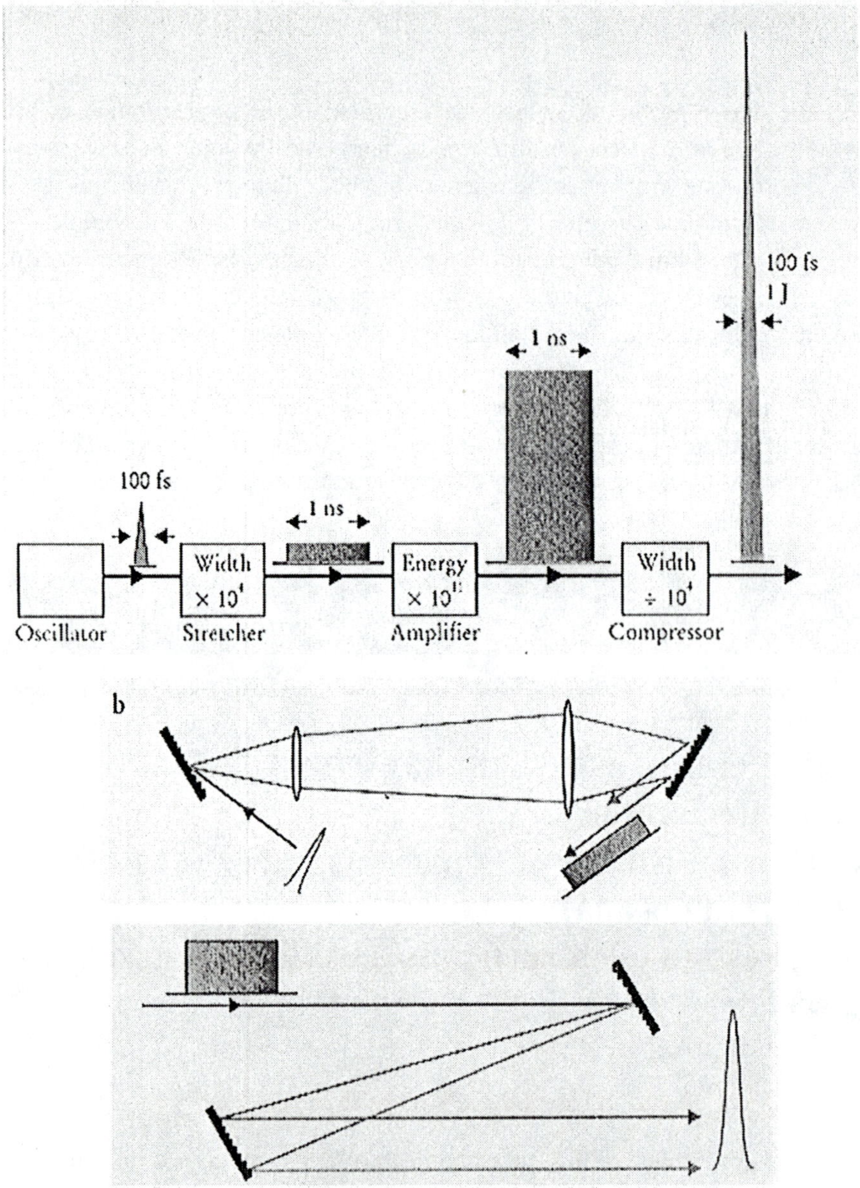

Fig. 1.2. CPA concept.

Figure 1.3 illustrates the progress made in the generation of maximum pulse intensities [5]. It is seen that the application of enhancement and compression has raised the peak power and intensity of chirped pulses by 6 orders of magnitude over the past ten years.

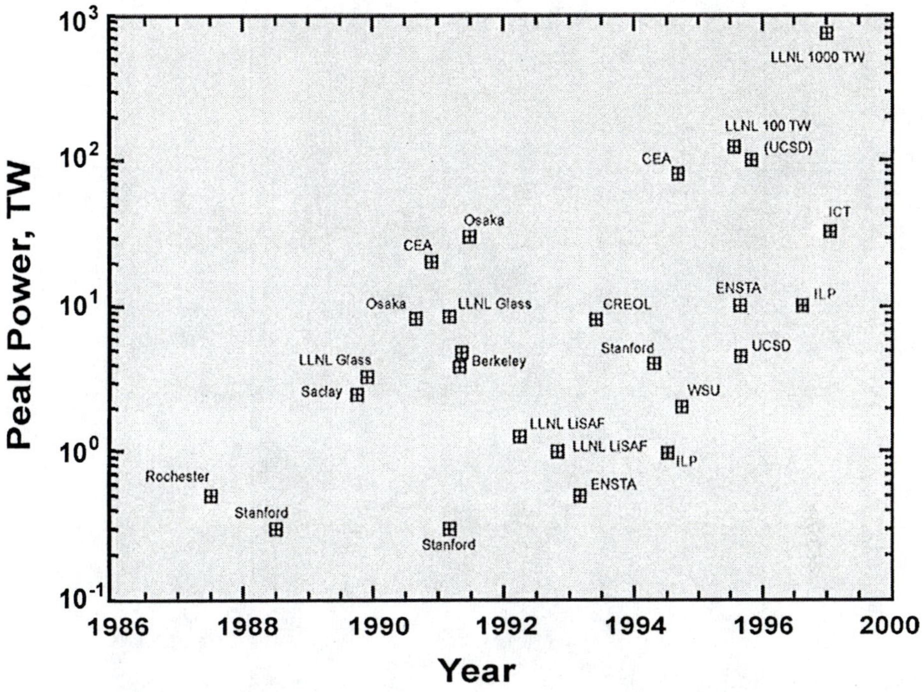

Fig. 1.3. Power lasers with compression of pulse.

The application of super-high laser fields has allowed observation of some new physical phenomena (**Figure 1.4**) to be discussed in detail in the Part 3. In particular, the electric field created in matter may be as high as that in the atom or even higher. Of special interest here are such effects as the ionization of matter by the electric field, the production of ions of high charge, and nonlinear un-stationary processes in rarefied and dense gases [1,5].

As the density of the laser energy flux is increased further, the next physical threshold is reached when the oscillation energy of an electron in the electromagnetic field becomes equal to its energy at rest. This trend of research is concerned with nonlinear relativistic optics [7], because the electron mass increases (the critical density rises) and the medium becomes transparent when the electron energy gained in the laser beam field exceeds one mega-electron volt [8].

The next physical threshold of laser intensity (and, hence, a new line of research) arises at future laser fields when the energy gained by an electron along the Compton length reaches its energy at rest. It becomes possible then to study nonlinear quantum electro-dynamic effects, such as the effective production of electron-positron pairs, and some other phenomena [1,6].

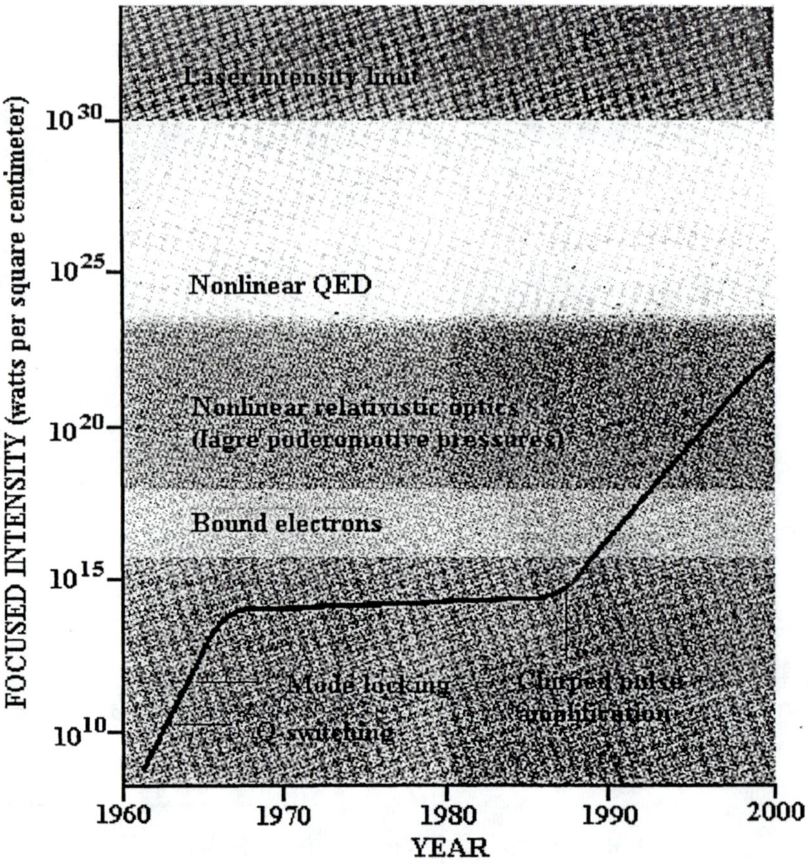

Fig. 1.4. Physics of ultra-high laser fields.

The aim of this book is to describe novel approaches to the generation of super-high laser fields, based on laser pulse compression in high power solid-state laser systems, and to discuss various applications of super-short laser pulses and the interaction between laser radiation and matter.

References for Part 1

1. Andreev A.A., Mak A.A., Yashin V.E. *Kvantowaja elektronika* v.24, p.99, 1997.
2. Mak A.A., Soms L.N., Fromzel' V.A., Yashin V.E. *Lazery na Neodimovom Stekle* (Neodymium Glass Lasers) Moscow: Nauka, 1990, p. 288.
3. Mak A.A., Ljubimov V.V., Serebrjakov V.A. et al. *Izw. AN SSSR, ser. Phizicheskaja* v.6, p.1858, 1982.

4. Maine P., Strickland D., Bado P., Pessot M., Mourou G. *IEEE J.Quant.Electron.* v. 24, p. 398, 1988.
5. Mourou G.A., Barty C.P.J., Perry M.D. *Physics Today*, v. 51, p. 22, 1998.
6. Perry M.D., Mourou G. *Science*, v. 264, p. 917, 1994.
7. Umstadter D et al. *Science* v. 273, p. 472, 1996.
8. Andreev A.A., Charuhchev A.V., Yashin V.E. *Usp. Fiz. Nauk*, v. 169, p. 72, 1999.

Chapter 2

SOLID-STATE LASERS FOR GENERATION OF ULTRA-SHORT HIGH-INTENSITY PULSES

INTRODUCTION

There are a variety of methods for generating high-energy ultra-short pulses. They can be divided arbitrarily into two types: direct amplification of short laser pulses and amplification of relatively long pulses followed by their compression. Generation of pulses with a high peak intensity in lasers with direct amplification is hindered by such nonlinear effects as optical damage threshold and self-interactions [1]. The optical damage threshold W of materials is approximately proportional to the square root of the pulse duration, right down to ~ 10 ps [2]. For this reason, generation of pulses with high output energies requires the use of amplifiers and other large-aperture components, which greatly increases the difficulties encountered in construction of suitable systems and, in the final analysis, their cost. Small-scale self-focusing of radiation in optical components of a laser system also has a strong influence on the characteristics of the system and limits the output energy, which is approximately proportional to the pulse duration.

2.1 LASER SYSTEMS WITH PULSE COMPRESSION

One of the most promising methods for the generation of short laser pulses free of many of the above-mentioned shortcomings is that based on the amplification of phase-modulated (chirped) relatively long optical pulses, followed by their compression in a dispersive delay line consisting of a pair of diffraction gratings [3,4]. Since the spectral width of the amplified (initial) pulse should be fairly large, it is necessary to use lasers with a wide spectral gain band, for example, neodymium glass or titanium-doped sapphire lasers. Comparative characteristics of various active media (stimulated emission cross section σ gain bandwidth $\Delta\nu$, wavelength λ_{max} corresponding to the maximum of the gain band, lifetime of the upper active level τ, and relative efficiency $\sigma\tau\Delta\nu$) suitable for the generation and amplification of subpicosecond laser pulses are listed in **Table 2.1**.

We can see that dye and crystal lasers have very wide gain bands and they are therefore capable of generating shorter (down to a few femtoseconds) pulses than neodymium glass lasers. On the other hand, neodymium glass lasers can generate fairly short (200-300 fs) pulses with a much higher energy because large-aperture glass active elements can be made. For these reasons, neodymium glass lasers are ahead of competition in tasks requiring high energies.

Table 2.1

Active media	10^{20} σ/cm^2	$\Delta\nu/\text{cm}^{-1}$	λ_{max} $/\mu m$	$\tau/\mu s$	$\sigma\tau\Delta\nu/cm\ s$
Dyes	2 x 10^4	1500	0.6	0.005	1.5
Nd : YLF	18	12	1.047-1.053	480	1
Neodymium glass	1.0-4.5	200-400	1.053-1.062	300-500	3
Ti:Al_20_3	30	3200	0.78	3.2	3.1
Cr : LiSAF	4.8	1900	0.83	70	6.1

2.1.1 Chirped Pulse Generation

Two approaches are used in the generation of phase- modulated (chirped) pulses. In the first approach a relatively long (20 - 50 ps) pulse is first generated in a laser with active or passive mode locking. This pulse is then directed to a single-mode fiber[5] where, because of the non-linearity of the refractive index of the fiber, the pulse becomes phase- modulated:

$$\Delta\varphi(t) = k\Delta nl, \tag{2.1}$$

where $k = 2\pi/\lambda$ is the wave number; l is the fiber length; $\Delta n = n_2|E|^2/2$ is the nonlinear phase shift per unit length of the fiber. A change in the phase with time alters the frequency in accordance with the law

$$\Delta\omega(t) = 0.5kln_2\frac{d|E|^2}{dt} \tag{2.2}$$

If the time dependence of the intensity is nearly quadratic, as is true at the peak of a Gaussian pulse, the frequency varies linearly with time and a pulse can be compressed temporarily in a dispersive delay line, which may consist, for example, of a pair of diffraction gratings [3].

Since the fiber diameter is small (approximately 10 μm), the pulse energy at its output end is also very small (~ $I0^{-9}$ J). This makes it necessary to use amplifiers with a very high gain. The pulse contrast in such systems is low because of super-luminescence and also because of the chirp non-linearity, as a result of a deviation of the temporal pulse profile from the ideal square form. Self-phase modulation may be in competition with other nonlinear effects and among them the greatest hazard is presented by stimulated Raman scattering (SRS) (which, however, can sometimes have a stabilizing influence on the chirp [6]) and by the fiber dispersion [7]. A rigorous control of the intensity of the radiation entering a fiber is essential to ensure that the chirp is linear, because a change in the frequency is proportional to a change in the intensity.

The second approach is largely free from these shortcomings: in this case, a mode-locked laser generates first a short pulse of the required duration and then this pulse is stretched by a pair of diffraction gratings separated by a telescope [8] (**Fig. 2.1**). In view of the linearity of the system, the chirp is purely linear and the pulse energy reaches $\sim 10^{-6}$ J, which is several orders of magnitude higher than the energy attainable by the first approach. Therefore, we shall concentrate on practical implementation of the second approach.

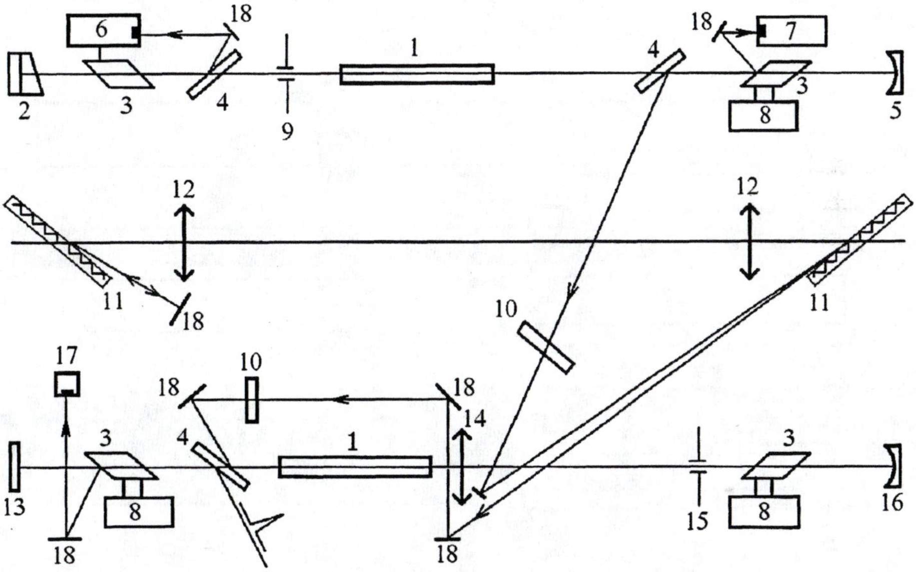

Fig. 2.1. Optical part of a picosecond start-up neodymium glass system: (1) laser heads; (2) cell with a bleachable dye placed against a non-transmitting mirror; (3) Pockels switches; (4) polarizers; (5, 16) non-transmitting mirrors; (6) negative feedback unit; (9, 15) apertures; (10) half-wave plates; (11) diffractions gratings; (12) telescope lenses; (13) non-transmitting mirror; (14) lens; (17) photo-detector; (18) aluminum mirrors.

A typical optical laser system intended for the generation of picosecond pulses, stretching these pulses (with the help of the chirp), and amplifying them in a regenerative amplifier is shown in Fig.2.1 [9] A train of pulses with a total duration 8 μs and with 13 ns separation between the pulses is generated in a laser in which the active element is made of GLS22.

Neodymum-doped phosphate glass, passive mode locking and a negative feedback loop are used in this laser. The loop enhances considerably the stability of the energy and duration of the pulses, compared with simple passive mode locking [10]. A single pulse of 1 ps duration is selected by an intra-cavity Pockels cell controlled by an electronic switch. The selected pulse is injected into a stretcher formed by a pair of holographic diffraction gratings (N = 1800 lines mm^{-1}) with a telescope between them. In this geometry the pulse is stretched to 450 ps, but the spectral width (about 1.8 nm at half amplitude) remains the same as for the short pulse and the pulse acquires linear phase modulation (chirp). Since the energy of this pulse (~ 10^{-6} J) is not high enough for efficient amplification in a high-power amplifier system, the pulse is injected into a regenerative amplifier with a phosphate glass rod (diameter 5 mm, length 100 mm). When a sufficiently high energy (2-5 mJ) is reached, the amplified pulse is coupled out of the amplifier by a Pockels cell.

Fig. 2.2 shows the optical system of a Ti: Al_2O_3 laser used to generate shorter pulses at the wavelength ~ 0.8 μm [11]. The main components are the same as in the laser described above, but there are several special features, which are discussed below. We shall now consider in greater detail the various components of the laser systems used to generate picosecond and femtosecond pulses, and we shall discuss their ultimate capabilities.

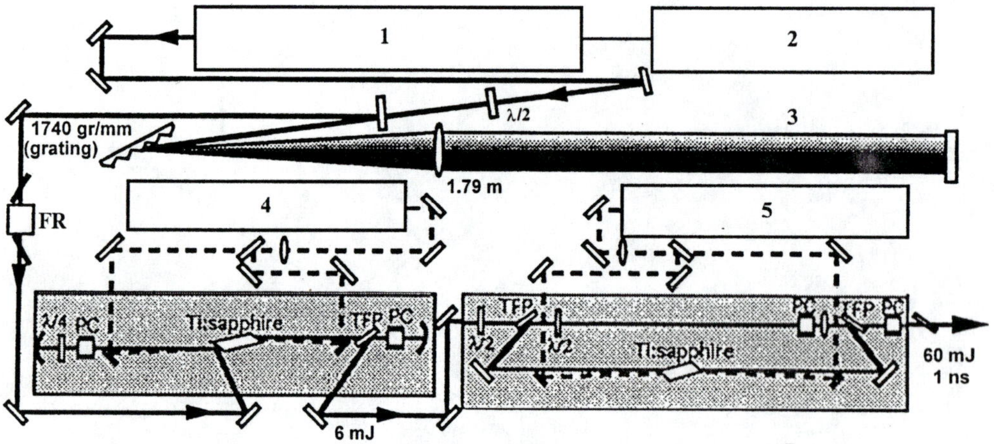

Fig. 2.2. Optical part of a picosecond start-up system on Ti: Al_2O_3 emitting output radiation of energy 60 mJ at the wavelength 1053 nm: (1) cw Ar laser providing pump power of 11 W; (1) Ti:Al_2O_3 mode-locked master oscillator, emitting single pulses of 100 fs duration with an average power of 400 mW; (3) pulse stretcher; (4, 5) Nd:YAG pump laser for regenerative amplifiers with the second-harmonic pulse energies of 340 and 850 mJ, respectively, and a pulse repetition frequency of 10Hz; (6, 7) linear and ring regenerative amplifiers.

2.1.2 Master Oscillator

The duration of the pulses generated by a neodymium phosphate glass laser with passive mode locking (performed by an organic dye) has a lower limit of ~ 1 ps because of the insufficiently wide gain band of the glass and because of the long relaxation time (several picoseconds) of the bleach-able dye. The use of broader-band neodymium-doped silicate glasses, or dyes with a faster response, or of self-phase modulation in the active element [12] makes it possible to reduce the pulse duration in such an oscillator to about 0.5 ps.

Shorter (~100 fs or less) pulses can be generated in a $Ti:Al_2O_3$ laser with Kerr-lens mode locking and pumped by a cw argon laser [13, 14]. A typical optical system of such a laser is shown in **Fig. 2.3**. Mode locking is then performed by the fast Kerr non-linearity and the dispersion of the refractive index of the optical components is compensated by a dispersive prism delay line. The gain band of this titanium laser partly overlaps the gain band of the neodymium glass lasers, so that the master oscillator considered here can be used in a neodymium glass laser amplifying system operating at the wavelength 1060 nm [11, 15, 16]. However, at this wavelength the gain of Ti: Al_2O_3 falls strongly compared with the gain at the maximum of the luminescence line, which means that the quality of $Ti:Al_2O_3$ crystals and the pump power have to be higher.

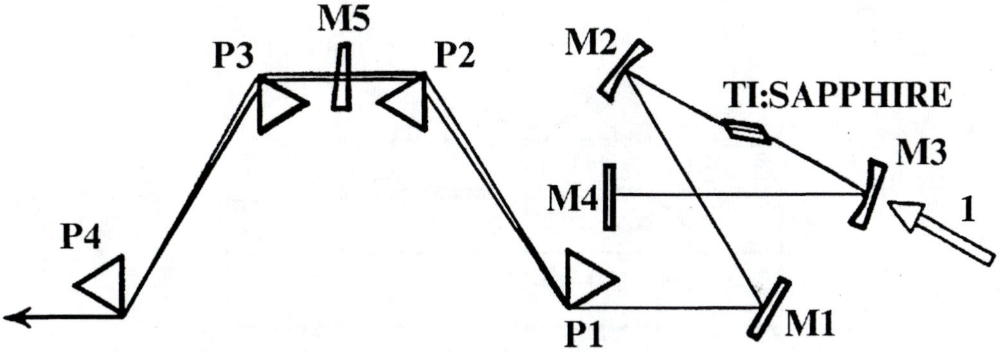

Fig. 2.3. Optical part of a femtosecond $Ti:Al_2O_3$ master oscillator: non-transmitting mirrors; intra-cavity prisms; output mirror; extra-cavity prisms; pump beam; output radiation.

In view of this, it is worth considering the feasibility of generation of 100 fs pulses directly in a neodymium glass laser with mode locking by a modulator based on quantum-well semiconductor structures [17, 18]. The optical system of this laser is shown in **Fig. 2.4**. A Fabry-Perot cavity is used to amplify the radiation field in the modulator and this ensures that the depth of modulation is of the order of a few percent. Stable operation of the laser under these conditions requires stable pumping of the active medium, which is performed by cw laser diodes. It is then possible to generate pulses of about 100 fs duration in neodymium phosphate and silicate glass lasers of this type [18], which makes these lasers fully competitive (compared with the $Ti:Al_2O_3$ laser) as master oscillators for neodymium glass laser systems.

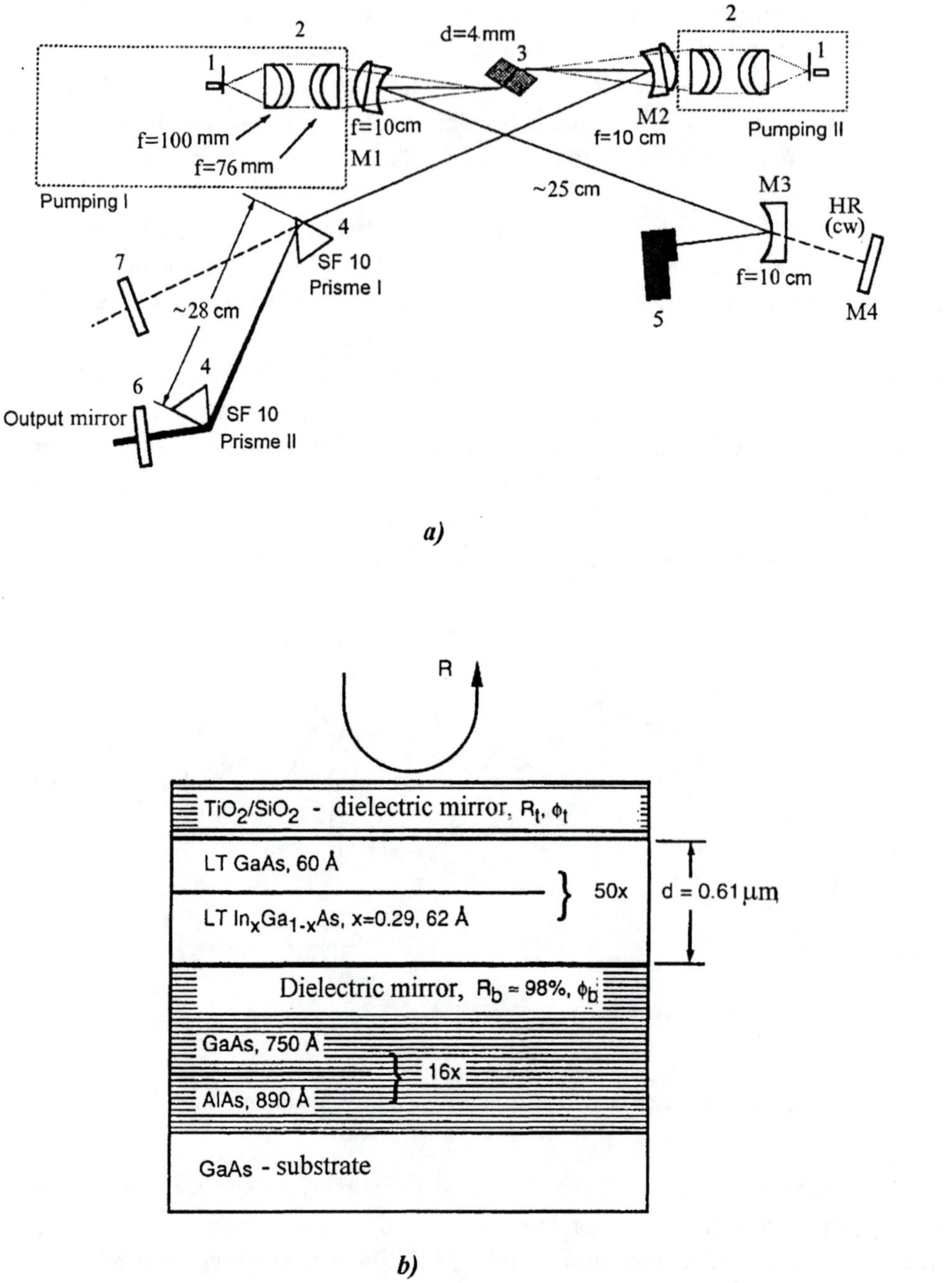

Fig. 2.4. Optical part of a femtosecond neodymium-glass master oscillator pumped by laser diodes: (1) laser diodes delivering pump radiation of 1.2 W power at $\lambda = 804$ nm; (2) optical focusing system; (3) neodymium-glass active element; (4) glass prisms;(5) anti-resonant saturable absorber; (6) output mirror; (7) mirror for a pump laser diode; spherical dichroic mirrors (f =10 cm).

Mode-locked lasers acting as sources of single short pulses suffer from a significant shortcoming, which is the need to select a single pulse from a train of pulses generated at a high repetition frequency $f = 2L/c$ (L is the cavity length). Such selection is performed by fast-response devices such as Pockels cells with a very high contrast. Two methods for generating high-contrast single pulses are worth noting. In the first method (**Fig. 2.5**) a single sub-picosecond pulse is generated by two-stage SRS pulse shortening in compressed SF_6 and H_2 gases [19]. These gases make it possible to convert a pulse of 20 ps duration, generated in a ruby master oscillator with mode locking, into a pulse of 0.5 ps duration at the wavelength 1056 nm, which lies within the gain band of neodymium glass. The SRS non-linearity makes it possible to select efficiently pre-pulses and to generate high-contrast short pulses. However, this non-linearity leads also to undesirable fluctuations of the energy of the scattered radiation and it makes it necessary to stabilize highly the energy and duration of the initial laser pulse.

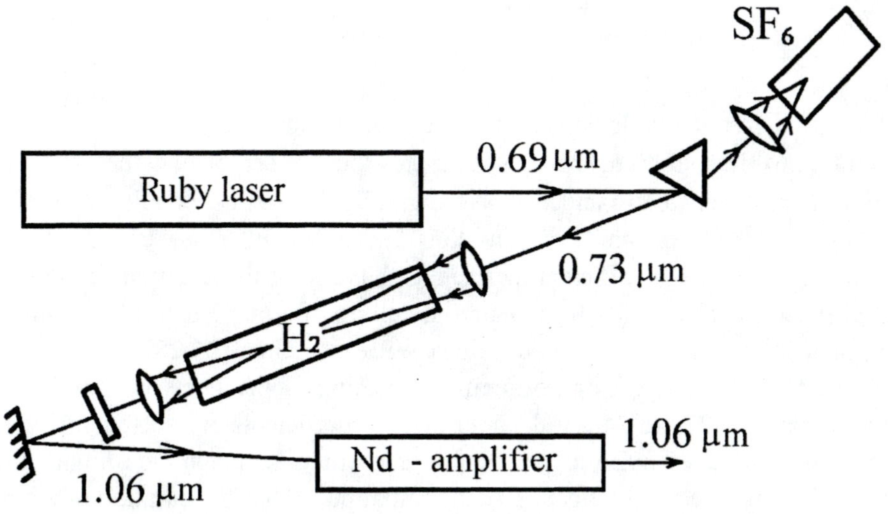

Fig. 2.5. Utilization of stimulated Raman scattering in a picosecond master oscillator.

The second method is based on a combination of a non-linear compression method with stimulated Brillouin scattering (SBS) and subsequent selection of a short pulse. A short pulse of 1 ps duration is generated in two stages. In the first stage, a master-oscillator pulse of 1-3 ns duration is compressed by SBS to about 100 ps [20]. Next, a short pulse of about 1 ps is selected from the longer pulse by a controlled silicon switch [21] of a high-resolution deflector system, which is capable of resolving about 100 elements. The feasibility of achieving such characteristics in the case of $LiNbO_3$ or $LiTaO_3$ deflectors was investigated quite a long time ago [22], although for longer laser pulses. A system of this kind can also be used to generate high-contrast single pulses without facing more stringent requirements in respect of the stability of the components of the system, because only one compression stage is employed. A fast travelling-wave electro-optical deflector, controlled by a silicon switch, can in principle generate also shorter pulses by selecting them from pulses of 1 ps duration [23].

2.1.3 Pulse Stretching System

The increase in the pulse duration in a diffraction stretcher should be as large as possible in order to suppress optical breakdown, and also to avoid self-focusing and self-modulation during amplification. The maximum pulse duration is governed by the dispersive delay time between the end points of a spectrum of width $\Delta\omega$:

$$t_L = \frac{\lambda \cdot L_d \cdot (\lambda n)^2 \cdot \Delta\omega}{2 \cdot \pi \cdot c^2 \cdot \left[1 - (\lambda \cdot n - \sin\Theta)^2\right]}, \tag{2.3}$$

where L_d is the effective length of the diffraction stretcher; λ is the wavelength corresponding to the middle of the spectrum; n is the number of lines per millimeter in the grating; Θ is the angle of incidence on the grating. Under real conditions, we have L_d = 2-3 m, n = 1800 lines mm^{-1}, the duration of a stretched pulse is 1-3 ns, and the maximum decompression coefficient, defined as the ratio of the duration of the stretched and initial pulses, is about 10^3-10^4. Greater stretching of a pulse requires elimination of the chromatic and spherical aberrations in a pulse stretcher [24-26] or new stretcher designs based on gratings with a non-uniform spacing of the lines or combinations of prisms and gratings [27, 28]. Such decompression coefficients are sufficiently large for the energy of a stretched pulse to be two or three orders of magnitude higher than the energy obtained by direct amplification of the initial pulse, but the duration of a stretched pulse is still insufficient for complete suppression of the self-interaction and optical breakdown effects. A further increase in the pulse duration even to 10 ns, would require either a stretcher and a compressor of excessive lengths or the use of multi-pass systems.

Since the energy of a pulse in a stretcher is low (from 10^{-9} to 10^{-6} J), the requirements in respect of the optical damage threshold and the reflection coefficient of diffraction gratings do not become more stringent. However, these gratings and other stretcher components should have a sufficiently large aperture to avoid limiting the spectrum by the aperture. When the width of the spectrum is 50 nm and the duration of a stretched pulse is about 1 ns, the aperture diameter should be about 10 cm. Spatial separation of this spectrum in a stretcher provides an opportunity for correction of the spectrum in order to improve the chirp linearity [29] or for control of the shape and duration of a compressed pulse by the use of masks or limiting apertures. The low energy makes it possible to employ different stretcher variants based on the same diffraction grating (see Fig. 2.2).

2.1.4 Amplifying System

The energy of a pulse at the stretcher output ranges, as already mentioned, from nanojoules to microjoules, depending on the nature of the master oscillator. Consequently, an output energy of just ~ 1 J requires an overall gain of 10^6-10^9. This is usually achieved by separating the amplifying channel into two parts: a preliminary amplification system with an overall gain of 10^3-10^6 and an output energy of ~ 1 - 10 mJ, and a main amplifying system with the desired output energy 1 - 1000 J. It is usual to employ regenerative amplifiers in the preliminary amplification system and the necessary number of passes is ensured by controlled electro-optical switches based on Pockels cells. Different variants of regenerative amplifiers with linear and ring cavities are shown in Figs 2.1 and 2.2 already. It is usual to employ photosensors (for example, those based on pin photodiodes) in a regenerative amplifier: they control a Pockels cell and ensure that the required output energy is reached quite accurately, which is achieved by altering the number of passes of a pulse through the amplifier in accordance with the input signal power.

When a pulse is amplified in a regenerative amplifier and also in the main amplifying system the pulse spectrum may be modified by a variety of effects: the limited width of the gain band of the active medium and of the pass bands of the various components of the amplifier, self-phase modulation resulting from the refractive index non-linearity, gain saturation, and temporal overlap of several master-oscillator pulses in the amplifier. The influence of modulation of the spectrum caused, for example, by interference effects in Pockels cells, lenses, polarizers, and mirrors, is enhanced by the multi-pass nature of the amplification in a regenerative amplifier. This makes it necessary to carefully select and develop the regenerative amplifier components, as well as to provide them with antireflection coatings, to tilt the surfaces of the components relative to the optic axis, and to avoid birefringence effects.

Small-scale modulation of the spectrum of a pulse, responsible for the low contrast after compression, may appear also when the contrast of separation of a single pulse from a master oscillator suffers from insufficient contrast. The main selected pulse and the pre-pulses or after pulses may then temporarily overlap in a regenerative amplifier, leading to modulation of the spectrum [23]. This effect can be avoided by careful selection of a single pulse with the help of optical de-couplers.

Narrowing of the spectrum of a signal in the course of its amplification in an active medium which has a Gaussian gain band of width $\Delta\lambda_{lum}$, is described by [15]

$$\Delta\lambda_{out} = \Delta\lambda_{in} \cdot \left[1 + \beta \cdot \ln(G) \cdot \left(\Delta\lambda_{in} / \Delta\lambda_{lum}\right)^2 \right]^{-1/2}, \tag{2.4}$$

where $\Delta\lambda_{out}$ and $\Delta\lambda_{in}$ are the widths of the spectra of the output and input pulses; G is the total gain experienced in the system; $\beta = 2[(\lambda_{in}-\lambda_{lum})/\Delta\lambda_{lum}]^2-1 \approx 1$. When the total gain in a neodymium phosphate glass laser is 10^9, the spectrum of an input pulse of 100 fs duration is reduced in width by a factor of 4-5 and the duration of a compressed pulse increases by the same factor.

Narrowing of the spectrum can be reduced by the use of active media with broader bands. In a regenerative amplifier with an active component that has a relatively small aperture, one can use successfully Ti: Al_2O_3 pumped by the second harmonic of a YAG laser [11, 15] or Cr:$LiSrAlF_6$ (Cr:LiSAF) pumped by a flash-lamp[16]. The gain band of a neodymium glass amplifier can be widened by the use of a frequency selector [30-32], as demonstrated in **Fig. 2.6**. This selector can be, for example, a Fabry-Perot etalon or an interference - polarization filter based on a crystalline quartz or Iceland spar plate in combination with a polarizer. A device of this type is best used in a regenerative or multi-pass amplifier.

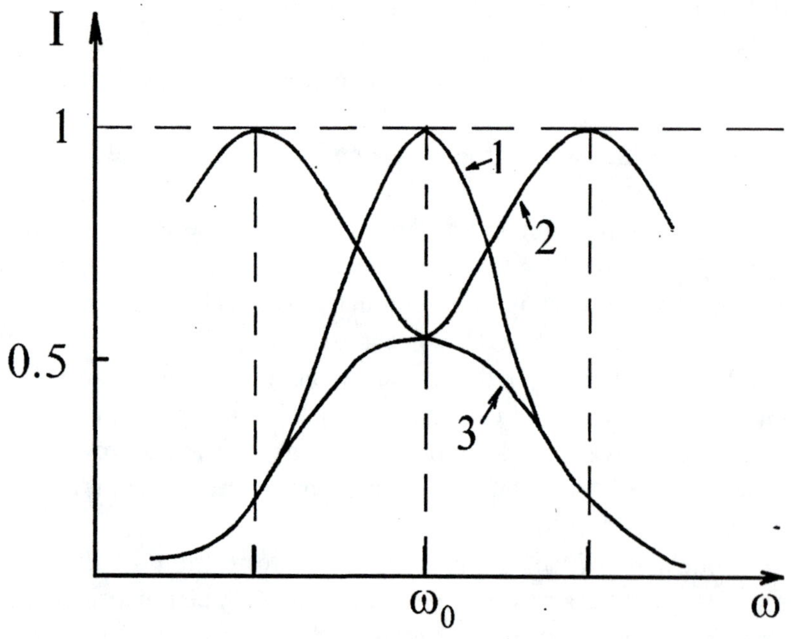

Fig. 2.6. Principle of operation of a frequency selector used to widen the gain band of neodymium glass: (1) luminescence spectrum of neodymium glass; (2) transmission curve of the selector; (3) modified gain profile of neodymium glass.

Neodymium glasses of different compositions, with different widths and positions of the maximum of the luminescence line, can be used in a regenerative amplifier or in a main amplifying system [15, 29]. This is shown in **Fig. 2.7**, which gives the luminescence spectra of neodymium phosphate and silicate glasses, and the spectrum of an amplifier in which a combination of these glasses is used. We can see that the use of such a combination of glasses makes it possible to widen the gain band of an amplifier by a factor of about 3. The gain spectrum can be widened still further by the use of a larger number of different glasses. Calculations show that the application of this method to neodymium glass laser systems should make it possible to generate pulses shorter than 300 fs with an energy in excess of 100 J [32, 33].

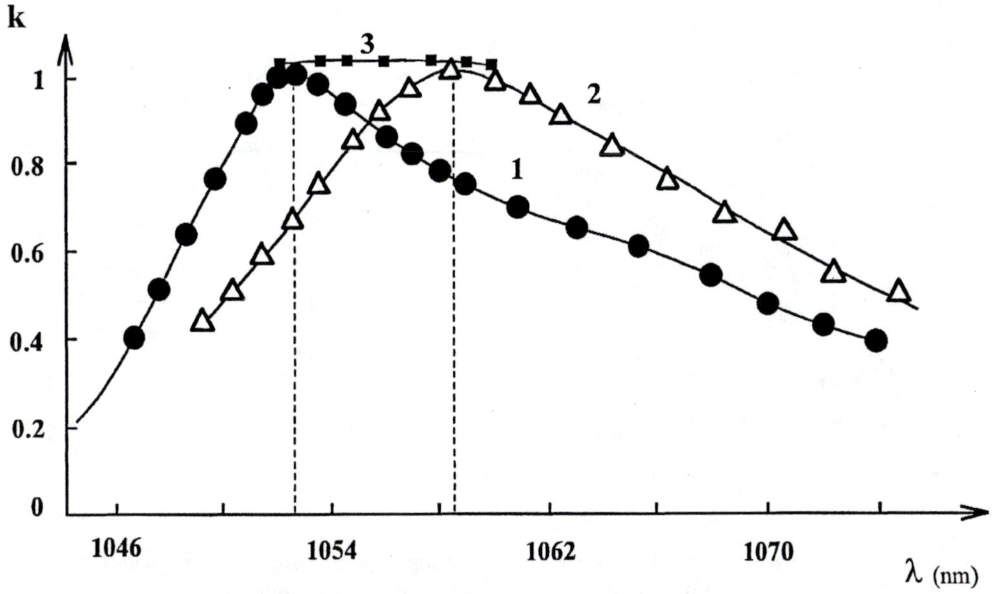

Fig. 2.7. Luminescence spectra of KGSS0180 neodymium phosphate glass (1) GLS6 silicate glass (2), and the combination gain profile of two amplifiers made of these glasses (3), the gains are equal at their maxim.

Gain saturation can also distort the pulse duration and profile [34, 35] since such saturation alters the profile of a stretched pulse in the course of its amplification: the temporal maximum of a pulse shifts to its leading edge and this alters the energy balance in a pulse for different parts of the chirp. The operation of this effect is demonstrated in **Fig. 2.8** which shows the change in the pulse profile for different values of the ratio of the energy density at the amplifier output to the saturation energy density $W_{sat} = h\nu/\sigma$. We can see that the pulse profile and duration change significantly only when the saturation energy density is exceeded significantly. For neodymium glass lasers with W_{sat} = 4-10 J cm^{-2} and pulses of 1-2 ns duration after a stretcher it is not possible to reach such an energy density because of limitations imposed by self-modulation and self-focusing (discussed below). Therefore, the influence of this effect (compared with the others) can be ignored.

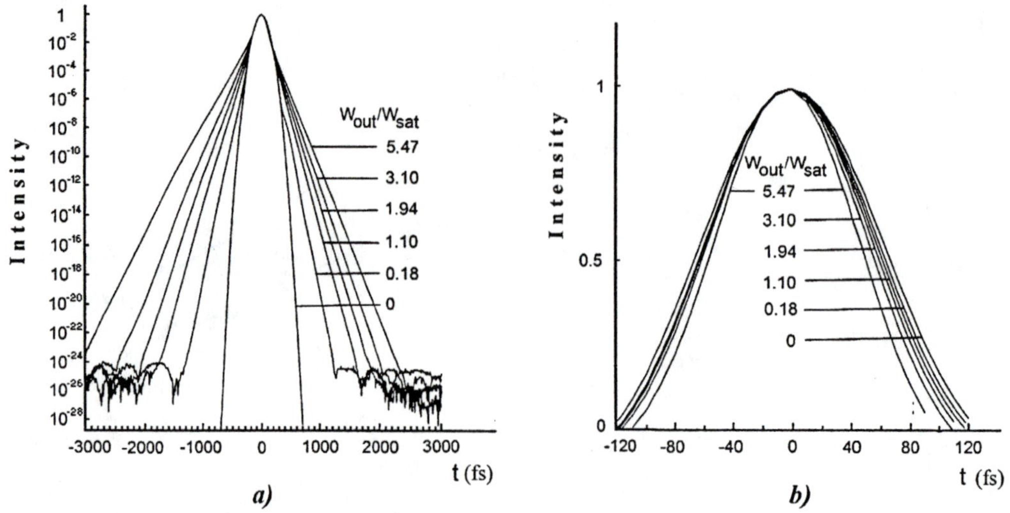

Fig. 2.8. Temporal profile of a laser pulse at an amplifier output plotted for a wide range of intensities (a) and at high intensities (b). The ratio of the output energy density to the saturation energy density is W_{out}/W_{sat} = 0 (1), 0.18 (2), 1.10 (3), 1.94 (4), 3.10 (5) and 5.47 (6).

The relatively short pulse duration after a stretcher (0.5 - 3 ns) means that it is not possible to avoid completely the undesirable influence of self-focusing and self-modulation during amplification. The influence of these effects can be described best by what is known as the breakup integral

$$B = \frac{8\pi^2 n_2}{\lambda c n_0} \int_0^{L_{nl}} I(z)\,dz, \tag{2.5}$$

where λ and c are the wavelength and velocity of light in vacuum; n_0 and n_2 are the linear and nonlinear refractive indices; I is the radiation intensity; L_{nl} is the length of the nonlinear medium. Numerous experiments have established that, as soon as B exceeds 2-3, a laser beam breaks up into filaments because of small-scale self-focusing and it is no longer possible to focus such a beam on a target to form a small spot.

The breakup integral B represents also the nonlinear phase shift, i.e. self-modulation. The effect is undesirable in the case under discussion because it distorts the initially linear chirp of a pulse [34, 36]. By way of example, **Fig. 2.9** demonstrates the dependence of the duration of a compressed pulse (initial duration 100 fs) for two values of the breakup integral of a regenerative amplifier [36]. We can see that even relatively low values, B = 1-2, which are not hazardous from the point of view of self-focusing, can increase the pulse duration and, what is more important, degrade significantly the contrast.

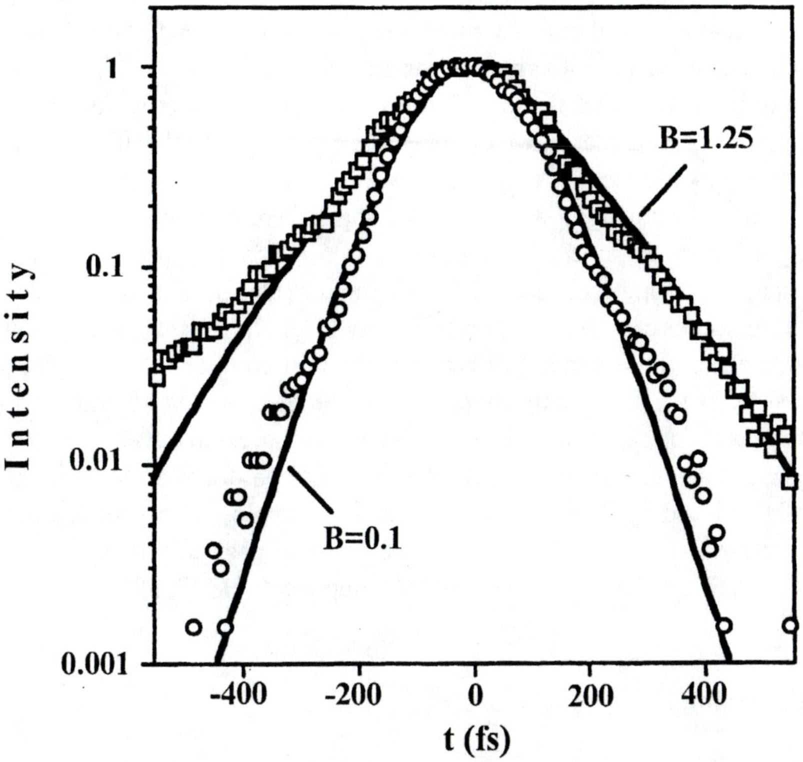

Fig. 2.9. Temporal profile of a compressed pulse at the out put of a regenerative amplifier plotted for two values of the breakup integral: (1) $B = 1.25$; (2) $B = 0.1$.

Large-diameter neodymium glass amplifiers should be used in further amplification of phase-modulated pulses. The output diameter of the amplifier is governed by two effects, which limit the peak power of the output radiation: the optical damage threshold and self-focusing [37]. Moreover, the amplification process should conserve also the linearity of phase modulation, which may be distorted under gain saturation conditions. We shall take account of these effects in estimating the required aperture diameters. We shall assume that the surface damage mechanism is thermal and we shall take account of the required service life of a laser and of the factor representing filling of the aperture with radiation characterized by a super-Gaussian distribution of the radiation over the cross section of the amplifying element. We can then use the following formula for the dependence of the energy density on the pulse duration:

$$W = at^{1/2} \tag{2.6}$$

where a = 3-10 J cm^{-2} ns$^{1/2}$ is a coefficient which depends on a number of factors: the depth of modulation of the radiation intensity in the laser system, the breakdown thresholds of materials and coatings used in the system, etc.

We can readily see that if the pulse duration is 0.5 – 1 ns, the permissible energy density is approximately 1 - 10 J cm^{-2} and energy of about 1000 J can therefore be obtained at the output of an amplifier with an aperture area $S \approx 1000\ \text{cm}^2$ (when the aperture diameter is about 30 cm). Amplifiers of this kind are readily available at present and are used widely in laser thermonuclear fusion studies [1, 38]. However, such an energy density in an amplifier is quite close to the saturation energy density (4-5 J cm^{-2} for neodymium phosphate glass), which may alter the chirp. On the other hand, lowering of the energy density reduces the output energy.

Small-scale self-focusing in a laser amplifying system can be suppressed by the use of a variety of devices or methods developed for this purpose [1, 38]: high-gain amplifiers, spatial filtering and beam re-translation, circular polarization of the radiation, etc. All these measures can raise the power density in amplifying stages to 5 - 6 GW cm^{-2}, which is quite close to the optical damage threshold of components. However, the total breakup integral for the whole amplifying system can then exceed 10 and, as pointed out already, this causes deterioration of the contrast of a compressed pulse.

The design of an amplifying system for short pulse production can be simplified and its cost can be reduced by the same methods as those employed in optimization of lasers for thermonuclear fusion. In particular, one can use multi-pass amplifying systems, which minimize the number of amplifiers and optical components [1, 38, 39].

2.1.5 Compression

A perfectly matched stretcher-compressor will have conjugate phase functions, $\psi_{str}(\omega)=-\psi_{com}(\omega)$. However, for the very large stretching and amplification required to produce terawatt pulses, the phase function of the compressor must account for the sum of the phase functions of the stretcher and the remainder of the laser system, $\psi_{sys}(\omega)$, including amplifiers, mirrors, and wave-plates. Writing the field of the pulse exiting the compressor as

$$E_{out}(\omega)=E_{in}(\omega)A(\omega)\, exp\{i[\psi_{str}(\omega)+\psi_{sys}(\omega)+\psi_{com}(\omega)]\}, \tag{2.7}$$

we see that a transform-limited pulse is achieved after compression only if $\psi_{com}(\omega)=-\psi_{sys}(\omega)-\psi_{str}(\omega)$. In this case, the amplified and compressed pulse has no residual chirp, $\delta=\psi_{com}(\omega)+\psi_{sys}(\omega)+\psi_{str}(\omega)=0$, and the temporal distribution of the field is just the Fourier transform of the spectrum of the field. In practice, the grating separation in the compressor is adjusted until the error δ is a minimum.

Effective compression of high-power chirped pulses is possible in single-pass and double-pass grating compressors. The geometric parameters of a compressor are selected on the basis of relationship (2.3) for the optical delay. A single- pass system, characterized by high energy efficiency, suffers from a significant shortcoming: the beam cross section at the output is elliptic, which increases the spot size and requires large-aperture optics for focusing of the beam. Therefore, two-pass compressors are used more widely: the radiation passes twice through the compressor and this compensates for the elliptically. The radiation is coupled out of a compressor by displacing the beam during the second pulse in a plane perpendicular to dispersion (this is done by a system of two mirrors or prisms).

Since a compressed pulse of 0.1 - 1 ps duration emerging from the output of a compressor occupies a fraction of a millimeter in space, a very high precision of the compressor alignment is required [40]. By way of example, **Fig. 2.10** gives the dependencies of the precision of setting the distance between gratings and of their parallel orientation on the angle of incidence of light when a pulse of 100 fs duration is generated. The precision can obviously be less for longer pulses.

Pulse compression increases the radiation intensity in a compressor in excess of 100 GW/cm^2, so that even the cubic non-linearity of air ($n_2 \approx 10^{-15}$ cgs esu) is sufficient for self- focusing ($B > 3$), which reduces strongly the focusing quality. This is why compressors in large laser systems are evacuated. The same hazard is the reason for the use of mirror systems in the focusing of radiation on a target.

Ruled and holographic diffraction gratings (of the reflection type with $n = 1500$-1800 lines mm^{-1}) operating in the first diffraction order are used in compressors. An increase in n makes it possible to increase the angle of incidence of light on a grating and reduce the distance between the gratings, so that a compressor can become more compact. However, this approach reduces the optical aperture of the gratings and makes even greater demands on the precision of their fabrication. For example, if $n = 1800$ lines mm^{-1} and the angle of incidence is about 70°, even a slight curvature of the grating surface is sufficient to reduce the intensity of radiation at the focus of a lens by a factor exceeding 100 because of astigmatism that appears when a fan of rays is incident on slightly bent tilted surfaces [41]. For this reason it is usual to select n in the range 1200-1740 lines mm^{-1}.

The multi-pass design of a compressor requires gratings of high diffraction efficiency. For example, an overall transmission coefficient of a two-pass compressor can exceed 60% provided the diffraction efficiency of a grating exceeds 90%. Such efficiencies are attainable for ruled and holographic gratings in the first diffraction order [42-44]. However, fabrication of large-aperture ruled gratings with the required value of n is difficult because of the cutting tool wear. It would therefore be of interest to see whether a high diffraction efficiency can be achieved in high (fifth to seventh) diffraction orders, because this would make it possible to reduce n to 200-300 lines mm^{-1} and to make large gratings without replacing the cutting tool. This is possible in the case of echelle gratings. For example, an echelle grating made at the Vavilov State Optical Institute enabled us to reach already a diffraction efficiency of about *80%* in the sixth diffraction order for $n = 300$ lines mm^{-1}.

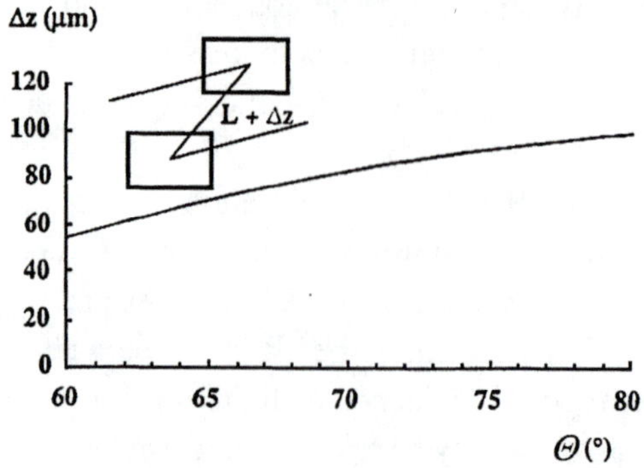

a)

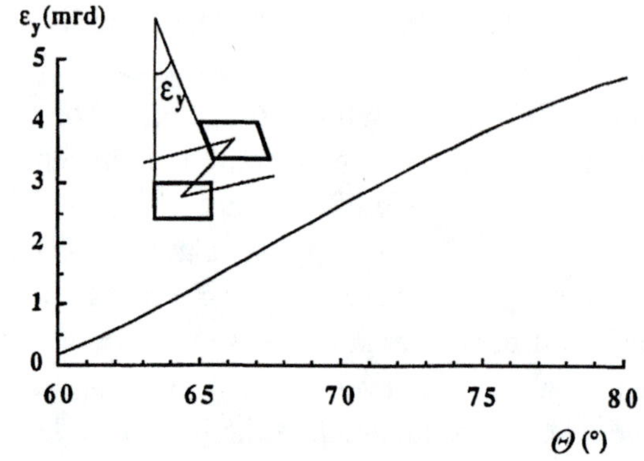

b)

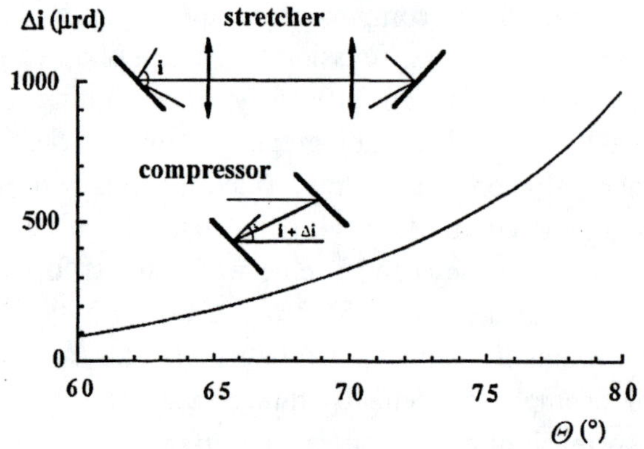

c)

Fig. 2.10. Precision of setting the distance between the diffraction gratings in a compressor (a), of the parallelism of the gratings (b), and of the mismatch of the angles of incidence in the stretcher and compressor (c), plotted as a function of the angle of incidence Θ on a grating for a Gaussian pulse of 100 fs duration at the $1/e^2$ level; compression coefficient 10000 and gratings with $n = 1740$ lines mm^{-1}.

The main requirements that diffraction gratings have to satisfy are high optical damage fluency. **Fig. 2.11** gives the dependence of the damage threshold of a holographic diffraction grating operating in the reflection mode on the thickness of its metal coating. We can see that for a pulse of 600 fs duration the damage threshold is less than 1 J cm^{-2} [44], which is over an order of magnitude less than the damage threshold of optical materials and coatings used in an amplifying system. For example, a laser with an output radiation of ~ 10^{18} W cm^2 power density, which is necessary (as shown below) for "fast ignition", should have gratings of meter size. The technology for the fabrication of such gratings is being developed successfully at the Lawrence Livermore National Laboratory [44]. At the Vavilov State Optical Institute we can make holographic diffraction gratings of up to 20 cm x 40 cm size for laser applications.

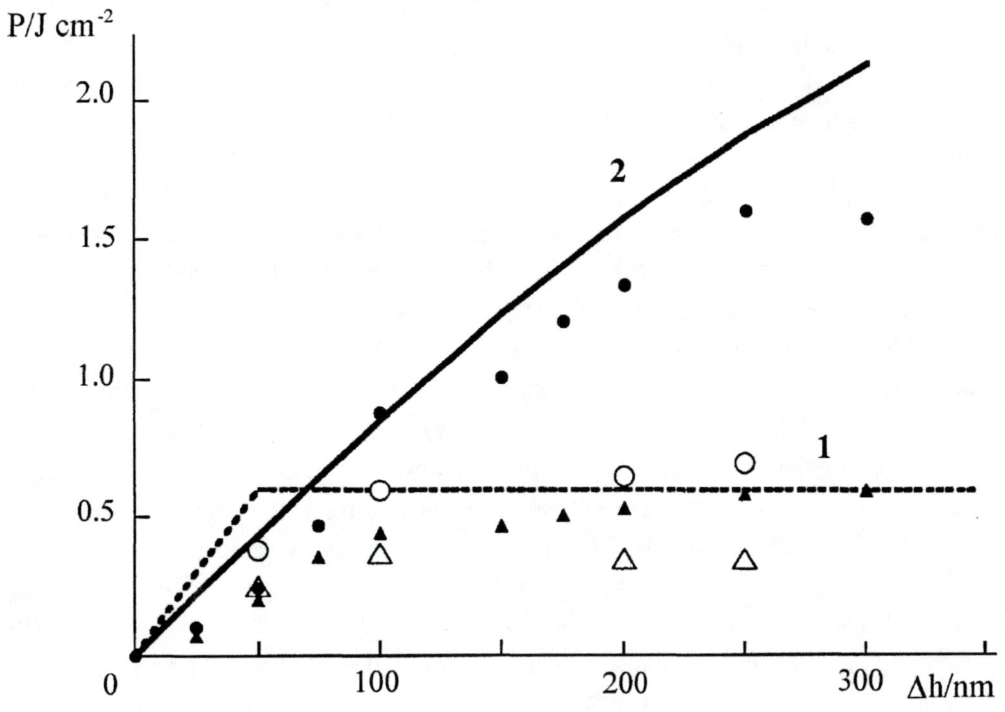

Fig. 2.11. Calculated (curves) and experimental (points) dependences of the damage threshold P of a diffraction grating on the thickness of a gold film Δh, obtained for laser pulses ($\lambda = 1053$ nm) of 600 fs (1) and 800 ps (2) duration.

It is thus clear that diffraction gratings are a weak link in a laser system. This has stimulated a search for the methods of fabricating stronger gratings. One of these methods involves formation of a diffraction structure in multi-layer dielectric coatings [45, 46], which are known to have a much higher damage threshold than metals. The published investigations [45] do indeed demonstrate that such gratings can be made and they can have a sufficiently high diffraction efficiency (80%-90%), although their optical damage threshold is enhanced approximately tenfold for nanosecond laser pulses, but only by a factor of ~ 1.5 for picosecond pulses (compared with gold-coated gratings).

This relatively modest increase in the optical damage threshold is the result of concentration of the electric field of an optical wave at the interfaces of layers in a multi-layer coating. Anyway the optimisation of such grating permits to get the enough high threshold for picosecond laser pulse [73].

2.1.6 The Example of CPA Laser System

The laser system to be described here [47] uses a specific way of producing a chirped pulse; it also has a different size of the amplifier stages and diffraction gratings of the compressor unit (**Fig. 2.12**). The system consists of four basic units: a triggering laser system, a high power amplifying unit based on a "Progress" amplifying channel, a compressor with holographic gratings, and a focusing system.

A chirped pulse is generated in the triggering unit by broadening the spectrum of the initial 30 ps pulse from a Nd:YLF master oscillator in a single mode fiber. Then the pulse is stretched in time to ~ 300 ps and amplified by a series of rod amplifiers on a phosphate neodymium glass of 85 mm diameter.

A single pulse is isolated and gated by a system of four Pockels cells controlled by synchronized high voltage electric pulse generators based on drift diodes with a fast restoration of the reverse voltage [47]. These generators are capable of producing electrical pulses of half-width duration ~ 1.5 - 5 ns and an amplitude up to 15 kV.

A 300 ps pulse appearing at the amplifying channel output is compressed to 1.5 ps duration in a single-pass compressor consisting of two holographic diffraction gratings with a 21 cm × 42 cm gold coating and a groove density of 1,600 lines/mm.

A beam is introduced into a target chamber through a high quality vacuum window made from a LiF crystal with a small nonlinear gain in the refractive index $n_2 \sim 0.35 \times 10^{-13}$ CGSE units (about 4 times less than in K-8 glass) to avoid small-scale self-focusing.

The focusing of a laser beam of a relatively low power (about 1 TW) onto a target was performed by an aspherical lens of 13 mm thick and 140 mm focal length, which concentrates 75% beam power into a diffraction-limited spot of 9.7 μm or by an off-axis parabolic mirror with a 120 mm focal length, which focuses 65% power of an ideal beam into a 7.2 μm spot. In the case of lens focusing, the lens serves as a vacuum window. These focusing systems were used in experiments with power densities up to 10^{18} W/cm^2. To focus more powerful beams (up to 30 TW), an axial parabola of 200 mm in diameter and 1:1 light power was used; it could focus 50% of the ideal beam power into a 5 μm spot, providing a focal intensity above 10^{19} W/cm^2.

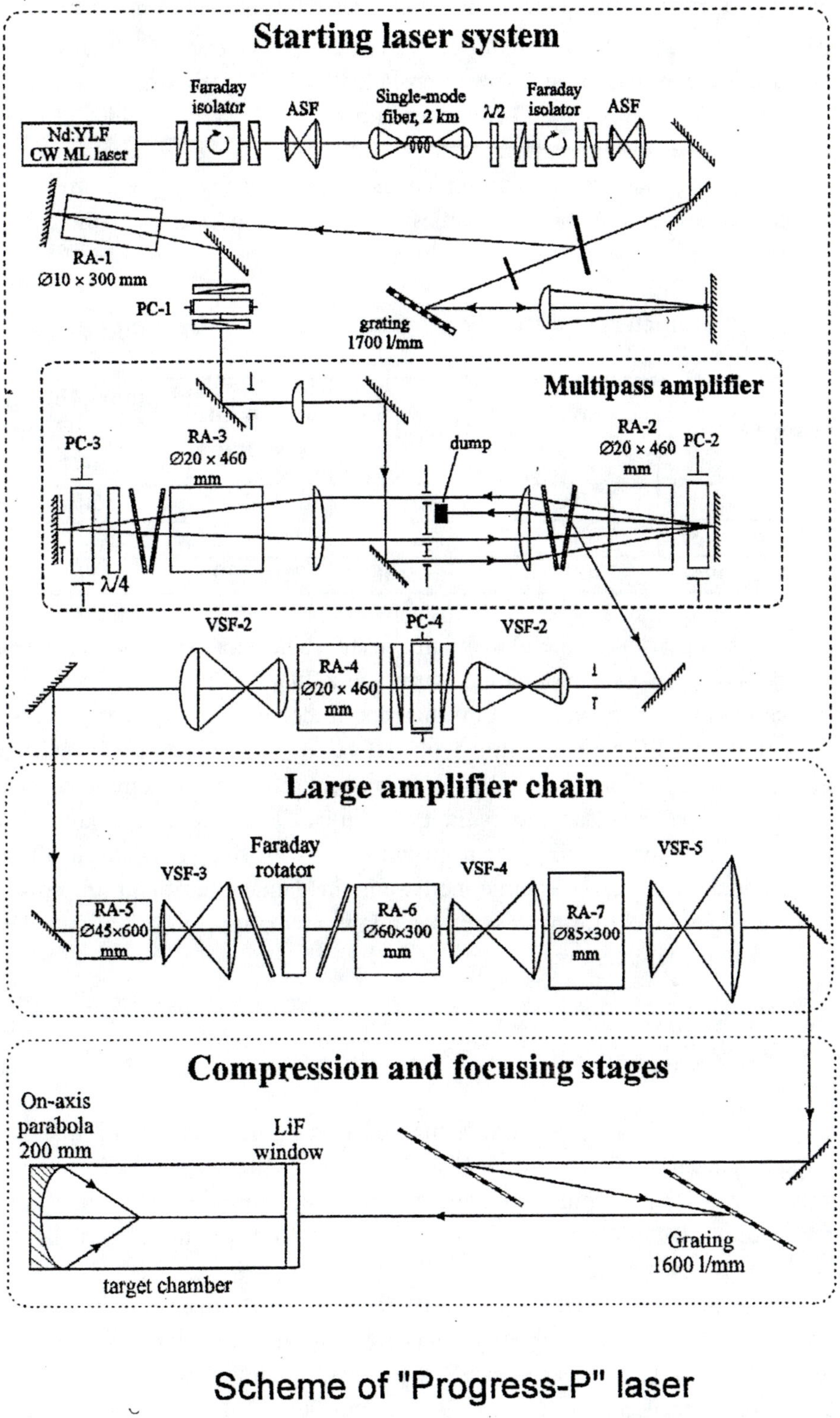

Fig. 2.12. The example of CPA laser system.

The energy of intense spontaneous radiation through an output compressor was less than 10 μJ for 5 ns prior to the maximum pulse. The energy contrast of a laser pulse was at least 10^5. High radiation intensities were provided by high quality optics of the laser systems and controlled by optical and X-ray techniques. To illustrate, **Table 2.2** presents the parameters of the key unit, i.e., of the diffraction gratings used for the pulse compression [48]. The characteristics presented show that they have a good safety factor, permitting an increase in radiation intensities several times greater than the available value.

Table 2.2. Parameters of diffraction gratings used for pulse compression

Grating type	Grooves per mm	Metal	Maximum size, mm	Damage threshold W_{th}, mJ/cm^2 (1ns/1 ps)
Cut grooves	1740	Au	170x170	360/240
Holographic grooves	1700	Au	210x420	320/100
Holographic grooves	1700	Au with coating	210x420	560/230

However, a comparison of the above damage thresholds for gratings and dielectric media [2, 49] shows that metallic gratings withstand a 30 - 40 times lower peak power. Of importance in this connection is the development of technique, which can increase the radiation stability of diffraction gratings [45]. One of such methods is to deposit additional dielectric layers on the metal [48]. It is clear from the last column of Table 1 that this method does increase the radiation stability. Numerical calculations from Maxwell's equations show [50,53] that it provides a 5 - 10-fold reduction in the field strength on the metal and nearly as large increase in the damage threshold. Improvement of this technique may give a further rise in peak power radiation, since it allows the use of gratings with a smaller aperture.

2.1.7 Radiation Contrast

In many investigations of the interaction of ultra-short pulses with matter, the contrast of such pulses should be sufficiently high to ensure that a plasma does not form before the arrival of the main pulse on a target. The plasma formation threshold is about 10^{10} W cm^{-2}, so that the contrast must be 10^8 if investigations are to be carried out at a power density of 10^{18} W cm^{-2}.

The contrast may be lowered by several factors, which manifest themselves in different ways at different time intervals measured from the maximum of the main compressed pulse. On the microsecond scale, the contrast may be reduced by luminescence or super-luminescence of laser amplifiers. However, in the case of neodymium glass with its long excited-state lifetime (300-500 μs) and a relatively low (compared with other media) small-signal gain, the luminescence power in separate amplifying stages is low and does not exceed several hundreds of watts in a wide solid angle. Super-luminescence, i.e. the luminescence amplified by successive stages, can be eliminated effectively by a variety of methods such as optical switches, spatial filters, and saturable absorbers. For example, spatial filters present in an amplifying system in order to control self-focusing also suppress effectively the luminescence [47]. The brightness contrast resulting from the noise in an amplifying system can be described by the following expression when the gain is unsaturated [19]:

$$K \approx \frac{10P_{in}}{I_s \lambda^2} \tag{2.8}$$

where P_{in} is the input signal power and I_s the saturation intensity of the active medium. It follows from this expression that the contrast can be increased by increasing the input pulse power.

A different type of noise grows in a few nanoseconds before the main pulse as a result of multi-pass amplification in a regenerative amplifier. The nonzero transmission of a switch that admits a pulse from a master-oscillator cavity into such an amplifier means that the main pulse is always preceded by pre-pulses of intensity that depends on the switch contrast, but is usually 10^{-2}-10^{-4} of the intensity of the main pulse. Such pre-pulses can be removed by fast optical switches, for example by Pockels cells. It is suggested in Ref. [21] that this purpose can be served by an electro-optical deflector controlled by a fast-response silicon switch. This device has a higher contrast than a conventional Pockels cell and, moreover, it responds fast.

Finally, the third group of factors that reduce the contrast is associated with distortion of the spectrum during the propagation of a beam through the optical components of an amplifying system. These factors degrade the contrast of the main pulse over picosecond and femtosecond time scales; they are discussed above together with the methods for their elimination. These methods reduce basically to removal of causes of spectral modulation, although time selection of pulses shortened in a compressor is possible (this can be done, for example, by saturable absorbers [48]). However, this is undesirable because of major energy losses and self- modulation caused by the non-linearity of such an absorber. Correct selection of the laser system parameters may ensure a contrast of 10^6-10^8 [48, 49].

The basic approach to improving the contrast of a pulse involves the use of some non-linear processes that depend strongly on the radiation intensity and are characterized by a shorter relaxation time. One of such processes may be, for example, SRS, which moreover can ensure matching of the master-oscillator and amplifying-system wavelengths. This use of SRS for increasing the contrast and matching the wavelengths of a ruby laser to a neodymium glass laser was demonstrated in Ref. [19]. In the case of a neodymium glass amplifying system it is also possible to use a master oscillator and regenerative amplifier with Ti:Al_2O_3 and/or Cr:LiSAF operating each at its own natural wavelength near 800 nm. This can be followed by frequency conversion to the gain band of neodymium glass (such conversion can be carried out in gaseous methane or nitrogen with a suitable frequency shift). However, when this approach is adopted, it is necessary to investigate how the spectrum of a chirped pulse is modified by SRS.

Another interesting non-linear process of practical importance is conversion of the frequency to the second and third harmonics in non-linear crystals. The feasibility of such conversion of picosecond pulses with an efficiency of about 70%- 80% has been demonstrated [50 - 52]. Moreover, under certain conditions it is possible to reduce also several fold the pulse duration [50, 52]. However, this reduction in the duration occurs only near the peak of a pulse but does not affect significantly its wings. The wings can be suppressed by degenerate parametric conversion [53], but this is possible only in beams of relatively small cross section and leads to a significant complication of the optical system of a laser. When shorter (femtosecond) pulses are converted, the efficiency of conversion to the second harmonic is limited to about 50% and this is primarily due to phase modulation in the non-linear crystal itself [54, 55]. Nevertheless, frequency conversion is of interest not only as means for increasing the contrast, but also for generation of shorter wavelengths that interact more strongly with plasmas.

2.2 Alternative Ways of Scaling of High-Power Laser System

We shall now consider some other ways of constructing high-power laser systems, which supplement or are alternative to those already considered. One such way is related to the observation that the optical damage threshold of transparent media is almost independent of the pulse duration in the range $t_L < 1$ - 5 ps [2] (**Fig. 2.13**). This makes it possible to use wide-aperture amplifiers with different active media (for example, excimers) for an additional increase of the energy of picosecond or subpicosecond pulses emitted by a neodymium laser and this could be done after pulse compression. The method used is tripling the neodymium laser frequency in a KDP crystal.

Another way of scaling-up the power involves the use of phase modulation of the radiation in wide-aperture non-linear (cubic) media [56, 57]. This approach is analogous to that based on spectral broadening in a fiber and is described by formulas (2.1) and (2.2). However, whereas in the case of single-mode propagation of the radiation in a fiber, self-focusing does not influence self-phase modulation, the situation in bulk media is different. Small-scale self-focusing limits the breakup integral (and, therefore, the spectral broadening) to B = 2-3, which is clearly insufficient for effective compression. When these values are exceeded, a beam splits into filaments, very strong spectral broadening occurs in an uncontrolled manner, and the angular divergence becomes much greater [1, 38]. For these reasons the attempts to broaden the spectrum in a non-linear medium followed by pulse compression have met with considerable difficulties and have proved to be ineffective [56, 57].

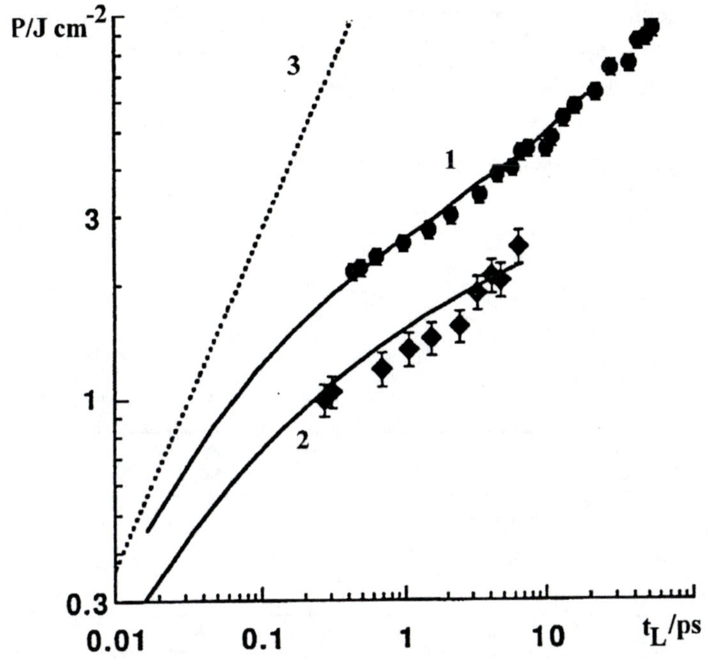

Fig. 2.13. Experimental (points) and calculated (curves) dependences of the surface energy density P causing damage to silica at $\lambda = 1053$ nm (1) and 526 (2), and of the multi-photon ionization threshold at $\lambda = 1053$ nm (3) on the laser pulse duration t_L.

Effective broadening of the spectrum by self-phase modulation requires also a uniform distribution of the intensity over the beam cross section. This condition is not satisfied by spectral broadening in oscillators or regenerative amplifiers reported in Refs. [12, 58, 59] and as a result the radiation contrast is poor. These difficulties can be overcome by adopting the well-developed methods for suppressing small-scale self-focusing [60]. One of the most effective methods is the use of spatial filters and optical repeaters, employed widely in neodymium glass lasers emitting high-brightness radiation [1, 38].

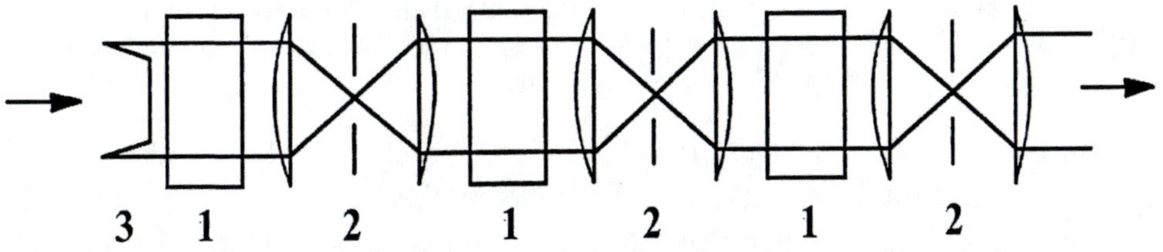

Fig. 2.14. Optical part of a system for broadening the spectrum of high-energy pulses, consisting of segments of a cubic nonlinear medium (1) and of spatial filters combined with imaging relays (2).

A spectral broadening system of this kind is shown schematically in **Fig. 2.14**. The total breakup integral, representing-in accordance with formula (2.2)-the spectral broadening is $B = NB_l$, where B_l is the breakup integral for one nonlinear component.

Simple passive, electro-optical or magneto-optical, de-couplers used in such a system make it possible to implement readily multi-pass or regenerative operation, which increases B by a factor m, where m is the number of passes. For example, even for just $N = 3$ and $m = 2$, the breakup integral can be increased to $B = 6B_I \approx 20\text{-}40$ and, consequently, the pulse can be compressed by the same factor.

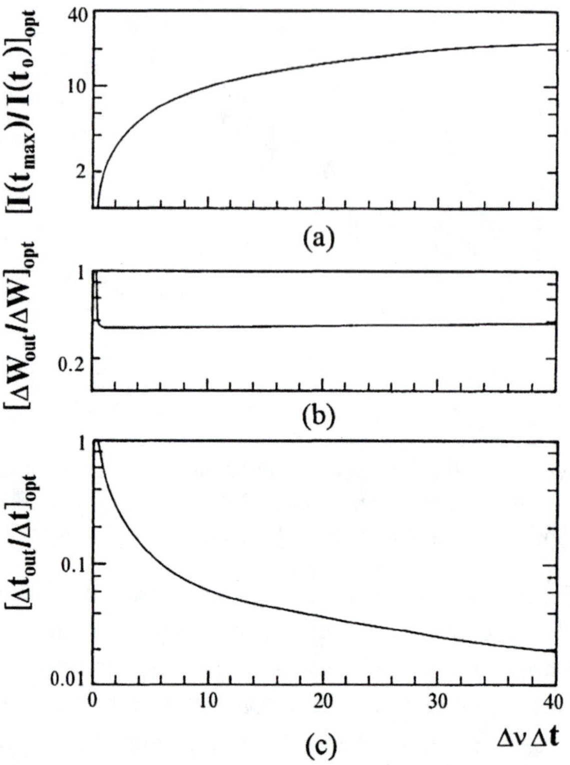

Fig. 2.15. Characteristic of compressed pulse, broadened in a nonlinear cubic medium, plotted as a function of the spectral broadening $\Delta\nu\Delta\tau$ for optimal distance between two gratings ($n = 600$ lines mm^{-1}) in a compressor: (a) ratio $\Delta t_{out}/\Delta t$ of the output duration of a pulse to the input duration measured at 0.5 of I_{max}; (b) ratio $W_{our}/\Delta W$ of the energies of the output and initial pulses contained in the center maximum, determined at half the maximum intensity; (c) ratio $I(t_{max})/I(t_0)$ of the maximum intensity to the initial intensity in the output pulse.

This is demonstrated in **Fig. 2.15** which gives the calculated dependence of the compression coefficient [60], defined as the ratio of the durations of the initial and compressed pulses measured as full width at half-amplitude, and which also gives the dependence of the intensity of a compressed pulse on $\Delta\nu\Delta\tau = B$ obtained for optimal compression conditions and a Gaussian temporal profile of the initial pulse.

Additional simulations [67] have shown that initial quality of a laser beam should be enough high to use this method.

Such a system with spectral broadening followed by pulse compression in diffraction gratings can be used for additional compression of picosecond pulses and for conversion of nanosecond into picosecond pulses (see **Fig.2.16**).

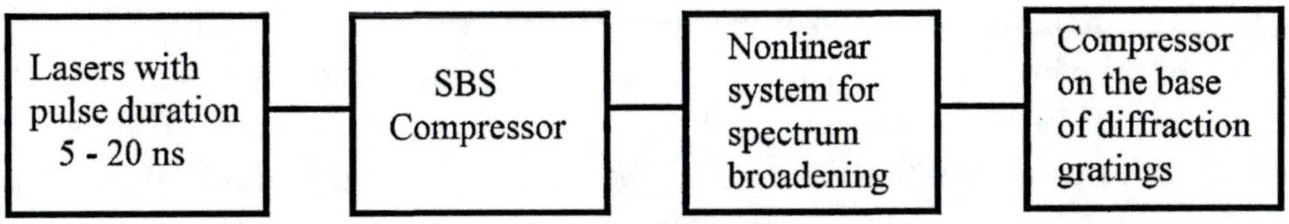

Fig. 2.16. The example of the laser system with nonlinear compression.

Let us estimate the requirements in respect of the parameters of the radiation, non-linear medium, and optical systems. The length L_{nl} of a non-linear component is found from expression (2.5) for the breakup integral. The intensity I is limited by the optical damage threshold of the medium, which means that it is possible to use optical glasses with $I \approx 5$ GW cm^{-2} for $t \approx 1$ ns [1, 38] and with $I \approx 30$ GW cm^{-2} for $t \approx 1$ ps [2]. Then, if the medium is silica ($n_0 = 1.46$, $n_2 = 0.9 \times 10^{-13}$ cgs esu), we obtain $L \approx 30$ cm for $t \approx 1$ ns and $L \approx 6$ cm for $t \approx 1$ ps. The length at one non-linear component can be reduced for glasses with a high ratio n_2/n_0, so that even lenses in spatial filters/repeaters can perform the role of a non-linear medium for picosecond pulses. When this is undesirable or if there are problems of any other kind (for example, when it is necessary to eliminate the chromatic aberration of the lenses in spatial filters which transmit short broad-band pulses), lenses can be replaced by mirrors. Another effect that limits the maximum value of B, as is true also of optical fibers, may be SRS with the threshold intensity for its excitation $I_{th} = \alpha / gL_{nl}$, where $\alpha = 30$ is the total threshold increment, g is the gain per unit length, and L_{nl} is the total length of the non-linear medium. Substitution of this formula into expression (2.5) for the breakup integral can yield the maximum value of B. For example, in the case of silica, we have $g \approx 10^{-2}$ cm GW^{-1} and $B \approx 70$. If necessary, SRS can be suppressed by dichroic mirrors, which reflect the Stokes radiation shifted considerably on the frequency scale relative to the pump radiation.

The phase shift can be distributed uniformly over the beam cross section only if the intensity distribution is uniform in the non-linear medium. This is usually achieved by an anodizing or 'hard' aperture, which produces a uniform beam profile, and a system of spatial filters/repeaters, which transmit the image of this aperture to consecutive parts of the non-linear medium. The angular pass band of the spatial filters $q = d/F$ (d is the aperture diameter and F is the focal length of the entry lens of a spatial filter) should be considerably greater than the diffraction-limit angular divergence $q_d = 2.44\lambda/D$ (D is the beam diameter on the lens). Nevertheless, in the case of short pulses it is possible to select $q = (4\text{-}5)q_d$, which is considerably less than for the longer pulses when the minimal aperture diameter is limited by the possibility that it becomes obstructed by a plasma [62]. For short pulses ($t \leq 0.1$ ns) such plasma obstruction is not a real hazard because it occurs after the passage of a pulse. A reduction in the pass band of a spatial filter makes it possible to eliminate more effectively the undesirable intensity peaks and it is then possible to increase the permissible breakup integral atone non-linear component. In such a spectral broadening system, as in the case when fibers are used, difficulties may be encountered in attainment of a high contrast of a compressed pulse when the temporal profile of the initial pulse differs from the ideal square shape. Calculations show (Fig. 2.9) that the time interval outside the main pulse width, measured at half-amplitude, may represent about 20 % of the energy [61]. However, in many tasks (including 'fast ignition') such a contrast is not required, but just the reverse, it is desirable to have a pulse with a long pre-pulse. However, when a high contrast is essential, it can be improved for nanosecond initial pulses by temporal profiling (for example, by electro-optical deflectors) and for picosecond pulses this can be done by spectral filtering in a stretcher and in a compressor [6, 29] or by frequency pre-conversion into harmonics.

In general, the chirp non-linearity in a compressor can be compensated by special diffraction gratings, for example, those with a non-uniform distribution of lines over the cross section. Gratings of this kind can be made, for example, by the holographic method and - by analogy with adaptive optics - this approach can be called spectral adaptive optics. It follows from calculations [63] that in this case one can achieve effective compression and a high contrast for pulses whose temporal profile differs from the square shape.

2.3 Limit Parameters of Pulse Compression Lasers

The laser systems described above have a peak power of 1 - 30 TW, providing radiation intensities on a target down 10^{19}W/cm. The maximum available peak power of 1.3 PW and the target intensity of ~ 10^{21} W/cm^2 were achieved in a PW laser system [68] at the Lawrence Livermor National Laboratory, using a 'Nova' laser channel.

Further concentration of laser radiation may become possible owing to a higher energy, shorter pulse duration, and smaller focusing spot. These potentialities will be discussed with reference to well-elaborated laser systems based on neodymium glass.

Table 2.3. Theoretical peak power

Laser type	Cross section (10^{-20} cm^2)	$\Delta\lambda$ (nm)	I_{th} (TW/cm^2)
Nd:glass phospate	4	22	60
Nd:glass silicate	2.3	28	100
Nd:glass combination	1.5	60	400
Ti:sapphire	30	120	120
Alexandrite	1	100	2000
Cr:LiSAF	3	50	300

The theoretical limit of laser peak power can be estimated as the ratio of the extractable energy (approximated by the saturation fluency) to the minimum pulse duration imposed by the gain bandwidth (neglecting gain narrowing). The peak power limit I_{th}, per unit area, assuming a time-bandwidth product $\Delta\nu\Delta\tau = 0.5$, is then estimated by

$$I_{th}=2F_{sat}\Delta\nu_a$$

This power also represents that needed to produce a Rabi oscillation in the amplifying medium. The theoretical peak power for different materials is estimated in Table 2.3. In principle, 100-TW pulses are achievable with compact systems, and even Exawatt (1 EW = 10^{18} W) pulses are possible with large-aperture (1 m) Nd-glass systems that contain a mixture of silicate and phosphate glasses to maximize the gain bandwidth.

The maximum output energy of a neodymium laser is limited by the damage threshold of the implication unit for relatively long chirped pulses and of the diffraction grating for short, subnanosecond pulses. The damage threshold for neodymium glass and coatings for 1 ns pulse is about 20 J/cm^2, providing a safe operation at the energy density of 10 J/cm^2 [69]. The transverse dimension of an active disk made from neodymium glass is limited to 30 cm by the excitation of spurious modes and by the fabrication technology. Active elements of this size are used in amplifiers of the most powerful lasers - of the operating Beamlet laser [69] and the NIF system [70] being constructed at the Lawrence Livermor National Laboratory (LLNL). The output energy of this amplifier is about 10 KJ, supported by experiments on the Beamlet at the LLNL [69].

It was pointed out above that the radiation stability of diffraction gratings is less than that of dielectric materials. However, there is a possibility to improve this parameter considerably and predict a value of about 2 J/cm^2 for femtosecond pulses with a stable operation at half the load. Under this condition, the grating area for the compression of a 10 KJ pulse must be 10,000 cm^2 and its size for a single-pass compressor must be 70 × 140 cm. The facilities necessary for the fabrication of gratings of this size are available today at the LLNL [71]. Segmented gratings made up of several smaller gratings can be employed, if necessary.

There are several ways of reducing the pulse duration in a neodymium glass laser [64]. One is due to an increased amplification band width provided by an anti-resonance filter in a generation amplifier and by laser amplifiers based on neodymium glass of various composition in the basic amplification unit as we mentioned above. Neodymium glasses, for example, silicate and phosphate glasses, have central luminescence peaks shifted toward the longer wavelengths, increasing the amplification spectral width by a factor of 3. These improvements reduce the duration of high power laser pulses to 200 - 400 fs, when using titanium-sapphire and neodymium master oscillators producing shorter pulses [68].

Further reduction in the pulse duration will be possible if the phase self-modulation in a cubic nonlinear medium is used. For example, this effect was applied in the glass light guide of the Progress-P laser [47] to produce the initial chirped pulse. A direct extension of this method to high power laser pulses is impossible because of the limit set by small-scale self-focusing. The application of gas-filled mirror light guides [72] increased the pulse power to the micro joule level, which is certainly insufficient for the method to be used in power beams. This method is especially effective, if nonlinear elements based on transparent large aperture glasses are separated by spatial filters [64], as we mentioned above. In this geometry, the nonlinear gain is attained gradually, from one element to another, using a regular filtration of small-scale perturbations. Calculations and experiments show that five nonlinear elements are sufficient for a 5-fold spectral broadening and for about as large increase in the peak power.

Ultrahigh laser intensity might be produced inside a plasma layer in a time short compare to the growth time for filamentation instability. Analysis of transient Raman backscattering in plasma as possible mechanism for such a fast compression of a laser pump indicates that a non-focused 10^{17} W/cm^2 intensity of 1 μm wavelength radiation might be achieved [74]. The corresponding fluences would be about 5 kJ/cm^2.

The above innovations allow the principal generation of laser pulses with an exowatt (10^{18} W) peak power, capable of producing a spot of 10 μm diameter and 10^{24} W/cm^2 intensity. Note that the spot size produced by modern focusing systems is still far from the limit defined by radiation diffraction. This is due to the remnant aberrations of these systems and to distortions introduced by the elements of the amplification and compression channels. To compensate for such distortions, linear and nonlinear adaptive optics seems quite promising [37]. In the latter case, it is necessary to design and study techniques for the wave front conversion of relatively short wide band chirped pulses. It should be stressed that such techniques can provide an automatic beam phasing in segmented diffraction gratings [37], whose application appears feasible at high energies. With these methods, the focusing spot size can be expected to decrease to a diffraction-defined value (about 1 μm for neodymium lasers). This may give two orders of magnitude gain in intensity, up to 10^{26} W/cm^2.

It is likely, that further intensity increase may be attained by combined focusing of time and space phase-matched laser pulses from different laser channels. For example, the use of a 192-channel laser of the NIF-type [14] can principally provide 10^{28} W/cm^2 intensity. It will be shown below that this is, in a sense, limit intensity, since then the vacuum itself is 'broken down'.

Of course, to produce such a high intensity is a very complicated and costly task. Its implementation would require world wide cooperative efforts, as is the case with thermonuclear fusion. On the other hand, the generation of more moderate intensities of 10^{22} - 10^{24} W/cm^2 is a task feasible in the near future. Below, we discuss the possibilities that arise from the study of these and lower laser fields and their interaction with matter.

REFERENCES FOR PART 2

1. Mak A.A., Soms L.N., Fromzel' V.A., Yashin V.E. *Lazery na Neodimovom Stekle* (Neodymium Glass Lasers) (Moscow: Nauka, 1990) p. 288.
2. Stuart B., Feit M.D., Perry M.D., Rubenchik A.M., Shore B.W. *Phys. Rev.B,* v.53, 1749, 1996.
3. Treacy E.B. *IEEE J. Quant. Electron.* QE-5, p.454, 1969.
4. Maine P., Strickland D., Bado P., Pessot M., Mourou G. *IEEE J.Quant.Electron.*, v.24, p.398, 1988.
5. Stolen R.H., Lin C. *Phys. Rev. A,* v.17, p.1448, 1978.
6. Heritage J.P., Weiner A.M., Hawkins R.J. et al. *Opt. Commun.*, v.67, p.367, 1988.
7. Grischkowsky D., Balant A.C. *Appl.Phys.Lett.*, v.41, p.1, 1982.
8. Martinez O.E. *IEEE J. Quant. Electron.* QE-23, p.59, 1987.
9. Van'kov A., Kozlov A., Chizhov S., Yashin V. *Proceeding of SPIE,* v.2095, p.87, 1994.
10. Burneika K., Grigonis R., Piskarskas A. et al. *Sov. J. Quant. Electron.*, v.18, p.1034, 1988.
11. Stuart B.C., Herman S., Perry M.D. *IEEE J.* QE-31, p.528, 1995.
12. Danielius R., Stabinis A., Valiulis G., Varanavicius A. *Opt. Commun.,* v.105, p.67, 1994.
13. Spence D.E , Kean P.N. and Sibbert W. *Opt. Lett.*, v.16, p.42, 1991.
14. Krausz F., Fermann M.E., Brabec T. et al. *IEEE J.* QE-28, p.2097, 1992.
15. Rouyer C., Mazataud E., Allais I. et al. *Opt. Lett.*, v.18, p.214, 1993.
16. Ditmire T., Nguen H., Perry M.D. *J. Opt. Soc. Am. B,* v.11, p.580, 1994.
17. Brovelli at al. *J.Opt. Soc. Am. B,* v.12, p.1, 1995.
18. Korf D, Kartner F.X., Weingarten K.J., Keller U. *Optics Lett.*, v.20, p.169, 1995.
19. Losev L.L., Soskov V.I. *Kvantovaja Elektron.,* v.22, p.531, 1995.
20. Buzelis R., Dement'ev A.S., Kosenko E.K., Murauskas E.K. *Kvantovaya Elektron.,* v.22, p.567, 1995.
21. Van'kov A.B., Kozlov A.A.,Chizhov S.A.,Yashin V.E. *Kvantovaja Elektron.,* v.22, p.583, 1995.
22. Kryzhanovskii V.I., Mak A.A., Sventitskaya I.N. et al. *Kfantovaya Elektron.,* v.4, p.345, 1977.
23. Van'kov A.B., Kozlov A.A., Chizov S.A. et al. *Technical Digest of Joint Symposium "Superintense Laser Fields" on 8th Laser Optics Conference* St-Petersburg, p.32, 1995.
24. Barty C.P.J., Gordon III C.L., Lemoff B.E. et al. *Proc. Soc. Photo-Opt. Instrum. Eng.*, v.2377, p.311, 1995.
25. Du D., Squier J., Kane S, Korn G *Optics Letters,* v.20, p.2114, 1995.
26. Sullivan A., White W.E. *Optics Letters,* v. 20, p.192, 1995.
27. Tournois P., *Electronic Letters,* v.29, p.1414, 1993.
28. Tournois P. *Opt. Commun.*, v.106, p.253, 1994.
29. Perry M.D., Patterson F.G., Weston *J. Opt. Lett.*, v.15, p.381, 1990.

30. McGeoch M.W. *Opt. Commun.*, v.7, p.116, 1973.
31. Heinz P., Laubereau A. *J. Opt. Soc. Am. B,* v.7, p.182, 1990.
32. Gorbunov V., Gogoleva N.. In *Technical Digest of Joint Symposium "Superintense Laser Fields" on 8th Laser Optics Conference* St-Petersburg, p.34 1995.
33. Blanchot N., Rouyer C., Sauteret, Migus A. *Opt. Lett.*, v.20, p.395, 1995.
34. Chuang Y.H., Zheng L., Meyerhofer D.D. *IEEE J. Quant. Electron.*, v.29, p.270, 1993.
35. Houliston J.R., Key M.H., Ross I.N. *Opt. Commun.*, v.108, p.111, 1994.
36. Perry M.D., Ditmire T., Stuart B.C. *Opt. Lett.*, v.19, p.2149, 1994.
37. Gamalii E.G., Dragila R., Tikhonchuk V. T. *Nature,* v.15, p.904, 1990.
38. Brown D.C. *High-Peak-Power Nd:Glass Laser Systems* Berlin: Springer, p.276, 1981.
39. ICF *Quartely Report, Special Issue: Beamlet Laser Project, 5, 1, Lawrence Livermore National Laboratory*, Livermore, CA, UCRL-LR-105821-95-1, p.1, 1994.
40. Fiorini C., Sauteret C., Rouyer C. et al. A. *IEEE J. Quant. Electron.*, v.30, p.1662, 1994.
41. Sauteret C., Husson D., Thiell G et al. *Optics Letters*, v.16, p.238, 1991.
42. Gerke R.R., Koreshev S.N., Semenov G.V. et al. *Opticheskiy Zh., no.*1, p.26, 1994.
43. Loewen E., Maystre D., Popov E., Tsonev L. *Appl. Optics*, v.34, p.1707, 1995.
44. Boyd R.D., Britten J.A., Decker D.E. et al. *Appl. Optics*, v.34, p.1697, 1995.
45. Perry M.D., Boyd R.D., Britten J.A. et al. *Opt. Letters*, v.20, p.940, 1995.
46. Svakhin A. S., Sychugov V. A., Tikhomirov A. E. *Kyantoyaya Elektron.,* v.21, p.250, 1994.
47. Borodin V.G., Komarov V.M., Charuhchev A.I. *Kyantoyaya Elektron.,* v.25, p.115, 1998.
48. Yashin V., Mak A., Bakh L. et al. *Proceedings of SPIE,* v.3291, p.199, 1998.
49. Kozlov A.A., Andreev A.A., Chizhov S.A. et al. *Proceedings of SPIE* v.3093, p.75-79, 1997.
50. Vinokurova V.D., Gerke R.R. *Optika i Spektroskopija,* v.83, p.990, 1997.
51. Bayanov V.I.,Vinokurov G.N., ZhulinV.I.,Yashin V.E. *Kvantovaya Elektron.,* v.16, p.253, 1989.
52. Chuang Y.H., Meyerhofer D.D., Augst S. et al. *J.Opt.Soc. Am. B,* v.8, p.1226, 1991.
53. Beaudin Y., Chien C.Y.,Coe J.S. *Opt. Lett.,* v.17, p.865, 1992.
54. Wang Y., Luther-Davies B. *Opt. Letts.*, v.17, p.1459, 1992.
55. Wang Y., Luther-Davies B., Chuang Y.-H. *Opt. Letts.*, v.16, p.1862, 1991.
56. Chien C.Y., Korn G., Coe J.S. *Opt. Lett.,* v.20, p.353, 1995.
57. Danielius R., Dubietis A., Valiulis G. *Optics Letters*, v.20, p.2225, 1995.
58. Rodriguez G., Roberts J.P., Taylor A.J. *Opt. Lett.*, v.19, p.1146, 1994.
59. Krylov V., Rebane A., Kalintsev A. *Opt. Lett.*, v.20, p.198, 1995.
60. Laubereau A. *Phys. Lett.* v.29A, p.539, 1969.
61. Lehmberg R.H., McMahon J.M. *Appl. Phys. Letts.*, v.28, p.204, 1976.
62. Yan L., Ling J.-D., Ho P.T. *IEEE J.Quant. Electron.*, v.24, p.418, 1988.
63. Scheidler W., Penzkofer A. *Opt. Commun.*, v.80, p.127, 1990.
64. Mak A.A., Jashin V.E. *Optika I spektroskopija,* v.70, p.3, 1991.

65. Penzkofer A. *Optical and Quantum Electronics*, v.23, p.685, 1991.
66. Auerbach J.M., Holmes N.C., Hunt J.T. *Appl. Opt.*, v.18, p.2495, 1979.
67. Andreev A.A., Sutyagin A.N. *Proceeding of SPIE,* v.2701, p.123, 1996.
68. Pennington D.M., Perry M.D., Stuart B.C. et al. *Proceedings of SPIE,* v.3047, p.490, 1997.
69. Van Wonterghem B., Murray J.R., Campbell J.H. et al. *Applied Optics,* v.36, p.4932, 1997.
70. NIF Laser System Performance Rating. /*Supplement to Proceedings of SPIE,* v.3492, 1998.
71. Boyd R.D., Britten J.A., Decker D.E. et al. *Appl. Optics,* v.34, p.1697, 1995.
72. Nisoli M., De Silvestri S., Svelto O. et al. *Optics Letters,* v.22, p.522, 1997.
73. Andreev A.A., Vinokurova V.D., Shatsev A.N., *Optics and Spectroscopy*, v.85, no.2, p.259, 1998.
74. Malkin V.M., Shvets G., Fish N.J., *Physics of Plasmas*, v.7, no.5, p.2232, 2000.

Chapter 3

APPLICATIONS OF ULTRA-STRONG LASER FIELD

INTRODUCTION

The use of super-intense laser radiation provides new opportunities for investigating the interaction of ultra-strong laser fields with matter and opens up new avenues in this branch of physics (see **Fig 3.1.1**). This applies particularly to the creation of such states of a medium that the relaxation times of any of its parameters exceed the duration of a laser pulse and, consequently, it is possible to investigate the fundamental properties of matter in such states. Moreover, it is now feasible to generate electric fields which have intensities in matter of the order of or even considerably greater than the intra-atomic intensity E_a. Such fields can be generated by a linearly polarized laser beam of intensity:

$$I_a = \frac{c|E_a|^2}{8\pi} = 3{,}4 \times 10^{16}\ W \cdot cm^{-2}.$$

The availability of laser beams of such intensity has led to a new topic in physics: nonlinear atomic (electron) physics. There is major interest in the ionization of matter by such fields and in the creation of ions with a high ionic charge [1,2], and also in studies of nonlinear transient processes in low- and high-density gases (generation of the radiation super-continuum and of higher harmonics [3-6], and changes in the frequency and pulse duration [7]) at intensities below the ionization threshold under conditions of fast (occurring within several periods of the field) ionization [8]. There is also interest in related topics of the special properties of low-density plasma created by field ionization [9]. One of the main applications of such a plasma is the formation of active media for x-ray lasers (with the collision excitation mechanism) [10-11].

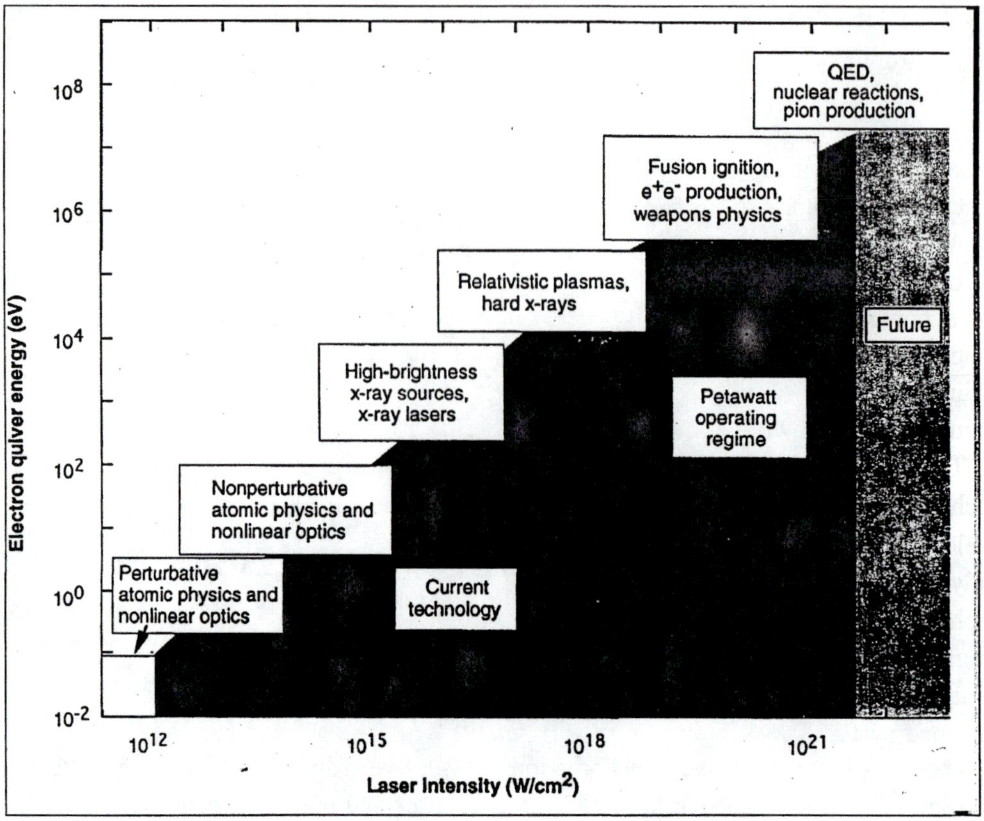

Fig. 3.1.1. Super-strong laser fields applications

At laser radiation intensities $I > 10^{17}$ W cm^{-2}, when the acquired velocity of the electron oscillations υ_E is higher than the thermal velocity υ_T, a new physical object can be created: this object is a high-temperature picosecond over dense laser plasma, subjected to high-contrast laser pulses, in which electrons do not manage to transfer any significant energy to ions during the plasma lifetime. Therefore, there is hardly any heating of ions or plasma expansion during a laser pulse and a hot plasma with a density of the same order as in solids can be generated (see, for example, Ref. [12,13]). Such an over dense plasma is of interest as a source of ultra-short pulses of fast particles and of x-ray radiation [14,15], and as a medium suitable for a recombination x-ray laser [16-19]. At such laser intensities it is possible to accelerate particles in an under dense plasma to ultrahigh energies [20-22] and the acceleration gradient can be ~ GeV m^{-1}, exceeding the corresponding gradient in conventional accelerators.

A further increase in the intensity above 10^{19} W cm^{-2} makes it possible to reach the next physical threshold when the energy of electron oscillations in the field of an electromagnetic wave becomes equal to the electron rest energy. This situation corresponds to the physics of relativistic (nuclear) laser plasma when the electron energy acquired from the laser beam exceeds 1MeV and the laser field can influence directly the state of the nuclei. This applies particularly to the nuclear reactions involving electrons [23].

There have been several recent investigations of new methods of laser thermonuclear fusion, which promise a considerable reduction in the laser energy needed to 'ignite' the fuel and to increase the thermonuclear gain of a target [24-26]. In all these methods it is necessary first to condition the thermonuclear fuel and then ignite rapidly a small proportion of the fuel. The improvement in respect of the energy can then be considerable ('fast ignition' method).

At still higher laser intensities ($I > 10^{20}$ W cm^{-2}) the processes of excitation of nuclei and of nuclear reactions by the direct action of a strong field become probable [27], so that a considerable number of excited nuclei can be created. Selection of the target composition and irradiation conditions can be used to form, from such excited nuclei, a medium with an inverted population of the nuclear levels, similar to that in a recombination x-ray laser [28].

The next physical threshold and, consequently, a new avenue in laser physics, is reached at intensities $I > I_Q = (c/4\pi)(mc^2/\lambda_c) = 10^{26}$ W cm^{-2} when the energy acquired by an electron in one Compton length λ_c exceeds its rest energy. At such intensities one can study the phenomena of nonlinear (laser) quantum electrodynamics, such as the scattering of photons by electrons in ultra-strong laser fields of intensity up to (10 - 100) E_a. The intensity of laser pulses required for such investigations can be reduced by utilizing collisions between a laser beam and a beam of ultra-relativistic electrons [29]. In a reference system in which an electron is at rest the electric field is then γ times higher than the field in the laboratory system (γ is the relativistic factor). Therefore, in collisions of 50 GeV ($\gamma = 10^5$) electrons with a laser beam characterized by 10^{19} W cm^{-2} ($E = 10^{11}$ V cm^{-1}), the electric field in the reference system in which the electron is at rest can reach the stability limit of vacuum (10^{16} V cm^{-1}) when the formation of electron - positron pairs and several other nonlinear quantum electrodynamic phenomena become possible, in accordance with theoretical predictions [30].

If it were possible to generate laser radiation of intensity in excess of 10^{23} W cm^{-2}, then collisions of beams of this kind could create a strong field that of itself could have a significant influence on the state of physical vacuum. The polarization of vacuum in an ultra-strong electromagnetic field converts it into a medium, which can be regarded as having a definite refractive index [31]. Then, for example, the passage of a fast electron through this medium would generate vacuum Cherenkov radiation and the interaction of three laser beams would result in four-wave scattering [32].

3.1 Absorption of High Intensity Laser Pulse in Over-Dense Plasma

We shall begin by reviewing the mechanisms of absorption of laser radiation in dense, hot plasma, as the transfer of laser energy to different channels of interaction (see **Fig. 3.1.2**) depends on this.

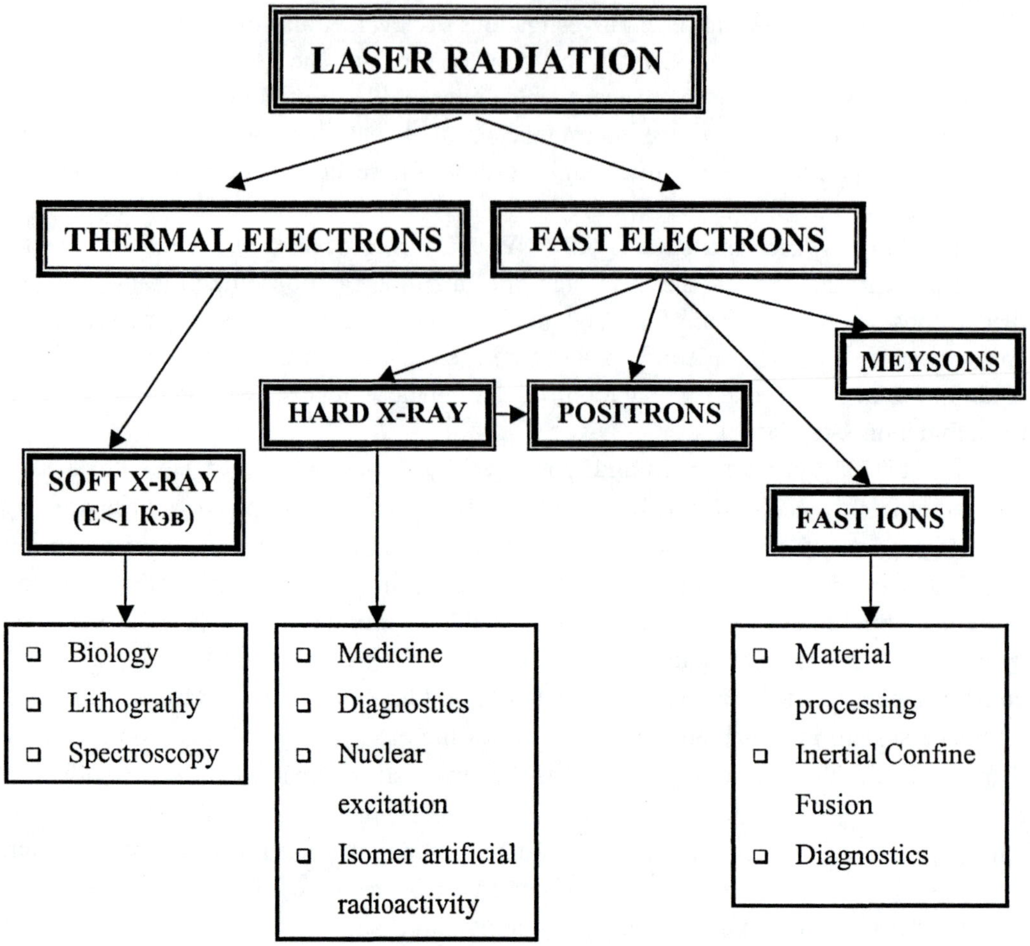

Fig. 3.1.2. Energy transformation of laser radiation

As is well known, for laser intensities I greater than 10^{15} W/cm^2, electronic temperature rises rather quickly $T_e \sim I^{1/2}\omega t$, so collision absorption becomes ineffective $\nu_{ei} \sim I^{-3/4}t^{-3/2}$ [28]. Besides, the oscillation velocity of electrons becomes comparable to their thermal velocity, which also reduces the effective collision frequency [34]:

$$\nu_{ef} \approx \nu_{ei}\frac{\upsilon_T^2}{(\upsilon_E^2+\upsilon_T^2)^{3/2}}$$

Thus, at intensities higher than 10^{16} W/cm^2, the collision-less mechanisms of absorption begin to be significant.

In our past experiments [35], with laser intensity less than 5x10^{17} W/cm^2, resonant absorption plays a basic role [36], modified by the mechanism of plasma wave breaking [37,38]. In this case, electrons - accelerated by a plasma wave field in vacuum - are reflected from an ambipolar barrier, accelerating the ions in vacuum, and returned to the target, heating it to a depth of their free path.

The absorption coefficient in this case is approximately [39]:

$$\eta_r \approx \frac{\sin^2\theta}{\cos\theta} kL,$$

here θ - angle of incidence to the target, $k = \omega/c$, L - scale of plasma density inhomogenity in vicinity of critical concentration - $n_c = \omega^2/4\pi e^2$.

In the case where the amplitude of electrons in the field of a laser wave exceed L $\upsilon_E/\omega > L$, (here $\upsilon_E = \frac{eE}{m\omega}$, oscillation velocity of electrons), we have Brunel absorption [40]:

$$\eta_b \approx a\frac{\sin^3\theta}{\cos\theta} kr_E \qquad a = 3 \text{ for } \eta_b << 1$$

$$a = 0.4 \text{ for } \eta_b \sim 1$$

For normal skin effect, an electron oscillates in a laser field and absorbs the energy of its electromagnetic field when it collides with an ion $\eta_n \sim kl_s$. This is possible, if $\frac{\upsilon_T}{\nu_{ei}} = l_{ei} < l_s$. When T_e increases, we obtain $l_{ei} > l_s : \upsilon_T/\omega > l_s$. Given these requirements, the laser field penetrates a length into plasma l_{sa}. Substituting $\nu_{ef} = \upsilon_T/l_{sa}$ into the equation for the length of the skin-layer (for $\nu_{ef} \geq \omega$) $l_s = \frac{c}{\omega_p}\left(\frac{\nu_{ef}}{\omega}\right)^{1/2}$, we deduce, that the field penetrates to the thickness of the anomal skin-layer: $l_{sa} = \frac{c}{\omega_p}\left(\frac{\upsilon_T\omega_p}{c\omega}\right)^{1/3}$. The absorption coefficient again:

$$\eta_a \sim k_0 l_{sa} \sim \alpha_a^{2/3}(\upsilon_T/c), \quad \alpha_a \equiv \frac{c\omega}{\upsilon_T\omega_p}.$$

For $\alpha > 1$, $l_s > \upsilon_T/\omega$ we have the regime of anomal skin effect at high frequency, in this case $\eta_{SIB} \sim \alpha_a^{-\gamma}$.

In the case of high temperatures and short laser pulses for the anomal skin effect conditions the absorption coefficient is as follows [41]:

$$\eta_a \approx \frac{2.8kl_s}{\cos\theta},$$

For the temperature of fast electrons we have the following approximation [42,43] (see **Fig.3.1.3a**): $T_h \approx 30(I_{17}\lambda^2_{\mu m})^{1/3}\left(T_{c,keV}\right)^{1/3} \quad [KeV]$.

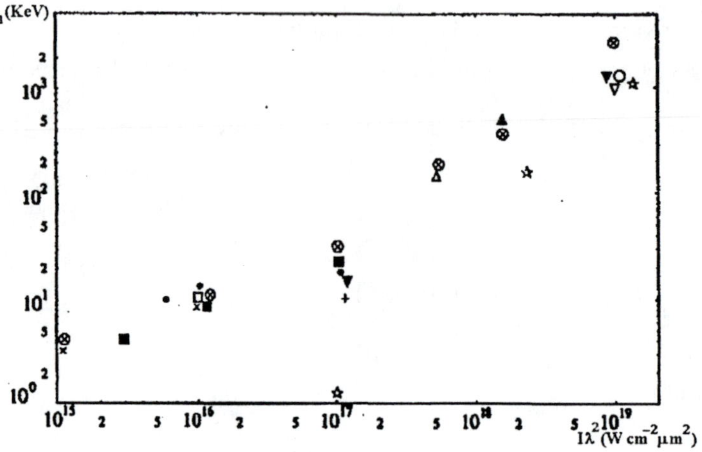

a)

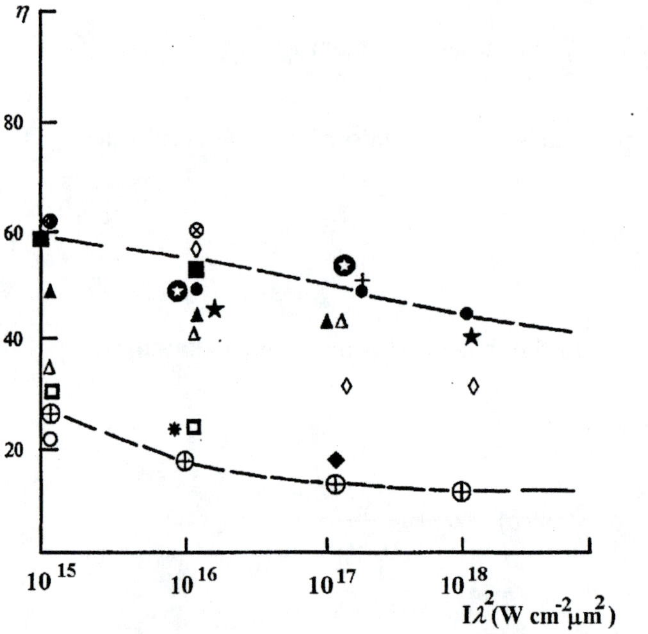

b)

Fig. 3.1.3. a) Dependence of the temperature of fast electrons on the product $I\lambda^2$.
b) Dependence of the absorption coefficient of laser radiation in Al plasma, on the product $I\lambda^2$, for P and *S*-polarized radiation.

The movement of ions must be taken into account at times $t > \frac{\upsilon_T}{\omega c_s}$ [44]. We should note, that even for subpicosecond laser pulses the plasma does not have a sharp boundary, because even laser pulses with high contrast ratio have enough energy in pre-pulse to create a plasma density gradient. Though the scale of plasma inhomogeneity in this case $L < \lambda$, however, for other scales it can be even less, the absorption is determined by L: $\eta_L \propto kL$.

At sub-relativistic intensities, when the difference in absorption for S- and P-polarized waves decreases, because of strong ponderomotive pressure, for an absorption coefficient it is possible to write the following expression [45]:

$\eta_c \approx 16\,\pi\, a_i^2\, x^3\, ln^2(1/2x)$, where $a_i = \upsilon_E/c < 1$, $x = kL < 1$

At ultra relativistic intensities $\upsilon_E > c$, the laser pulse acts on the plasma, like a piston accelerating ions by pressure $P_r = (I/c)\,(1 + R)$.

When ions are moving under the action of light pressure the expression for the velocity of the ionic front $\upsilon_i = 2\; 10^{-3}\; c\; I_{18}^{1/2}$ can be written from the law of conservation of mass [26]. Here I_{18} laser intensity has units of 10^{18} W/cm^2. In this case absorption coefficient $\eta_i = 3\; 10^{-3}\; (n_c/n_e)^{1/2} k_0 r_E \propto \upsilon_E/\omega$, as well as in the case of Brunel absorption.

These approximations were tested by simulation using one-dimensional kinetic code "KINET" [46], and the results of calculation and scaling agreed with experimental data [47,48] (see **Fig.3.1.3b**), although there were some difference between calculated and experimental data, qualitatively they agreed well. We will discuss experimental and simulation results for absorption coefficient in detail later.

The important effects breaking the one-dimensional pattern, are "hole boring" [26] and rifling of a surface [49]. Both these effects arise from cross inhomogeneity of a laser beam at relativistic intensities and were discovered in simulations using 2D and 3D PIC codes. "Hole boring" arises from the pressure of light $P_r = 2I/c = 0.6 I_{18}$ Gbar can exceed pressure of plasma $P_l = 160(n_e/10^{23}\,cm^{-3})T_e$ Mbar, and the plasma will move inside the target. As the radial ponderomotive force (from transversal laser beam spatial shape) pushes out electrons from the center of the beam, because of the ambipolar field the ions follow the electrons and thus there is a "hole" in the plasma. We should also note that, at such intensities, length of skin layer increases from relativistic magnification of electron mass and decreasing the plasma frequency $\omega_{pe}^2 = \omega_p^2/\gamma$, where $\gamma = (1+\upsilon_E^2/2c^2)^{1/2}$ - the relativistic factor.

As a curved surface is formed, absorption and temperature of fast electrons increase, since the density gradient formed in parallel with the laser electric field. We should also note the generation of magnetic fields in such plasma because of density gradients and temperature $\nabla n_e \times \nabla T_e$ [50], and also current originating at a density gradient, because of temporal modification in a ponderomotive force [51] and surface currents of fast electrons [52]. A strong magnetic field changes requirements of absorption, connected with direct acceleration of electrons, excitation of plasma waves and curls [53,54].

We should note, that for obliquely incident P-polarized laser pulse, electronic jets are generated, and not twice per laser pulse period, as in case of normal incidence, but only one time because of the magnetic field, which arises from current spread along the surface of the plasma [55]. This magnetic field can change the resonant properties of plasma and consequently the dependence of absorption coefficient on the angle of incidence does not have a maximum, such as in the case of small intensities [41,56].

3.1.1 Theoretical Model for Hydro Simulations of Absorption

In the case of laser intensity less then relativistic one the laser plasma dynamics is studied using a one-fluid two-temperature Lagrangian hydro-code SKIN [57-59] including the electron and ion thermal conductivities, both natural and artificial ion viscosities, and the ponderomotive force impact on the plasma motion. The set of hydrodynamics equations including the mass, momentum and energy conservation laws is written with the Lagrangian mass coordinate $s = \int_{-\infty}^{x} \rho(x')dx'$ as follows [60]:

$$\begin{aligned}
&\frac{\partial}{\partial t}(1/\rho) = \frac{\partial}{\partial_s} U \\
&\frac{\partial U}{\partial t} + \frac{\partial}{\partial_s}(P_e + P_i + \mu_e + \mu_i) + \frac{\mathrm{Re}(\varepsilon)-1}{16\pi}\frac{\partial |E|^2}{\partial s} = 0 \\
&\frac{\partial x}{\partial t} = U \\
&\frac{\partial\left(\mathcal{E}^e + \mathcal{E}^{ion}\right)}{\partial t} + (P_e + \mu_e)\frac{\partial}{\partial s}U + \frac{\partial}{\partial s}q^e = (1/\rho)\left(-Q_{ei} - Q_r + Q_L + Q_f\right) \\
&\frac{\partial \mathcal{E}^i}{\partial t} + (P_i + \mu_i)\frac{\partial}{\partial s}U + \frac{\partial}{\partial s}q^i = (1/\rho)Q_{ei}
\end{aligned} \qquad (3.1.1)$$

where ρ and U are the plasma density and velocity; $P_{e,i}$ and $\mu_{e,i}$ denote the electron and ion pressures and viscosities; $\mathcal{E}^{e,i}$ are the electron and ion thermal energies per unit mass; and $q^{e,i}$ stands for the electron and ion heat fluxes. The symbol Q_{ei} represents electron-ion relaxation; $\mathcal{E}^{ion}$ is the energy utilized for the plasma ionization; E is the amplitude of the laser electric field and ε is a complex plasma dielectric constant; Q_r is the energy loss due to x-ray emission; Q_f is the energy deposed by fast electrons. The collision absorption Q_L of the laser radiation is expressed as

$$Q_L = \frac{\omega}{8\pi}\mathrm{Im}(\varepsilon)\left(|E_l|^2 + |E_\perp|^2\right),$$

where the indices l and $\perp$ denote the components of the laser electric field, parallel and normal to the target normal, and ω_o is the laser frequency. A one-dimensional planar description is usually sufficient for short pulse interactions with a solid target, as the diameter of the laser focus is much longer than the scale length of the plasma expansion during the laser pulse.

A simplified model of atomic physics is included in order to calculate the mean ion charge Z and the averaged ion charge squared $\overline{Z}^2$. The populations $N^{(z)}$ of the charge states Z in the plasma are described by a set of atomic rate equations:

$$\frac{\partial}{\partial t}\left(N^{(z)}/n_i\right)=\frac{\partial}{\partial t}y^{(z)}=\Gamma^{(z)} \qquad Z=1,2,...Z_{nucl} \tag{3.1.2}$$

where the overall rates

$$\Gamma^{(z)}=s^{(z-1)}y^{(z-1)}-\left(s^{(z)}+\alpha^{(z-1)}\right)y^{(z)}+\alpha^{(z)}y^{(z+1)}$$

$$\Gamma^{(z_{nucl})}=s^{(z_{nucl}-1)}y^{(z_{nucl}-1)}-\alpha^{(z_{nucl}-1)}y^{(z_{nucl})}$$

are expressed as the ionization $s^{(z)}$ and recombination $a^{(z)}$ rates for an ion in the charge state Z. The charge state of a completely stripped ion is Z_{nucl}. The population of neutrals is denoted by $N^{(0)}$ and expressed in terms of the normalization condition

$$n_i=N^{(0)}+\sum_{i=1}^{z_{nucl}}N^{(i)},$$

where n_i is the total concentration of ionized and neutral atoms (ion density). The rates of the collision ionization, the radiate and three-body recombination, taken from [61], include the depression of the ionization potential in dense plasma. The theory ADK [62] is used for the calculation of the rate of tunneling ionization by laser radiation. The populations of the ion charge states are used explicitly in the description of fast electron transport into the target. For a solid Al target, the tunneling ionization is important only in a very thin layer (a few Å) on the target expanding so fast that the collision ionization is inefficient. The plasma recombination is important for the simulation of experiments when a pre-pulse is present. Also, it has a serious effect on the x-ray emission by the target.

The radiation energy loss Q_r consists of the bremsstrahlung and recombination emission multiplied by the escape factor, so that only the part of the emission that reaches the plasma-vacuum boundary is included in the energy conservation.

In the present model, Maxwell's equations are solved for both S- and P-polarized laser light. As the description of the S-polarized laser radiation absorbed only collision mechanism is straightforward, we will concentrate on the P-polarized oblique radiation. When it is incident on inhomogeneous plasma, some of its energy is collision absorbed and some is transformed to a longitudinal electron plasma wave. The energy of the longitudinal wave is totally absorbed by the plasma, either collision or by Landau damping, or by wave-breaking in a nonlinear way. While the collision absorption of transverse and longitudinal waves heats thermal electrons, the remainder of the longitudinal wave energy is transferred to a group of electrons accelerated to high energies (fast electrons). The laser electromagnetic fields and absorption are calculated numerically by solving Maxwell's equations for hot plasma:

$$\frac{d^2H}{dx^2}-\frac{d\ln\varepsilon}{dx}\frac{dH}{dx}+\frac{\omega^2}{c^2}\left(\varepsilon-\sin^2\theta\right)H=-\frac{\omega^2}{c^2}\sin\theta\left(\varepsilon E_l+\sin\theta\, H\right)$$
$$3\beta^2\frac{\omega_{pe}^2}{\omega^2}\frac{d^2E_l}{dx^2}+2i\frac{\omega}{c^2}\hat{\Gamma}E_l+\frac{\omega^2}{c^2}\varepsilon E_l=-\frac{\omega^2}{c^2}\sin\theta\, H, \qquad (3.1.3)$$

where the plasma dielectric constant is $\varepsilon = \varepsilon(\omega, k=0) = 1-\omega_p^2/\omega(\omega+i\nu)$, ω_p ν are the plasma frequency and the effective electron-ion collision frequency, and $\beta_T^2 = k_B T_e/mc^2 \equiv \upsilon_T^2/c^2$ is a dispersion factor. Instead of the laser magnetic field B_z, we have used the quantity H, introduced by the following substitution

$$H(x)=B_z(x)+3\beta_T^2\sin\theta\int_{-\infty}^{x}dx'\frac{\omega_p^2}{\omega^2}\frac{dE_l(x')}{dx'}$$

These equations are solved together with four boundary conditions: two in the over-dense plasma and two at the plasma-vacuum boundary.

The methodology developed in [63] is used to calculate the collision frequency ν, which describes the collision fraction of the laser light absorption. Landau damping of longitudinal plasma wave, due to the acceleration of resonance electrons, is described precisely by integral operator:

$$\hat{\Gamma}E_l(x)=\int\gamma_L(k)F_k e^{ikx}dk=\int dk\,\gamma_L(k)\int\frac{dx'}{2\pi}E_l(x')e^{ik(x-x')},$$

where the damping rate $\gamma_L(k)$ is expressed by the electron distribution function $f(\upsilon)$.

When the plasma density is assumed to be linear and the effective collision frequency ν to be spatially constant, an analytical approximation for the longitudinal electric field E_l can be found. The field E_l is expressed through the field H at the critical surface x_c

$$E_l=-i\sin\theta\frac{L_c\omega}{\upsilon_T}H(x_c)\int_0^{\infty}d\tau\exp\left\{-L_c\omega\left[i\frac{x\tau}{\upsilon_T L_c}+\frac{i\tau^3}{\upsilon_T}+\frac{\nu}{\omega}\frac{\tau}{\upsilon_T}+\frac{\pi}{n_c}f_c\left(\frac{\upsilon_T}{\tau}\right)\right]\right\}$$

for any electron distribution f_c at the critical surface; the critical density is n_c. The approximate formula is in good agreement with numerical results for relatively long density profiles with the density scale lengths $L_c \geq 0.1\ \lambda$.

However, the solution of a set of integral differential equations in each time step of the hydrodynamic code is extremely time consuming. The complex approximation proposed in [64] has been used in the hydro-code instead of the integral operator for Landau damping. Here, an imaginary part is added to the dispersion coefficient β_T^2 to account for Landau damping

$$\beta_T^2 = \frac{k_B T_e}{mc^2} \frac{1}{1+i\nu_L}, \quad \nu_L = \frac{\xi_L}{4}\left(\frac{n_c}{n_e} - 1\right)$$

The Landau damping rate ν_L is zero for electron densities $n_e > n_c$ and it is high in a low density plasma. The coefficient ξ_L can be modified to ensure that the longitudinal wave is damped out entirely before it reaches the plasma-vacuum boundary. We have verified numerically that the errors introduced in the absorption efficiency and laser fields are insignificant, at least for the typical profiles of plasma parameters encountered here. The absorption efficiency scaling for density scale lengths less than the laser wavelength differs considerably [65] from the well-known scaling for an extensive plasma expansion.

The plasma resonance is limited nonlinearly by so-called wave-breaking [66] for high laser intensities at

$$\frac{\upsilon_E}{\upsilon_{T_e}} \geq \frac{9}{\Phi(q)} \left(\frac{L_c \upsilon_{T_e}^2}{\lambda^3 \omega_{pe}^2}\right)^{1/6}$$

Here, υ_{Te} and υ_E are the electron thermal velocity and the oscillation velocity of an electron in the laser wave in vacuum, respectively, $q = (\omega L_c/c)^{2/3} \sin^2\theta$ and $\Phi(q)$ is Ginzburg function. A detailed description of the wave-breaking mechanism is possible only in terms of a kinetic approximation, when the resonance is formed and destroyed in a small number of laser periods. The limitation of the resonance field is then described phenomenologically, permitting the calculation of effective resonance laser fields from the stationary wave equation. Wave-breaking leads to a nonlinear limitation of the electric laser field amplitude in the resonance.

When the resonance absorption is treated with no spatial dispersion, an effective collision frequency is introduced into the dielectric constant to account for the wave-breaking [67]. However, this should be modified by introducing an effective damping rate:

$$\gamma_B = \omega \frac{\upsilon_c}{c} \left[\left(1 + \frac{\omega L_c}{2c} \frac{\upsilon_c}{c}\right)^2 - 1 \right]^{-1/2}, \quad \upsilon_c = \frac{eH(x_c)}{m\omega} \sin\theta$$

into the equation for the longitudinal component E_l of a high frequency electric field.

The impact of wave-breaking on the laser fields is demonstrated in **Figure 3.1.4**. The vacuum heating [34] has been ignored here, as the plasma density scale length is large enough to meet the condition $\omega L_c \geq 0.3\,\upsilon_T$. Laser-induced surface waves may enhance the radiation absorption [68], but its contribution to the absorption of short laser pulses is unclear. Corrugations and imperfections of the target surface may dramatically enhance the absorption of laser radiation at normal incidence, when the absorption efficiency of a target with a perfect surface is rather low, ~ 10% [69]. The possible enhancement of laser absorption is much less important for *p*-polarized light because the absorption efficiencies are generally much higher (~ 50%).

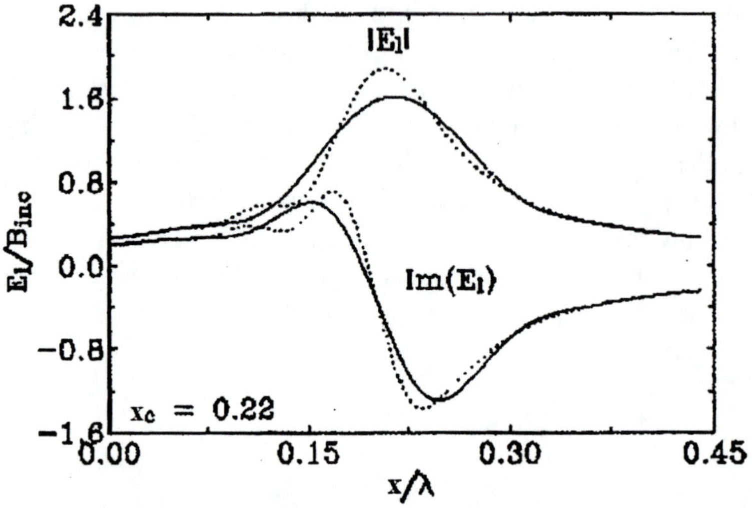

Fig. 3.1.4. Profiles of the longitudinal electric field E_l (the absolute and imaginary components) with wavebreaking (solid line) and without it (dashed line). The exponential plasma density profile with the scale length $L_c = 0.1\ \lambda$ and $T_e = 0.5$ keV; P-polarized Nd-laser pulse of intensity 8 10^{15} W/cm^2 is incident at 45°. The critical surface is located at x_c,

The acceleration of electrons in resonance absorption is treated in each time step in terms of a stationary electron diffusion [70] in the velocity space. The average diffusion coefficient $D(\upsilon)$ is expressed through the electron acceleration by the Fourier resonance component F_k of the longitudinal electric field

$$\upsilon\frac{\partial f}{\partial x}=\frac{\partial}{\partial \upsilon}\left[D(\upsilon)\frac{\partial f}{\partial \upsilon}\right], \tag{3.1.4}$$

where

$$D(\upsilon)=\xi_D\frac{e^2}{2m^2}\frac{1}{L\upsilon}\left|F_k\right|^2_{k=\omega/\upsilon}$$

and F_k is the Fourier transform of the longitudinal electric field. Electrons are assumed to be accelerated in a region of width $L \sim (L_c r^2_{De})^{1/3}$, approximately equal to that of the plasma resonance region, with r_{De} as the Debye electron radius. The factor $\xi_D \sim 1$ is iterated to achieve the exact matching between the laser radiation flux absorbed by fast electrons and the energy flux of the fast electrons.

When wave-breaking is absent and the Landau damping is described via the integral operator, we have exactly $\xi_D = 1$. The diffusion coefficient $D(\upsilon)$ for linear plasma density profiles can then be expressed as

$$D(\upsilon) = \frac{e^2}{2m^2}\frac{L_c^2}{L\upsilon}\left|H(x_c)\right|^2 \sin^2\theta \exp\left\{-\frac{L_c}{r_{D_e}}\left[\frac{2\pi\upsilon_T}{n_c}f(\upsilon)+\frac{\nu(\upsilon)r_{D_e}}{\upsilon}\right]\right\}$$

through the laser field H at the critical surface x_c.

Electrons are accelerated preferentially in the direction to the under-dense plasma. The energy of fast electrons matches the difference between the overall laser absorption and the integrated local collision absorption, so that energy conservation remains valid. We assume here a Maxwell distribution of electrons entering the critical region from over-dense plasmas with the local temperature T_e and concentration $n_e = \alpha_f n_c$. Typically, the coefficient is taken to be $\alpha_{fm} \sim 1.5$. The characteristic electron distribution function $f_A(\upsilon)$ of accelerated electrons at the vacuum side of the acceleration region is plotted in **Figure 3.1.5** for the maximum laser pulse.

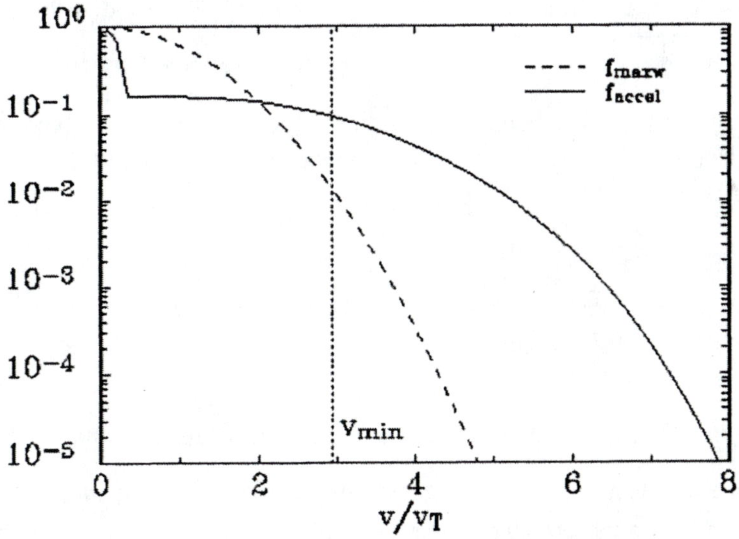

Fig. 3.1.5. The velocity spectrum f_A of electrons accelerated by resonance absorption and the Maxwell spectrum f_M of electrons incident from the over-dense plasma. The spectra are normalized, $f(\upsilon=0)=1$. Velocity υ_{min} is the minimum velocity of the distribution function δf_A for fast electrons. The electron spectra are for the maximum P-polarized Gaussian 400 fs FWHM Nd-laser pulse with peak intensity 10^{16} W/cm^2.

The electron distribution is split into the region of thermal electrons and that of fast electrons to provide a suitable description of the energy transfer to the target. Electrons are assumed to be fast if they provide the net gain in the energy flux density due to the collision-less absorption (Landau damping and wave-breaking). When the distribution f_M of the electrons incident onto the acceleration region is compared with the distribution f_A of the electrons leaving this region, there is a certain velocity υ_{min}, denoted by a vertical dotted line in Figure 3.1.5. This means that the total energy density flux of the electrons with $\upsilon \leq \upsilon_{min}$ in the distribution function f_A is equal to that in the original Maxwell distribution f_M. Thus, the distribution function δf_A of fast electrons is defined as

$$\delta f_A = 0 \quad for \quad \upsilon \prec \upsilon_{\min};$$
$$\delta f_A = f_A - f_M \quad for \quad \upsilon \succ \upsilon_{\min}$$

When fast electrons are reflected by the plasma-vacuum boundary, a fraction $\tilde{\eta}_i$ of their energy is lost for the acceleration of ions by the ambipolar electric field. For simplicity, $\tilde{\eta}_i$ is assumed to be independent of the electron velocity. The quantity $\tilde{\eta}_i$ is defined by the requirement of the energy conservation, and the total energy lost by fast electrons must be equal to the energy gained by fast ions.

The flight of fast electrons is a means of energy transport over relatively large distances, but no net particle flux is induced, as the return current substitutes fast electrons by thermal ones everywhere to ensure quasi-neutrality. The fast electron transport is described just as continuous slowing down, and the time of flight of fast electrons inside the simulation box is neglected. The distribution function $f_f(x_c,\upsilon) = \sqrt{1-\tilde{\eta}_i}\,\delta f_A\left(\upsilon/\sqrt{1-\tilde{\eta}_i}\right)$ of fast electrons reflected by the plasma-vacuum boundary is then used as a boundary condition for solving the problem of energetic electrons transport to the dense target. The stopping power of fast electrons is written in the Bethe-Bloch form [71]:

$$\frac{\partial \mathcal{E}_f}{\partial s} = \frac{1}{\rho}\frac{\partial \mathcal{E}_f}{\partial s} = -\frac{2\pi e^4}{\rho \mathcal{E}_f}\left[\sum_{z=0}^{z_{nucl}-1} Z_\upsilon^{(z)} N^{(z)} \ln\left(\frac{1.16\,\mathcal{E}_f}{\chi^{(z)}}\right) + n_e \ln\left(\frac{1.16\,\mathcal{E}_f}{\chi_o}\right)\right], \quad (3.1.5)$$

where $\mathcal{E}_f = m_e\upsilon^2/2$ is the kinetic energy of a fast electron; $Z_\upsilon^{(z)}$ and $\chi^{(z)}$ represent the number of valence electrons and the ionization potential of an ion in the charge state z, respectively; and χ_0 is the effective potential related to the stopping power by collective processes in the plasma. The distribution function f_f of fast electrons is calculated in each cell, and the electrons slowed down to and below the minimum speed υ_m (usually set to be equal to the local thermal speed) are assumed to be stopped within one spatial cell. The power Q_f per unit mass of thermal electrons heated by fast electron energy dissipation in a spatial cell is expressed as

$$Q_f = \frac{dI_f}{ds} = -\frac{d}{ds}\int_{\upsilon_m}^{\infty} \upsilon \mathcal{E}_f(s,\upsilon)\,d\upsilon = -\int_{\upsilon_m}^{\infty} \frac{d\mathcal{E}_f}{ds}\upsilon f_f(s,\upsilon)\,d\upsilon - \int_{\upsilon_m}^{\infty} \mathcal{E}_f \frac{d}{ds}\left[\upsilon f_f(s,\upsilon)\right]d\upsilon. \tag{3.1.6}$$

Here, I_f is the energy flux created by fast electrons and $f_f(x, \upsilon)$ is the distribution function of these electrons. The first term in the right-hand side of the equation represents the energy lost by fast electrons during their flight through a given cell, while the second term is the energy of fast electrons stopped there. The power Q_f, dissipated by fast electrons gives rise to a precursor of the thermal wave that heats the target. The energy of the fastest electrons which have escaped the simulation region into the target bulk is lost to the simulation. This fact is relevant to thick solid targets which do not practically reflect fast electrons back into the interaction region. The time of flight of fast electrons inside the simulation region is generally shorter than the laser pulse duration of ~1 ps, since the width of the simulation box is l_s<< 10 μm and the approximation used is that of instantaneous fast electron transport. The small portion of the fast electron energy utilized for K_α radiation has been neglected. The examples of calculations of the fields and density distributions have shown on **Fig. 3.1.6a,b**.

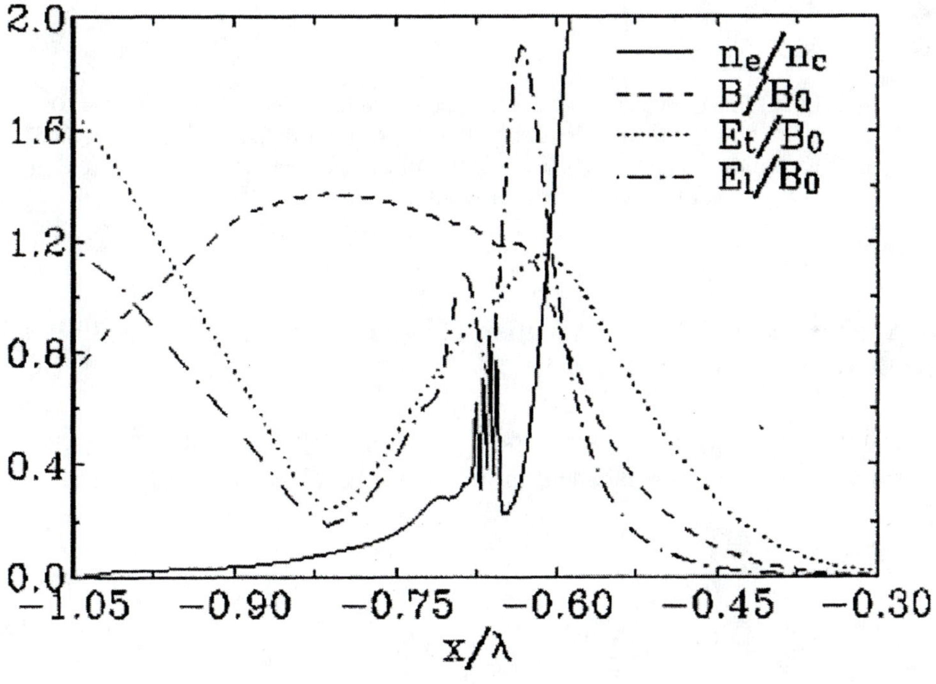

a)

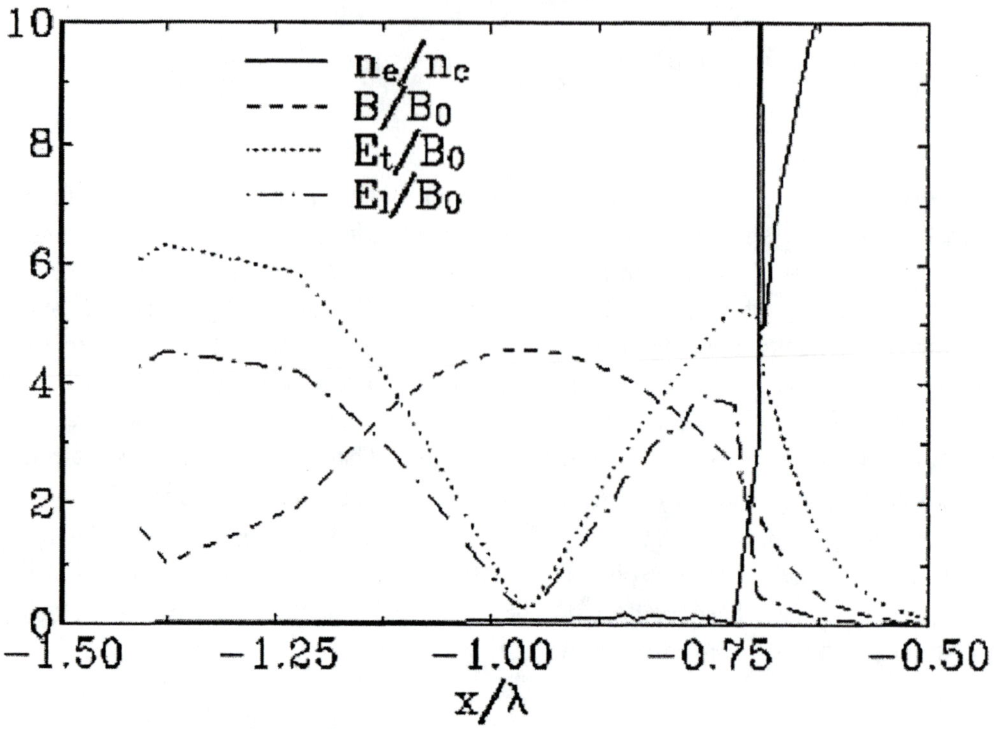

b)

Fig. 3.1.6. a) Profiles of the laser density and of the laser fields at the peak $I = 10^{16}$ W/cm^2 of a 1.5 ps Gaussian pulse of p-polarized Nd-laser radiation, incident at a 45° angle on a solid Al target. B_0 is the amplitude of the incident laser magnetic field in vacuum; b) the same for laser fields at the peak $I = 10^{17}$ W/cm^2.

3.1.2 Analytical Model for Absorption of P-Polarized Short Laser Pulses in Strongly Inhomogeneous Plasmas

For simplified analysis of laser absorption we developed next analytical model. Let us consider a plasma, occupying a half-infinite space (x≥0), with a linear electron density profile :

$$\eta_e(x) = n_e(x)/n_0 = \begin{cases} 1, & x \geq 0 \\ 1 + x/(LN), & -LN \leq x \leq 0 \end{cases} \tag{3.1.7}$$

where $N = n_0/n_{cr} = \omega_p^2/\omega^2 \geq 1$. The electromagnetic field is given by the following expressions:

$$E = (E_x, E_y, 0),\ H = (0,\ 0,\ H),\ E,H \sim \exp(-i\omega t + i\omega y \sin\theta / c)$$

The dielectric constant of the plasma is written in the familiar way as

$$\varepsilon = 1 - N\eta_e(x)/[1 + i\beta N\eta_e(x)]^{-1}, \tag{3.1.8}$$

where $\beta = \nu_c/\omega$ - is the electron collision frequency at critical density normalized to ω. As usual, we consider the magnetic component equation

$$\frac{d^2H}{dx^2} - \frac{1}{\varepsilon}\left(\frac{d\varepsilon}{dx}\right)\left(\frac{dH}{dx}\right) + \frac{\omega^2}{c^2}\left(\varepsilon - \sin^2\theta\right)H = 0. \tag{3.1.9}$$

Introducing a new variable $\xi = x/L + N + 1$ and substituting (3.1.8) into (3.1.9), we obtain :

$$\frac{d^2H}{d\xi^2} - \frac{dH/d\xi}{[\xi - i\beta(1+\xi)][1 + i\beta(1+\xi)]} - \alpha H\left[\frac{\xi - i\beta(1+\xi)}{1 + i\beta(1+\xi)} + S\right] = 0 \tag{3.1.10}$$

where $\alpha = (\omega L/c)^2$, $S = \sin^2\theta$.

We will seek for the solution of (3.1.10) on the assumptions $\alpha \ll 1$ and $\beta \ll 1$. It should be noted that one cannot expand H in terms of α directly in Eq.(3.1.10), since the last term in (3.1.10), proportional to α, may become rather large at $\xi \to \infty$. So we will solve Eq.(3.1.10) for the edge region ($\xi \cong 1$) and in the plasma bulk ($\xi \gg 1$) separately and then will perform the matching of the expansions obtained. Knowing the magnetic field H, one can easily obtain the electrical component E_y and the absorption efficiency η_p:

$$E_y = i\varepsilon^{-1}a^{-1/2}dH/d\xi,$$

$$\eta_p = 1 - \left|(\zeta - \cos\vartheta)/(\zeta + \cos\vartheta)\right|^2 \tag{3.1.11}$$

where $\zeta = (E_y/H)|_{\xi=-1}$ is the surface impedance.

Let us consider the region $-1 \le \xi \le \alpha^{-1/3}$, where the last term in Eq.(3.1.10) is small as compared with the other terms. Setting $\alpha=0$, we obtain a zeroth-order approximation to the "edge"-solution, which is valid for all values of β:

$$H_0 = 2D\left\{\frac{i}{\beta^2}arctg[\beta(1+\xi)] - \frac{\xi(i+\beta)}{\beta} + \frac{1}{2\beta^2}\ln\left[1 + \beta^2(1+\xi)^2\right]\right\} + B \tag{3.1.12}$$

where D and B are undefined constants. Treating β as a small parameter in the same way as α and substituting the expansion

$$H = H_0(\xi) + \alpha H_\alpha(\xi) + i\beta H_\beta(\xi) + ... \tag{3.1.13}$$

into (3.1.10), one can obtain the following result:

$$H_0 = D\xi^2 + B, \tag{3.1.14}$$

$$H_\alpha = B_1 - D/10 + DS/8 - B/6 + BS/4 + \xi^2(D_1 + D/6 - DS/4 + B/2 -$$

$$BS/4) + (BS/2)\xi^2 ln\xi + (B/3)\xi^3 + (DS/8)\xi^4 + (D/15)\xi^5 + \xi^2 \begin{cases} 0, & \xi \le 0, \\ i\pi BS/2, & \xi \ge 0 \end{cases} \tag{3.1.15}$$

$$H_\beta = B_2 - 2D\xi + D_2\xi - (2/3)\, D\xi^3, \tag{3.1.16}$$

where D_1, D_2, B_1 , B_2 are constants to be defined by matching the "edge" and "bulk" solutions. The capability of the matching procedure is provided by the assumption $\alpha \ll 1$, because in this case the edge region can overlap the bulk one. Substituting (3.1.15)-(3.1.16) into (3.1.11), we obtain the expression for the impedance :

$$\zeta = 2i\alpha^{-1/2}(D + \alpha D_1 + i\beta D_2)(D + B + \alpha(D_1 + B_1) + i\beta D_2)^{-1} \tag{3.1.17}$$

Turning to the solution of Eq.(3.1.10) in the limit $\xi \gg 1$, we start with the case , when the relations:

$$\alpha \gg \beta^3 \tag{3.1.18}$$

and

$$\alpha \gg N^3 \tag{3.1.19}$$

are valid.

The last condition implies, as will be shown later, that the laser wave amplitude vanishes within the region of inceasing density. Introducing a new variable $x = \alpha^{1/3}\xi$ and taking into account (3.1.19), one can rewrite Eq.(3.1.10) as

$$x(d^2H/dx^2) - dH/dx - x^2H = \alpha^{1/3}SxH - i\beta\alpha^{-1/3}x(dH/dx + x^2H), \tag{3.1.20}$$

Substituting the expansion

$$H = h_0(x) + \alpha^{1/3}Sh_\alpha(x) + i\beta\alpha^{-1/3}h_\beta(x) + ...,$$

where $\alpha^{1/3}$ and $\beta\alpha^{-1/3}$ are new small parameters, into Eq.(3.21) and using the boundary condition $H(x=\infty)=0$, we obtain

$h_0 = xK_{2/3}(u),$ (3.1.21)

$$h_\alpha = (2/3)^{2/3} x\left\{K_{2/3}(u)\left[C_1 - \int_0^u t^{1/3} I_{2/3}(t)K_{2/3}(t)dt\right] + I_{2/3}(u)\int_\infty^u t^{1/3} K_{2/3}^2(t)dt\right\},$$

(3.1.22)

$$h_\beta = (2/3)^{4/3} x\left\{K_{2/3}(u)\left[C_2 + \int_0^u t^{-1/3} I_{2/3}(t)\left((3/2)tK_{2/3}(t) + \left(1+9t^2/4\right)K_{2/3}(t)\right)dt\right] + \right.$$
$$\left. + I_{2/3}(u)\left[\int_u^\infty t^{-1/3}\left(1/2+9t^2/4\right)K_{2/3}^2(t)dt - 3u_{2/3}^2(u)/4\right],\right.$$

(3.1.23)

where $u=(2/3)x^{3/2}$, C_1, C_2 are constants. Evidently, condition for (3.1.22)-(3.1.24) is $x(z=0) > 0$, which gives the above relation (3.1.20). Expanding expressions (3.1.22)-(3.1.24) in the limit $x \ll 1$ and matching the obtained series with the "edge"-solution (3.1.15)-(3.1.17) by the use of relations

$$H_0(\xi \gg 1) = h_0(x \ll 1),\ H_{\alpha(\beta)}(x \ll 1),$$

we obtain the desired values of the constants :

$$D = -(3^{1/3}/4)\Gamma(1/3)\alpha^{2/3},\ B = (3^{3/2}/2)\Gamma(2/3),\ C_1 = C_2 = B_2 = 0,$$

$$D_1 = -\frac{B(1+i\pi S)}{2} - S\left[\frac{q_0}{6^{1/3}}\Gamma(2/3) + \frac{\Gamma(2/3)\left(\gamma + \ln\left(2\alpha^{1/2}/3\right)\right)}{2\cdot 3^{1/3}}\right],$$

$$B_1 = \frac{B(1-3S/2)}{6},$$

$$D_2 = \frac{\alpha^{1/3}\left(3\Gamma(1/3)/40 + q_1/\left(3\cdot 2^{2/3}\right)\right)}{\Gamma(2/3)},$$

where $\gamma=0.577$ is the Euler constant ,

$$q_0 = \int_0^\infty t^{1/3}\left[K_{2/3}^2(t) - \Gamma^2(2/3)\left(2t^2\right)^{-2/3}\exp(-t)\right]dt \cong -0.31,$$

$$q_1 = \int_0^\infty t^{-1/3}\left[K_{2/3}^2(t) - \Gamma^2(2/3)\left(2t^2\right)^{-2/3}\right]dt \cong -4.29.$$

in view of the above written expressions, relation (3.1.17) leads to

$$\zeta \approx 2i\alpha^{-1/2}\frac{D}{B}\left[1+\frac{\alpha D_1 + i\beta D_2}{D}\right] \approx -1.38i\alpha^{1/6}\left[1+\alpha^{1/3}\left(0,72-\right.\right.$$
$$\left.\left.-S\left(0.21+0.24\ln\alpha\right)\right)\right]+\pi\alpha^{1/2}S+0.38\beta\alpha^{-1/6}. \tag{3.1.24}$$

The real terms of (3.1.24) are responsible for resonant and collision parts of the total absorption efficiency. The absense of the angular dependence in the last term of (3.1.24) implies the equality of collisional absorption efficiency for the S- and P-polarized waves in the considered approximation. The expression for Reζ at $\theta = 0$ agrees with the result η = 0.5$\beta(L/\lambda)^{-1/3}$, empirically obtained [72]. Using (3.1.12) and (3.1.25) we can write a simplified expression for the resonant absorption efficiency:

$$\eta_r \cong 8\pi^2(L/\lambda)\sin^2\theta\cos\theta\left[\cos^2\theta + 6.4(L/\lambda)^{-2/3}\right]^{-1} \tag{3.1.25}$$

The maximum value of η_r occurs at $\theta_m = \arccos\left[2.5(L/\lambda)^{1/3}\right]$and is roughly equal to 15.5(L/λ)2/3. Let us note, for the sake of comparison, that for a long scale length of the density profile, the asymptotic solution yields: $\theta_m = \arcsin\left[0.35(L/\lambda)^{-1/3}\right]$ and $\eta_r \infty (L/\lambda)^{2/3}\sin^2\theta$ in the limit $(L/\lambda)^{1/3}\sin\theta \to 0$.

The angular dependence of the absorption coefficient given by (3.1.12), (3.1.25) is illustrated in **Fig. 3.1.7** for N=200, β=0.04, L/λ=0.01 and compared with the corresponding numerical result. One can see that the agreement between them is quite reasonable. The absorption efficiency as a function of L/λ is depicted in **Fig. 3.1.9a** for the same values of N and β. It is notable that in this specific case the theoretical curve is consistent with the numerical one even up to $L/\lambda \cong 1$.

Now we consider the case when the major part of the collisionally absorbed energy deposition occurs in the uniform lasma density region (x≥0). This leads to the following validity condition (opposite to (3.1.20):

α<< N-3.

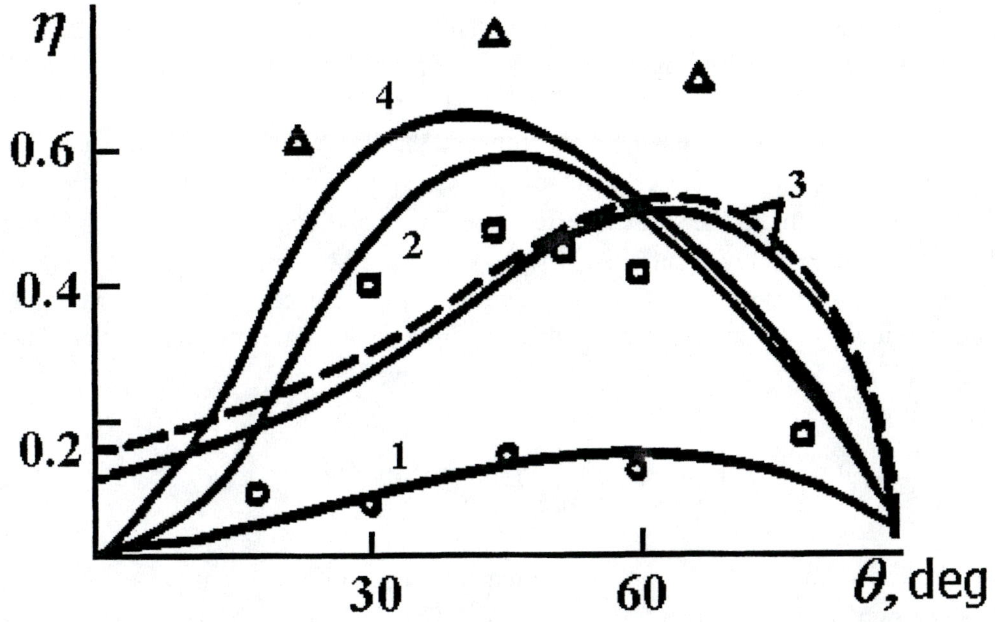

Fig. 3.1.7. Laser absorption versus laser incidence angle θ. Unshaded circles (L/λ=0.01) and squares (L/λ = 0.07) show the results of the simulations for N=2 and β = 0. Curves 1 and 2 display the results of the corresponding analytical calculations. Triangles represent the experimental results for N 50 and L/λ 0.2. Curve 4 shows the result simulations corresponding to the experimental data. Curve 3 display the results of analytical (the solid line) and numerical (the dashed line) calculations of absorption for N = 200, β =0.04 and L/λ=0.01.

Equating the solution of Eq.(3.1.9) in the region x≥0 to the "edge"-solution (3.1.15)-(3.1.17), we obtain :

$$\zeta \approx \pi\alpha^{1/2}S + \frac{\beta N^2(N-\cos 2\theta))}{2(N-1)^{-2}(N-\cos^2\theta)^{-1/2}} - i\frac{(N-\cos^2\theta)^{1/2}}{N-1}\cdot$$
$$\left\{1+\alpha^{1/2}\left[\frac{(N-\cos^2\theta)^{1/2}}{2(N-1)} + \frac{(N-1)(N+\cos^2\theta)}{2(N-\cos^2\theta)^{1/2}} + \frac{(N-1)S\ln(N-1)}{(N-\cos^2\theta)^{1/2}}\right]\right\}. \quad (3.1.26)$$

At $\alpha \to 0$, this expression approaches a weakly collisional limit of the Fresnel absorption coefficient. The resonant part of the laser-plasma coupling efficiency now reads as

$$\eta_r = 8\pi^2(L/\lambda)\sin^2\theta\cos\theta\left[\cos^2\theta + (N-\cos^2\theta)(N-1)^{-2}\right]^{-1} \quad (3.1.27)$$

One can see that η_r is proportional here to L/λ at all values of incidence angle in contrast to that obtained for $\alpha >> N-3$ (see Eq.(3.1.26)). Maximum value of η_r when $N>>1$ (but $N << \beta - 1$) occurs at $\theta \cong \arccos N^{1/2}$ and is roughtly equal to $4\pi^2(L/\lambda)N^{1/2}$.

The θ and L/λ dependences of the absorption efficiency, given by (3.12) and (3.27), are plotted in Fig.3.1.7 for the parameters of the numerical simulation [73] (N=2, β=0). One can see that thetheoretical curves are in reasonable agreement with the data of [73] for $L/\lambda \leq 0.1$.

The numerical results qualitatively confirm the above behavior of theabsorption coefficient, in particular, the estimate of θ ($\theta_m \cong \pi/4 = \arccos N^{-1/2}$). Note that the laser-plasma coupling due to the Brunel's effect under the conditions of [62] ($\omega L/c \cong .08$) can play a dominant role only at sufficiently small values of L/λ, namely at $L/\lambda \leq 0.01$.

Similarly to the previous section, we first consider the case when the wave field vanishs within the region of varying density. Introducing a new variable, which now reads $x=(\alpha/\beta)^{1/2}\xi$, we can rewrite Eq.(3.1.10) as follows:

$$\frac{d^2H}{dx^2}+iH=(\alpha/\beta^3)^{1/2}(H/x-i\frac{dH}{dx}/x^2)-i\beta(1-S)H. \tag{3.1.28}$$

The condition x >>1, which eliminates the uniform density region from consideration, now takes the form $\alpha >> bN^2$. The neglect of the right-hand side of Eq.(3.1.28) gives us a zeroth-order solution in terms of β and (α/β^3), which appears to be identical to that for uniform density. In order to obtain the required "edge"-solution, one has to expand expression (3.1.14) in term of ($\beta\xi$)>>1 to give : $H_0 \cong -2iD\,\xi + B$.

Matching the appropriate zeroth-order solutions, we obtain, as could be expected, the same expression for the surface impedance as in the case of step-like density profile:

$$\zeta \cong (1-i)(\beta/2)^{1/2}. \tag{3.1.29}$$

This results from the cancellation of the functions $\eta_e(x)$ from the numerator and denominator of the expression (3.1.8) for ε in the collisional regime considered. The resonant part of the laser-plasma coupling efficiency is now a small correction of the order of (α/β)1/2 for the Fresnel absorption coefficient and can be written as follows:

$$\eta_r \cong 8\pi^2(L/\lambda)\sin^2\theta\cos\theta\left[\cos^2\theta+(2\beta)^{1/2}\cos\theta+\beta\right]^{-1}, \tag{3.1.30}$$

$$\eta_r(\theta_m) \cong 8\pi^2(L/\lambda)\beta^{-1/2}/(2+\sqrt{2}),\ \theta_m=\arccos\beta^{1/2}.$$

These formulas are, evidently, valid for the near step-like density profile case, with the condition $\alpha << bN^2$.

The parameter regions corresponding to the above studied absorption regimes, are depicted in **Fig. 3.1.8** in the plane of parameters ($\beta,\alpha^{1/2}$) for the case N = 40 given as an example. It is notable, that the regime of a weakly collisional step-like density profile is essentially limited to very short density scale lengths $L << \lambda N^{-3/2}$.

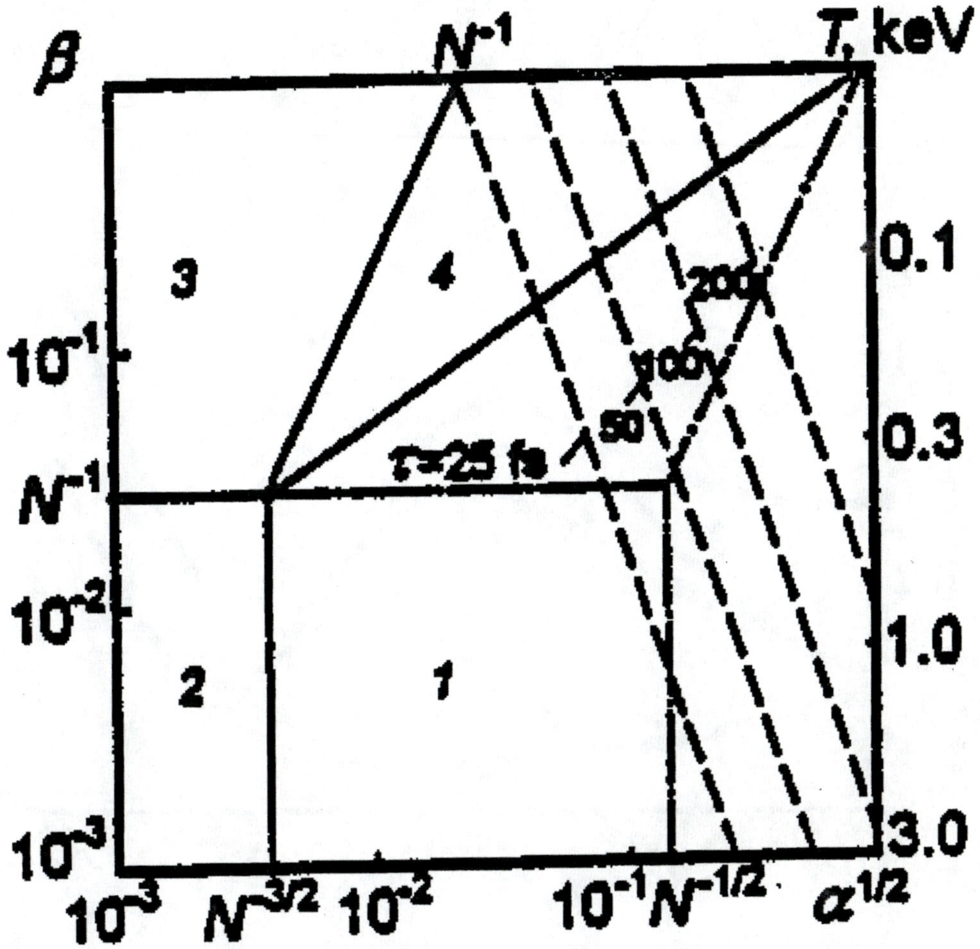

Fig. 3.1.8. Various absorption regimes of in parameter plane ($\beta,\alpha^{1/2}$) for the illustrative example N=40.

This is a specific feature of the linear density assumption. In the case of the exponential profile, the corresponding restriction is not so rigid and reads as $L/\lambda << N^{1/2}$. This is illustrated in **Fig. 3.1.9b**, where the L/λ absorption dependence for the exponential profile, numerically obtained in [72,73], is also shown. It should be noted that one cannot describe properly the absorption coefficient behaviour in the regime A and, simultaneously, in the regimes C and D (see Fig.3.1.8) with expression (3.1.8) for ε. When the major energy deposition occurs in the dense plasma region with $\beta N >> 1$, the effective collision frequency will be $\nu_c N(3\pi/32)$ instead of $\nu_c N$. This results from the more general expression for a dielectric constant, which, in the case of Maxwellian electron distribution and $Z >> 1$, reads [74]:

$$\varepsilon = 1 - (1/3)(2/\pi)^{1/2} N\eta(z) \int_0^\infty \frac{u^4 \exp(-u^2/2)du}{1 + 3(\pi/2)^{1/2} i\beta N\eta(z) u^{-3}},$$

where $\nu_c = (4/3)(2\pi)^{1/2} n_{cr} e^4 Z \ln\Lambda m_e^{-1/2} T^{-3/2}$

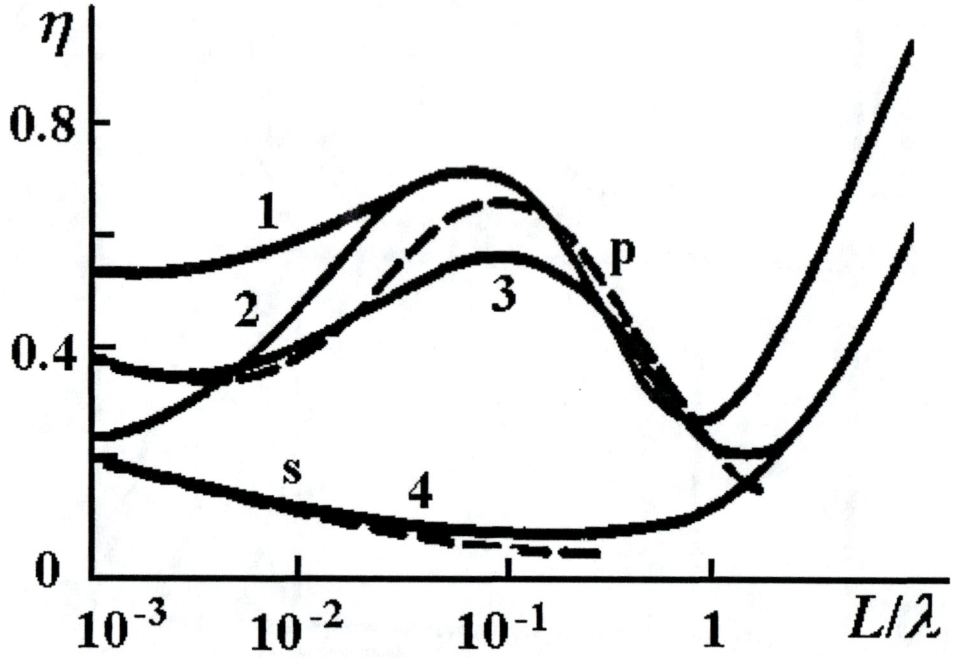

Fig. 3.1.9. Dependence of the absorption coefficient on L/λ at θ=45° for P polarized (1-3) and S-polarized (4) waves in media with the exponential (1 and 2) and linear (3) profiles of the density. The dashed lines show the results of calculations performed by using approximate formulas. The solid lines display the results of numerical simulations.

3.1.3 Experimental Results of Picosecond Laser Pulse Absorption in Hot Dense Plasma

Our experiments were carried out with a terrawatt laser setup in which a phase-modulated pulse was amplified and then its duration was reduced in a diffraction-grating compressor [75]. The maximum energy of a pulse compressed to 1.5 ps and with a contrast of 10^6 was limited to ~0.5 J by the optical strength and dimensions (110 mm x 110 mm) of the gratings. The angular divergence of the laser radiation was 10^{-4} rad., which made it possible to reach intensities of $q < 10^{17}$ W cm^{-2} on the target. A flat Al target was placed at the center of a vacuum chamber where the pressure was 10^{-4} Torr. The angle of incidence of the laser radiation on this target was 45° (**Fig. 3.1.10**). The experiments were carried out with S- and P-polarized laser radiation. This radiation was focused on the target by an aspherical objective (diameter 50 mm, focal length 120 mm). The thickness of this objective was 12 mm, so that it was possible to avoid nonlinear distortions. Consequently, 70% of the laser energy E_L reached a spot ~ 12 μm in diameter. The reflected optical energy was measured along the specular direction with an objective, characterized by $D/f = 1$ and located at a distance of 40 mm from the target, and with an IMO-2M calorimeter. The scattering pattern was investigated with light-sensitive photographic paper located on a circle of radius 30 mm around the target (see Fig. 3.1.10).

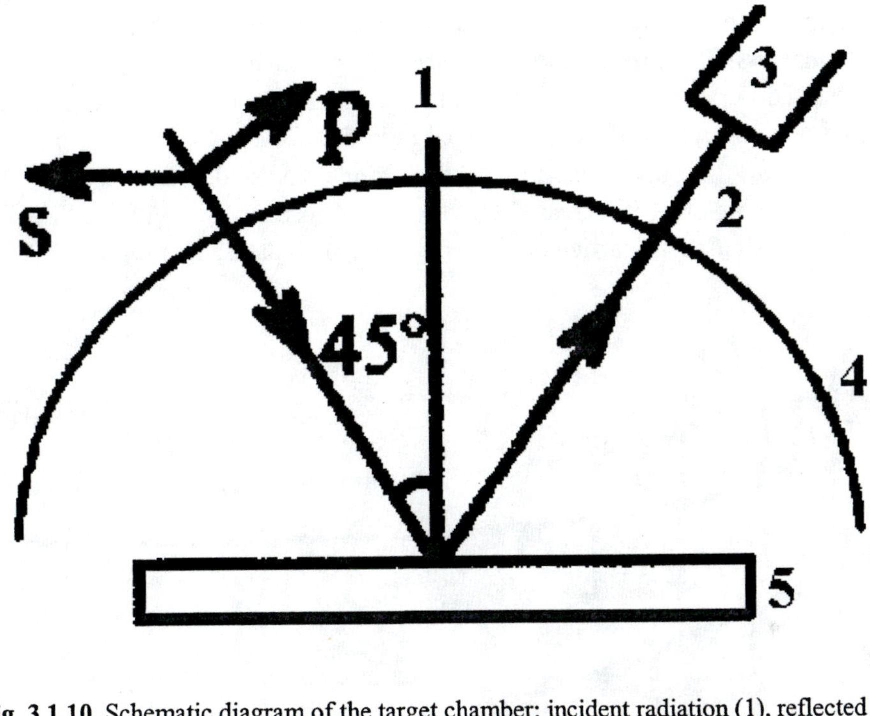

Fig. 3.1.10. Schematic diagram of the target chamber: incident radiation (1), reflected radiation (2) calorimeter or ion collector (or two-channel analyzer) (3), photographic film (4) target (5).

Under our conditions the oscillatory velocity of electrons inside dense plasma was less than their thermal velocity because of a steep plasma density gradient [22,23]. Therefore, the absorption was determined from an expression for the local permittivity. The time scales of changes in such hydrodynamic characteristics as the electron temperature (T_e) and density (n_e) were greater than for the electro-dynamic characteristics. Therefore, the absorption of laser radiation was calculated employing steady-state wave equations with given values of T_e and n_e. Since the length of the plasma formed in this way was in our case less than the diameter of the laser beam spot, we used the one-dimensional approximation realized in code SKIN. The results of our determination of the scattering diagram of the laser radiation incident at $\theta = 45°$ on a flat target revealed a considerable broadening of the specularly reflected component of this radiation in the plane of incidence on the target, which could be due to distortion of the reflection surface in the plasma caused by the ponderomotive pressure exerted by the intense laser radiation.

The results of determination of the reflection coefficient of the S- and P-polarized radiation by the target plasma are represented by the continuous curves in **Fig. 3.1.11**. We did not measure accurately the energy of the laser radiation scattered by the plasma into the angle 2π in front of the target. However, estimates indicated that the fraction of the energy of the diffusely reflected laser radiation did not exceed 2% of the energy incident on the target. The dependences of the reflection coefficients on the incident laser energy (Fig. 3.1.11) showed that the absorbed energy increased with increase in the incident energy of either polarization. However, the reflection coefficient for the S-polarized component was considerably greater than for the P-polarized radiation. The results of calculations by SKIN of the absorption under our experimental conditions are represented by the dashed curves in Fig. 3.1.11. Clearly, an increase in the laser energy increased somewhat the absorption of the P-polarized radiation. This was the result of an increase in the slope of the plasma density profile from 0.25 to 0.1 μm resulting from the ponderomotive pressure: it increased the resonant absorption (see **Fig. 3.1.9**, which gives the results of calculations carried out in accordance with the SKIN program).

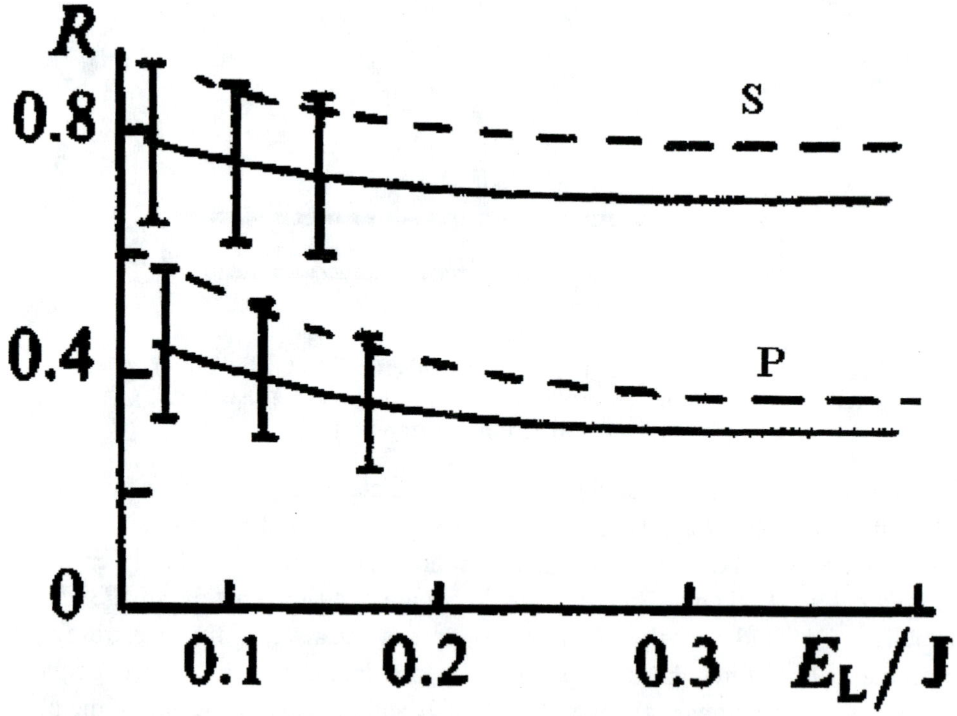

Fig. 3.1.11. Dependences of the experimental reflection coefficient of S- and P-polarized radiation (dashed curves are theoretical ones) from incident laser energy.

Calculations carried out for the S-polarized laser radiation, absorbed usually only as a result of bremsstrahlung, yielded the value of the absorption coefficient considerably less than the experimental one. However, under our experimental conditions the ponderomotive pressure could create a non-uniform distribution of the plasma density over the laser spot, which would give rise to a significant fraction of the P-polarized component in the initially S-polarized laser beam. Numerical calculations for the S-polarized radiation indicated that, when the energy of this radiation was 0.1 J, the absorption coefficient was 0.1 and the contribution of bremsstrahlung was ~0.01 in agreement with the experimental results.

A redistribution of the energy flux of the P-polarized radiation can be described as follows. In the case of a laser pulse of 100 mJ energy, about 60 mJ was absorbed in accordance with the resonant mechanism and only 2 mJ could be attributed to bremsstrahlung. Moreover, about 50% of the energy of fast electrons was converted into fast ions and the rest was lost in collision heating of thermal electrons (~ 30% of the laser energy) and in the emission of X rays (less than 2%).

We assumed that the distortion of the plasma surface by the S-polarized laser beam transferred about 20% of the energy to the P-polarized component and in this case our simulation results are closed to experiment ones (see Fig. 3.1.11). This energy was sufficient for the absorption of 12% of the laser energy, in accordance with the resonant mechanism, when the absorption coefficient was the same as before. Then, since 2% of the energy was absorbed because of bremsstrahlung and 12% in accordance with the resonant mechanism, the thermal electrons should receive ~ 8% of the laser energy. This was 4 times less than for the P-polarized laser radiation. Therefore, the thermal group of electrons was greater for the P-polarized laser radiation than for the S-polarized component, but in our experiments the difference was even greater. We also assumed that under our conditions there was a possibility of the excitation of surface wave, which would carry away the missing part of the laser energy.

3.1.4 Skin Effect for a P-Polarized Electromagnetic Wave in Inhomogeneous Plasma

Here we consider a typical situation when the hydrodynamic time of plasma expansion is greater than the laser pulse duration. In this case, the plasma density is the same as the density of irradiated matter, and the plasma ions can be treated as being immobile during the interaction. This section is concerned with the theory of the skin effect in inhomogeneous plasma, in particular, with the dependence of the absorption coefficient of laser radiation on the degree of anisotropy of the plasma temperature and the characteristic scale of plasma inhomogeneity for a wide range of this scale.

Basic equations

Consider an oblique incidence (at an angle θ to the plasma surface) of a P-polarized electromagnetic wave. The plasma is assumed to be inhomogeneous along the x-axis. The electric field of the wave, E, lies in the (y,x)-plane. The interaction is described by a set of equations including the Boltzmann equation for the distribution function $f(t,x,y,p_x,p_y)$

$$\frac{\partial f}{\partial t}+\mathrm{v}_x\frac{\partial f}{\partial x}+\mathrm{v}_y\frac{\partial y}{\partial y}+\left[e\left(E_a+E_x\right)-\frac{e}{c}\mathrm{v}_y B_z\right]\frac{\partial f}{\partial p_x}+ \\ +e\left(E_y+\frac{\mathrm{v}_x}{c}B_z\right)\frac{\partial f}{\partial p_y}=-\hat{\nu}\left(f-f_0\right) \tag{3.1.31}$$

and the Maxwell equations for the components of the fields E_x, E_y and B_z. These fields are the components of the external electromagnetic wave field and the ambipolar field E_a, which is related to the inhomogeneous profile of the ion density $n_i(x)$. We assume that the ion charge number is $Z > 1$. In this case, the frequency of electron-ion collisions is greater than that of electron-electron collisions by a factor of Z, and for smooth variations of $f(p)$, the collision integral $\hat{\nu}$ can be introduced in a form corresponding to the τ-approximation [74]. We consider a nonlinear interaction of a P-polarized wave with inhomogeneous plasma for the cases of high frequency anomalous and normal skin effects. The case when both regimes are important simultaneously is considered in [33,76-78] for the normal incidence of laser radiation on a sharp plasma boundary. The conditions under which these regimes can occur can be estimated as in [19,33,78]. For the anomalous skin effect, we obtain a limitation of $I \leq 60(A/Z)t_L^2\lambda^{-2}\ 10^{12}$ W/cm^2 on the laser radiation intensity and $t_L < 30(A/Z)^{1/2}\lambda$ fs on the laser pulse duration. Here, A is the atomic weight of an ion and λ is the radiation wavelength in microns. It follows from these estimates that this regime is possible at $I \leq 10^{17}$ W/cm^2, $t_L \leq 40$ fs, and $\lambda = 1$ µm, or after the laser pulse maximum. Similarly, for a high frequency normal skin effect, we obtain

$$I \leq (0.1Z)^{7/3}(n_i/6 \times 10^{22}\ \mathrm{cm}^{-3})^{5/3}\ \lambda^{2/3}[10^{18}\ \mathrm{W/cm}^2],\ \lambda \leq 0.5\ \mu\mathrm{m}.$$

In this case a laser with a shorter wavelength must be used but the pulse duration can be $t \sim 100$ fs.

We assume that the space (along the y-coordinate) and the time dependent on the laser field have the form $\exp(ik_y y - \iota\omega t)$. By linearizing the initial set of equations, one gets the following expressions for determining the unperturbed distribution function f_0 and its perturbation δf:

$$\mathrm{v}_x \frac{\partial f_0}{\partial x} + eE_a \frac{\partial f_0}{\partial p_x} = 0,$$

$$\frac{dE_a}{dx} = 4\pi e\left[\int f_0 dp - n_i(x)\right],$$

$$\left(-i\omega + \nu_{ei} + ik_y \mathrm{v}_y\right)\partial f + \mathrm{v}_x \frac{\partial \delta f}{\partial x} + eE_a \frac{\partial \delta f}{\partial p_x} =$$

$$= -\left(eE_x + \frac{e}{c}\mathrm{v}_y B_z\right)\frac{\partial f_0}{\partial p_x} - \left(eE_y + \frac{e}{c}\mathrm{v}_x B_z\right)\frac{\partial f_0}{\partial p_y},$$

$$\begin{aligned} &\frac{\partial^2 E_y}{\partial x^2} + \frac{\omega^2 E_y}{c^2} - i\frac{\omega}{c}\sin\theta \frac{\partial E_x}{\partial x} = -\frac{4\pi \;\; i\omega \;\; e}{c^2}\int \mathrm{v}_y \delta f d\mathbf{p}, \\ &-i\frac{\omega}{c}\sin\theta \frac{\partial E_y}{\partial x} + \frac{\omega^2}{c^2}\cos^2\theta \;\; E_x = -\frac{4\pi \;\; i\omega \;\; e}{c^2}\int \mathrm{v}_x \delta f d\mathbf{p} \end{aligned} \qquad (3.1.32)$$

The solution to the set of equations (3.1.32) is the Maxwell-Boltzman distribution with potential $E_a = -d\varphi/dx$ satisfying the Poisson equation:

$$\frac{d^2\varphi}{dx^2} = -4\pi e n_i(\infty)\left[\exp\left(\frac{e\varphi}{T_\parallel}\right) - \frac{n_i(x)}{n_i(\infty)}\right], \qquad (3.1.33)$$

with $n_i(\infty) = n_i(x)\,|_{z\to\infty}$. For convenience, our further analysis will deal with ion density profiles that satisfy equation (3.1.33); in other words, the ambipolar potential will be assumed to be the solution to this equation. Consequently, in the zero-order approximation, we find

$$f_0\left(x, p_y, p_x\right) = \frac{n_i(\infty)}{2\pi m\left(T_\perp T_\parallel\right)^{1/2}}\exp\left(-\frac{e\varphi(x)}{T_\parallel} - \frac{p_x^2}{2mT_\parallel} - \frac{p_y^2}{2mT_\perp}\right). \qquad (3.1.34)$$

There are the following reasons for choosing the distribution function. In the case of interest, the isotropization of the distribution function occurs over a time of about the characteristic time of electron-ion collisions $t_{ei}(n_c)$, where n_c is the critical density of ~ 30 fs. We consider laser pulses with durations that include $t_L < t_{ei}(n_c)$, which is natural because terawatt pulses with a duration of $t_L < 20$ fs are presently available [36]. Under these conditions, the distribution function may be slightly an-isotropic, since the particles can gain more energy by being accelerated along the electric field (than the energy gained along the transverse velocity component). Consequently, we have chosen the two-temperature Maxwell distribution function f_0, assuming $(T_\parallel - T_\perp)/T_\parallel$ 1. Note that the numerical modeling of a similar problem of laser pulse absorption [41,76] revealed an certain anisotropy of the distribution function under the conditions corresponding to the anomalous skin effect (see **Fig. 3.1.12**).

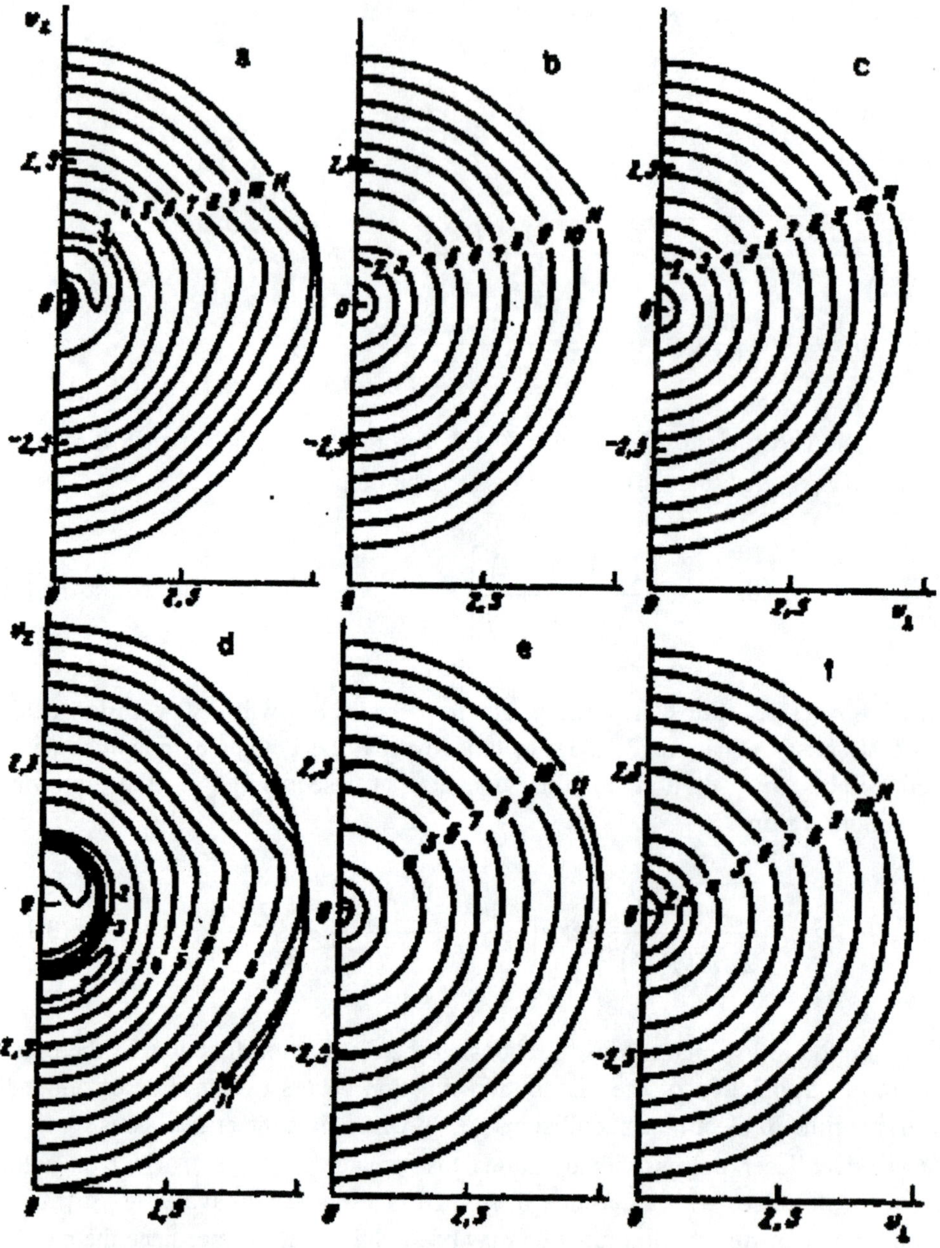

Fig. 3.1.12. Contour plots of the electron distribution function: a, d - near the plasma boundary, $x \approx 0$; b,c – at a distance $x/x_0 = 3$; e, f- at a distance $x \approx 6x_0$. a, b, c) at the time $t = 10t_0$; d, e, f) – 40 t_0.

The general solution to the equation for δf can be derived for an arbitrary profile $\varphi(x)$ using the method of characteristics. In a standard approach [74], we introduce δf as the sum of two functions, $\delta f = \delta f^+$ for $v_x > 0$ and $\delta f = \delta f^-(|v_x|)$ for $v_x < 0$. The potential is a smooth function satisfying the boundary conditions $\varphi(\infty) = 0$ and $\varphi(-\infty) = +\infty$ An electron moving in the field defined by this potential is rejected at the turning point x*, which is described by the equation

$$e\varphi(x^*) = e\varphi(x) + \frac{m v_x^2}{2}. \tag{3.1.35}$$

The first boundary condition for δf is obvious: $\delta f^- = \delta f^+|_{x=x^*}$. The second one can by found from the condition that the field does not penetrate deep into the plasma: $\delta f^-|_{x\to\infty}=0$. The kinetic equation for the function δf which satisfies the boundary conditions, specified above, has a single-valued solution. This solution allows us to express the plasma currents through the components of the electric field E_y and E_x as

$$\frac{\partial^2 E_y}{\partial z^2} + k^2 E_y - ik\sin\theta \frac{\partial E_z}{\partial x} =$$

$$= i\frac{\omega_p^2 k}{cn_0}\left\{\int_x^{\infty} dx' \int_{-\infty}^{+\infty} dp_y^2 \int_0^{\infty} \frac{dp_x}{\wp} e^{-\Phi(x,x')} \cosh\Phi(x',x^*) \times \left(E_y \frac{\partial}{\partial p_y} - \hat{L}\right) f_0(x',p_y,p_x) - \right. \tag{3.1.36}$$

$$\left. - \int_{-\infty}^{x} dx' \int_{-\infty}^{+\infty} dp_y^2 \int_{\wp_0}^{+\infty} \frac{dp_x}{\wp} e^{-\Phi(x,x')} \left[\cosh\Phi(x',x^*) E_y \frac{\partial}{\partial p_y} - \sinh\Phi(x',x^*)\hat{L}\right] f_0(x',p_y,p_x)\right\},$$

$$-ik\sin\theta \frac{\partial E_y}{\partial x} + k^2\cos^2\theta E_x = -i\frac{\omega_p^2}{cn_0} \times$$

$$\times\left\{\int_x^{\infty} dx' \int_{-\infty}^{+\infty} dp_y^2 \int_0^{\infty} \frac{dp_x}{\wp} e^{-\Phi(x,x')} \cosh\Phi(x',x^*) \times \left(E_y \frac{\partial}{\partial p_y} - \hat{L}\right) f_0(x',p_y,p_x) + \right.$$

$$\left. + \int_{-\infty}^{x} dx' \int_{-\infty}^{+\infty} dp_y^2 \int_{\wp_0}^{+\infty} \frac{dp_x}{\wp} e^{-\Phi(x,x')} \left[\cosh\Phi(x',x^*) E_y(x') \frac{\partial}{\partial p_y} - \sinh\Phi(x',x^*)\hat{L}\right] f_0(x',p_y,p_x)\right\}.$$

Here,

$$\hat{L} = E_x(x')\frac{\partial}{\partial p_x} + \left(\frac{i}{\omega}\frac{\partial E_y}{\partial x'} + \frac{\sin\theta}{c}E_x(x')\right)\left(v_x \frac{\partial}{\partial p_y} - v_y\frac{\partial}{\partial p_x}\right), \quad \omega_p^2 = \frac{4\pi e^2 n_0}{m},$$

$$\wp^2 = \frac{m v_x^2}{2} + e\left[\varphi(x) - \varphi(x')\right], \quad \wp_0^2 = e\left[\varphi(x') - \varphi(x)\right],$$

with $k = \omega/c$ and the function $\Phi(x_1,x_2)$ defined by the integral

$$\Phi(x_1,x_2) = \int_{x_1}^{x_2} \frac{i\omega\left(1+\frac{i\nu_{ei}}{\omega}-\frac{\mathrm{v}_y}{c}\sin\theta\right)}{\sqrt{\mathrm{v}_x^2+\frac{2e}{m}\left[\varphi(x)-\varphi(x'')\right]}}dx''. \tag{3.1.37}$$

By solving the set of integral differential equations (3.1.36), we can find the solution to the problem of electromagnetic wave reflection from a plasma with an arbitrary density profile for arbitrary ratios of the plasma inhomogeneity scales L and skin depth l_s, spatial dispersion $\mathrm{v}_{T\|}/\omega$, and the wavelength of the electromagnetic wave in vacuum $2\pi c/\omega$. Since, we consider collision-less absorption in the following analysis, we set $\nu_{ei} = 0$. This is possible because we discuss the case ν_{ei} ω. Collision absorption can be easily incorporated by using (3.1.36) and (3.1.37).

Electromagnetic Wave Absorption by a Plasma with a Sharp Boundary

We turn now to the dependence of the absorption coefficient η_a on the inhomogeneity scale length. Let us start with the case of L smaller than the skin depth l_s c/ω_p. For such inhomogeneity scales, the limiting case of a sharp plasma boundary, which is most frequently dealt with [41,79], corresponds to the potential $\varphi = 0$ for $x > 0$ and $\varphi = \infty$ for $x < 0$. This step function cannot be treated as an exact solution to the self-consistent equation (3.1.36), which assumes the step ion density profile $n_i(x)$ because the scale L cannot, in fact, be less than the Debye length $\mathrm{v}_{T\|}/\omega_p$. In their numerical simulations, Yang et al. [76] showed that if this is ignored, the error in the absorption coefficient value can be as large as 40% in a plasma with a sharp boundary. In our calculations, we will adopt a linearly varying potential (with the typical spatial scale L):

$$\varphi = 0, \quad x > L; \quad \varphi = \varphi_0\left(1-\frac{x}{L}\right), \quad x \le L \tag{3.1.38}$$

This corresponds to an exponentially varying plasma density and is close to the real situation.

The set of equations (3.1.36) and (3.1.37), in which the potential is defined by (3.1.38), contains the parameter $\alpha = \omega c/\omega_p \mathrm{v}_{T\|}$. In order of magnitude, this parameter equals the ratio of the skin depth and the scale of the spatial dispersion. The limiting cases α 1 and α 1 correspond to the anomalous and normal skin effects, respectively. In the zero-*th* approximation of the parameter L/l_s, the kernels of the integrals depend on the combination of variables $x - x'$, and the solution to the equations (3.1.36) can be found by using the Fourier transform. With the neglect of the small quantities, $\beta_{\|,\perp} = \mathrm{v}_{T\|,\perp}/c$. The absorption coefficient η_a can be expressed through the impedance ζ for any arbitrary α:

$$\eta_a = 1 - \left| \frac{\zeta - \cos\theta}{\zeta + \cos\theta} \right|^2, \tag{3.1.39}$$

$$\zeta = \frac{i\Omega}{\pi\alpha} 2 \int_0^\infty \left\{ 1 - \frac{\xi^3}{\alpha^2} \left[Z(\zeta) - \Delta\left(\frac{1}{\xi} + Z(\xi) \right) \right] + \frac{\sin^2\theta}{\cos^2\theta + \frac{2}{\Omega^2}\xi^2 \left(1 + \xi Z(\xi)\right)} \right\}^{-1} d\xi, \tag{3.1.40}$$

where

$$\Delta = 1 - \frac{T_\perp}{T_\parallel}, \ \Omega = \frac{\omega}{\omega_p}, \ \mathrm{v}_{T_{\parallel,\perp}} = \sqrt{\frac{2T_{\parallel,\perp}}{m}}, \ \xi = \frac{\omega_p x}{c},$$

and $Z(\xi) = i\sqrt{\pi}e^{-\xi^2} - 2e^{-\xi^2}\int_0^\infty e^{t^2}$ is the dispersion function of the plasma. For $T_\parallel = T_\perp$, expression (3.1.40) coincides with that derived in [79]. **Figure 3.1.13** demonstrates the dependence $\eta(\alpha)$, which means that the absorption increases with increasing incidence angle θ and anisotropy parameter $\Delta=(T_\perp - T_\parallel)/T_\parallel$. By expanding integrals in L/l_s , we can obtain the corrections to (3.1.39). Since the expressions for these corrections are too cumbersome, we will not present them here and only note that the inner correction to η_a for α 1 is positive, and the absorption decreases in the limiting case of α 1.

Figure 3.1.13 shows also the variation of the laser absorption coefficient at $\theta = 0$ with the skin depth parameter $\alpha = c\omega/\omega_p \mathrm{v}_T$ and anisotropy parameter Δ. In this case, the laser absorption is mainly due to the Landau damping, as $\nu/\omega = 0.01$. It is evident from Fig. 3.1.13 that the plasma anisotropy causes an increase in the absorption coefficient [41].

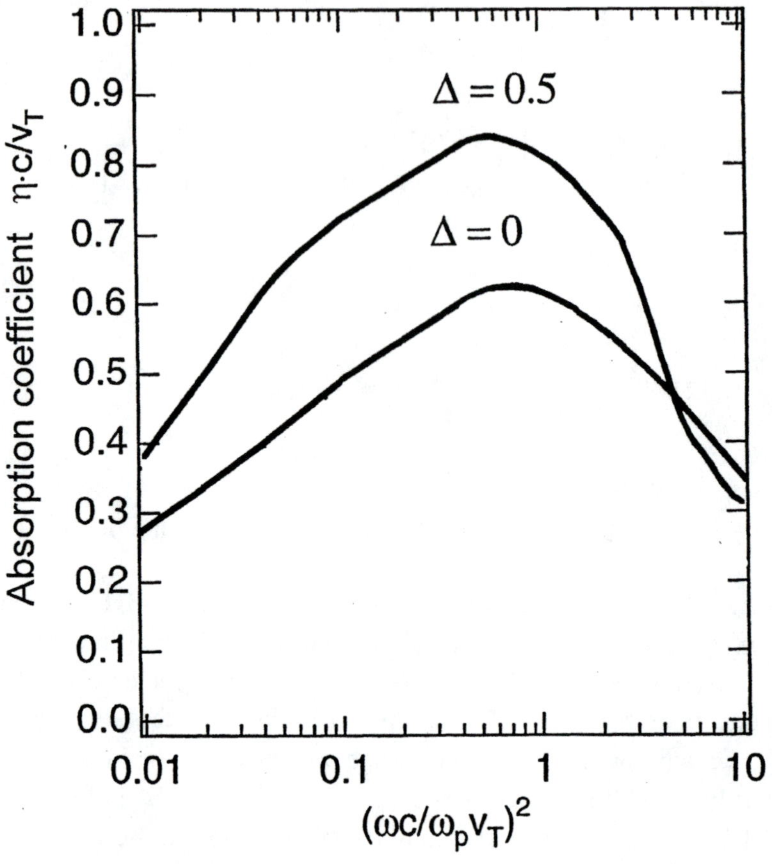

Fig. 3.1.13. Variation of the absorption coefficient with the skin depth parameter $\left(\frac{\omega c}{\omega_p v_T}\right)^2$ for two values of the anisotropy parameter $\Delta = \frac{T_\perp}{T_\parallel} - 1$ and $\theta = 0$, $\delta = 0.01$.

Absorption by a Plasma with a Smooth Boundary

Now, we will confider the scale range of L corresponding to $L > l_s$. This case is most interesting when the skin effect is anomalous, $l_s < v_{T\parallel}/\omega$. This is so because η decreases at $L > l_s$, when the opposite inequality is satisfied (the normal skin effect). In the limiting case of the anomalous skin effect (strong spatial dispersion), we have $v_{T\parallel}/\omega > L > l_s$, the phases Φ are small, and the equations of (3.1.36) are reduced to one equation for E_y:

$$\frac{\partial^2 E_y}{\partial \xi^2} + \kappa^2 \cos^2\theta \;\; E_y = -\frac{iA}{\sqrt{\pi}} \left\{ \int_{-\infty}^{+\infty} d\xi' E_y(\xi') K_0\left(\frac{|\tilde{\varphi}(\xi) - \tilde{\varphi}(\xi')|}{2} \right) \times \right.$$
$$\left. \times e^{\frac{\tilde{\varphi}(\xi) - 3\tilde{\varphi}(\xi')}{2}} \cos^2\theta + \frac{i\Delta\beta_{11}}{\kappa} \int_{\xi}^{\infty} d\xi' \frac{\partial E_y}{\partial \xi'} \times \mathrm{erfc}\left(\sqrt{\tilde{\varphi}(\xi) - \tilde{\varphi}(\xi')} \right) e^{(\tilde{\varphi}(\xi) - 2\tilde{\varphi}(\xi'))} \right\}, \tag{3.1.41}$$

where the dimensionless quantities are defined as follows: $\xi = x/L$, $\kappa = \frac{\omega L}{c}$, $A = \omega\omega_p^2 L^3/(v_{T\|}c^2)$, $\tilde{\varphi} = e\varphi/T$ and erfc(x) is a complementary error function. Equation (3.1.41) is almost identical to the field equation for a S-polarized incident electromagnetic wave interacting with the plasma [77]. Therefore, there is no significant difference between the absorption of the S- and P-polarized waves in the range of L under consideration. The only exception is that the corresponding angular distributions are different (this results in the appearance of a numerical coefficient of the order of unity in the absorption coefficient). For a linearly varying potential $\varphi(\xi) = -\xi$ (exponentially varying plasma density) and A= 0, the method of solving equation (3.1.41) is described in [77]. Using this method, we have obtained the absorption coefficient, $\eta_a \;\; 2\pi L\omega/c\cos\theta$, which increases linearly with $v_T/\omega > L > l_s$. The "resonant" mechanism (which causes a substantial difference in the absorption of the S- and P-polarized waves) does not contribute to the absorption in this case, because the wavelength of the longitudinal oscillations $\propto v_{T\|}/\omega$ exceeds the characteristic inhomogeneity scale L, and the longitudinal oscillations, which may lead to an additional absorption, cannot the excited.

For the linear potential and $\theta = 0$, which we consider now, the parameter α becomes small, and the latter equation can be written in a simpler form:

$$\frac{d^2 E_y}{d\xi^2} + \kappa^2 E_y = -\frac{iA}{\sqrt{\pi}} C \int_{\xi-1}^{\xi+\alpha^{-2}} d\xi' e^{\xi'} E_y(\xi') \tag{3.1.42}$$

where $C \sim 1$ is an unknown numerical constant. For small α, the upper integration limit in Eq.(3.1.42) tends to infinity. By using the variable $\mu = \xi - 1$, it is easy then to reduce this equation to the third-order linear differential equation:

$$\frac{d^3 E_y}{d\mu^3} + \kappa^2 \frac{\partial E_y}{\partial \mu} - \frac{iA}{\sqrt{\pi}} C e^{\mu} E_y(\mu) = 0 \;, \tag{3.1.43}$$

After the change of the variable $e^{\mu} = x$, Eq.(3.1.43) reduces to the standard form of the generalized hyper-geometric equation. The solution that matches the incident and reflected waves at $\xi \to -\infty$ ($x \to 0$) is

$$E_y(\xi) = C_1 e^{i\kappa(\xi-1)}\,{}_0F_2\left(1+i\kappa;1+2i\kappa;\frac{iAC}{\sqrt{\pi}}e^{\xi-1}\right) +$$

$$C_2\, e^{-i\kappa(\xi-1)}\,{}_0F_2\left(1-i\kappa;1-2i\kappa;\frac{iAC}{\sqrt{\pi}}e^{\xi-1}\right), \qquad (3.1.44)$$

where ${}_0F_2(a;b;z)$ is the generalized hyper-geometric function [80].

The arbitrary coefficients C_1 and C_2 can be found by imposing $E_y \to 0$ at $\xi \to \infty$, and from the known amplitude of the incident wave. Using the asymptotic behavior of ${}_0F_2$ at large values of the argument modulus in [80], we can find the following expression for the amplitude reflection coefficient R:

$$R = e^{-\pi\kappa}\left(\frac{AC}{\sqrt{\pi}}\right)^{2i\kappa}\frac{\Gamma(1+i\kappa)\Gamma(1+2i\kappa)}{\Gamma(1-i\kappa)\Gamma(1-2i\kappa)}. \qquad (3.1.45)$$

In the limiting case under study, the parameter $\kappa = \frac{\omega L}{c}$ is quite small, because $\frac{\omega L}{v_{T_\parallel}} << 1$. Consequently, $\kappa \le \frac{v_{T_\parallel}}{c}$, and Eq.(3.1.45) can be expanded into series in κ

$$R = 1 - \pi\kappa - 2i\kappa\left(\ln\frac{AC}{\sqrt{\pi}} + 3\widetilde{C}\right), \qquad (3.1.46)$$

where $\widetilde{C} = 0.577$ is the Euler constant.

Note that the unknown constant C appearing in Eq. (3.1.46) influences only the imaginary part of the reflection coefficient. So, the absorption coefficient η_a $2\pi\kappa << 1$ is the first order in κ. A result different from Eq. (3.1.46) only by the numeric factor under the logarithmic sign was obtained in [81] from the analysis of a functional equation.

Similar calculations can be made for the potential $\varphi(x)$ of the form $-e\varphi(x)/T = ln\xi$, corresponding to a linear variation in the electron plasma density. Introducing a new variable $\varsigma = ln\xi$ and approximating the kernel, we again arrive at the hypergeometric type of equation. In spite of the other expression $E_y(\xi)$, the absorption coefficient is again proportional to the plasma inhomogeneity scale.

Fig. 3.1.14 shows the plot $\eta(L)$ denoted by a dotted line, which was found for $\omega_p^2/\omega^2=60$, $v_T/c = 0.1$, to compare it with the results of [79] denoted by the symbol ×. As $c\omega/v_T\omega_p = 1.3$, it means that we are in a transient area between ASE and SIB; but even in this case, when our analytical formula is not quite good, we see a fairly good agreement between the analytical and numerical results. There are also experimental points from the paper [69] denoted by the symbol • , where the experimental plasma parameters were close to our ASE region. In this case we also observe an agreement.

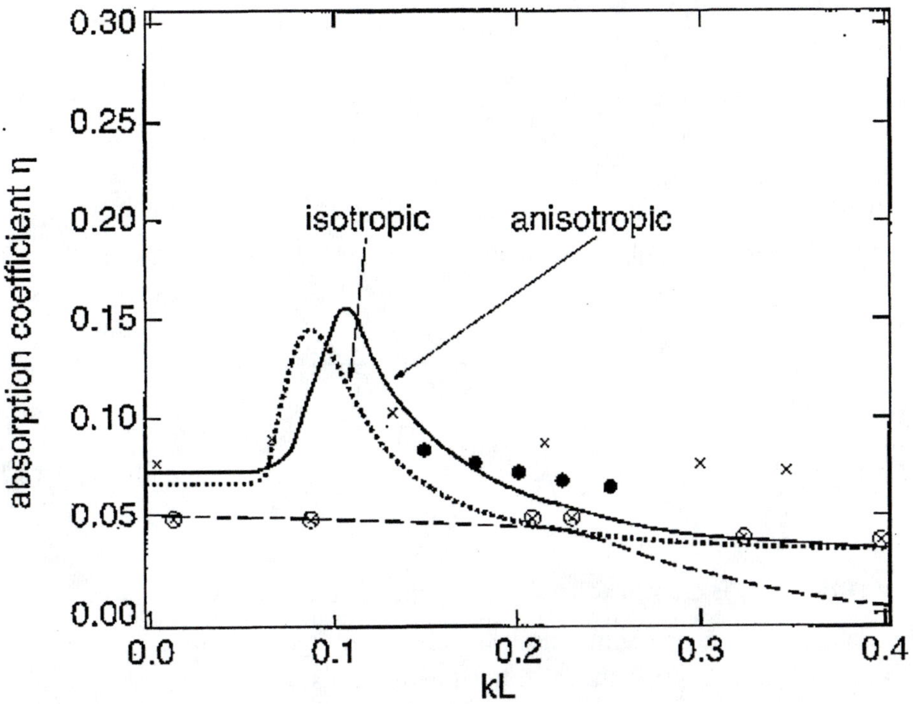

Fig. 3.1.14. Variation of the absorption coefficient as a function of the scale of plasma inhomogeneity for an isotropic (Δ=0) and anisotropic (Δ=0.5) distribution. Parameters: (ASE) $v_T/c=0.1$, $\omega_p^2/\omega^2=60$ (Δ=0 - dotted line, Δ=0.5 - solid line, × - simulation), • - experiment. (SIB) $v_T/c=0.1$, $\omega_p^2/\omega^2=20$ (dashed line, ⊗ - simulation).

Let us now consider the solution to the field equation, taking into account the temperature anisotropy. Using the reduced variables defined above, Eq. (3.1.41) can be written as

$$\frac{d^2 E_y}{d\xi^2} + \kappa^2 E_y =$$

$$-\frac{iA}{\sqrt{\pi}} \int_{-\infty}^{+\infty} d\xi' \left\{ K_0\left(\left|\frac{\xi - \xi'}{2}\right|\right) \exp\left(\frac{\xi' - \xi}{2}\right) E_y(\xi') + \right.$$

$$\left. \frac{i}{\alpha} \frac{T_\perp - T_\parallel}{T_\parallel} \operatorname{erfc}\left(\sqrt{\xi' - \xi}\right) H(\xi' - \xi) \frac{\partial E_y}{\partial \xi'} e^{\xi' - \xi} \right\} e^{\xi'} . \qquad (3.1.47)$$

It is difficult to solve this equation if the relative magnitude of the first and the second terms under the integral is arbitrary. We consider the case of $\Delta >> \alpha$, leaving only the second term under the integral. Estimation of the kernel of the integral equation gives the following approximation for Eq.(3.1.47):

$$\frac{d^2E_y}{d\xi^2}+\kappa^2E_y=\frac{AC}{\alpha\sqrt{\pi}}\frac{T_\perp-T_\parallel}{T_\parallel}\int_\xi^\infty\frac{\partial E_y}{\partial\xi'}e^{\xi'}d\xi' , \tag{3.1.48}$$

where C is a numerical constant of the order of unity.

The solution to Eq.(3.1.48) for an incident and reflected wave is

$$E_y(\xi)=C_1\,e^{i\kappa\xi}\,{}_1F_2\left(1;1+i\kappa;1+2i\kappa;-\frac{AC}{\sqrt{\pi}\alpha}\frac{T_\perp-T_\parallel}{T_\parallel}\right)+$$
$$C_2\,e^{-i\kappa\xi}\,{}_1F_2\left(1;1-i\kappa;1-2i\kappa;-\frac{AC}{\sqrt{\pi}\alpha}\frac{T_\perp-T_\parallel}{T_\parallel}\right), \tag{3.1.49}$$

where ${}_1F_2(a;b;c;z)$ is the generalized hypergeometric function [80].

The condition of a negligible electric field in the target bulk yields the relation between C_1 and C_2. As a result, the reflection coefficient is

$$R=\frac{C_2}{C_1}=\ e^{-3\pi\kappa}\left(\frac{AC}{\sqrt{\pi}\alpha}\frac{T_\perp-T_\parallel}{T_\parallel}\right)^{3i\kappa}\ \approx\ 1-\ 3\pi\kappa\ +3i\kappa\ \ln\frac{AC}{\sqrt{\pi}\alpha}\frac{T_\perp-T_\parallel}{T_\parallel};\ T_\perp\succ T_\parallel .$$

In the opposite case of $\frac{T_\perp-T_\parallel}{\alpha T_\parallel}<<1$, the anisotropic correction to the absorption coefficient can be estimated from the perturbation theory:

$$\Delta\eta=\frac{A(T_\perp-T_\parallel)}{\sqrt{\pi}\kappa\alpha T_\parallel}\cdot\frac{\int_{-\infty}^{+\infty}d\xi\int_\xi^\infty d\xi'\,\mathrm{Im}\frac{\partial E_y}{\partial\xi'}\,\mathrm{Re}\,E_y(\xi)\,\mathrm{erfc}\left(\sqrt{\xi'-\xi}\right)\exp(2\xi'-\xi)}{\left[\mathrm{Re}\,E_y(\xi=-\infty)\right]^2},$$

where E_y is determined by Eq. (3.1.47). Estimations for the sign of $\Delta\eta$ give $\Delta\eta>0$ for $T_\parallel<T_\perp$, and $\Delta\eta<0$ for $T_\parallel>T_\perp$.

Thus, in the limit of great anisotropy, $T_\parallel<T_\perp$, the absorption coefficient increases by a factor of three (in case of a weak anisotropy, the difference in the absorption is not so large, see Fig. 3.1.15). At $T_\parallel>T_\perp$, a similar analysis gives a lower absorption coefficient than in the isotropic case.

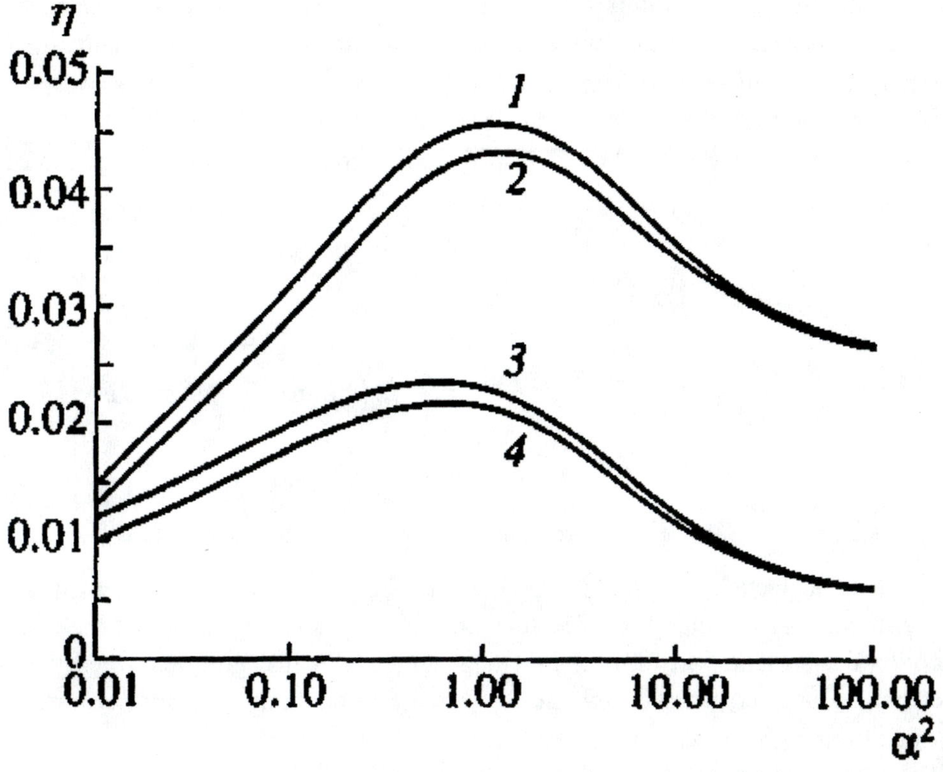

Fig. 3.1.15. Dependence of the absorption coefficient η on α for a sharp plasma boundary for different parameters of anisotropy and incident angles: (1) $\Delta = -0.1$ and $\theta = 60°$; (2) $\Delta = 0$ and $\theta = 60°$; (3) $\Delta = -0.1$ and $\theta = 30°$; and (4) $\Delta = 0$ and $\theta = 30°$.

The function $\eta_a(L)$ substantially depends on the type of skin-effect ($\alpha << 1$ for the anomalous skin-effect and $\alpha \gg 1$ for the normal one). At $\alpha << 1$ (ASE for a sharp boundary) with increasing L, the absorption varies as follows: at $0 < L < l_S$, η_a is determined by the expression for a well-defined boundary. At $l_s < L < v_T/\omega$, the absorption grows linearly with increasing L, so that $\eta_a \approx 2\pi\omega\,(L+l_s)/c$ in these two length ranges. At $L \approx v_T/\omega$, the absorption reaches its maximum. With further increase, $\alpha > v_T/\omega$, the anomalous skin-effect is replaced by the normal one (the integral differential equation for the field is reduced to the ordinary differential one). Then η_a decreases with increasing L. Thus, for the S-polarized electromagnetic wave propagating in the regime of an anomalous skin-effect, there is a maximum absorption on the inhomogeneity scales L of the order of the scale of spatial dispersion v_T. If $\alpha >> 1$, we observe the normal skin-effect regime even at $L \approx 0$. In the range $0 < L < c/\omega_p$, the absorption is then described by the formula for a well-defined boundary, and at $c/\omega_p < L$ it decreases with increasing L. If we optimize the absorption with respect to the parameter α and the scale L, it will become evident that the maximum will occur at $\alpha = 1$ and $\omega L/v_T \approx 1$.

The next range of the inhomogeneity scales, $L > v_{T\|}/\omega$, corresponds to the case of a weak spatial dispersion. For this range, the phases Φ are much larger than unity, and we can expand the rapidly oscillating integrals in expression (3.1.36) in the parameter $v_{T\|}/L\omega < 1$ by using the perturbation technique [81]. As a result, we get a set of differential equations for the fields $E_x(\xi)$ and $B_z(\xi)$, in which the permittivity is a local function of the coordinates

$$\frac{3}{\Omega^2}\beta_\|^2\frac{d^2E_x}{d\xi^2}+k^2\varepsilon(\xi)E_x=k^2\sin\theta\ \ B_z,$$
$$\frac{d^2B_z}{d\xi^2}-\frac{1}{\xi}\frac{d\varepsilon}{d\xi}\frac{dB_z}{d\xi}+k^2\left(\varepsilon(\xi)-\sin^2\theta\right)B_z=-k^2\varepsilon(\xi)\sin\theta E_x-k^2\sin\theta B_z, \tag{3.1.50}$$

where $\varepsilon(\xi)=1-\frac{\beta_\|A}{k^3}e^{-\tilde{\varphi}(\xi)}$ is the permittivity of an inhomogeneous plasma if the spatial dispersion is neglected. The solution to the set of equations (3.1.50) obtained by the Wentzel-Kramers-Brillouin (WKB) method and the respective treatment of the resonant absorption are presented in [82] for inhomogeneity scales that greatly exceed the laser wavelength. However, we will not use the WKB approximation because we are more interested in studying the case with $L < \lambda$.

The other approximate approach to solving (3.1.50) is to introduce the permittivity of an inhomogeneous plasma with allowance for the spatial dispersion in the form $\varepsilon(q(\xi),\xi)$ [74]. With this approach and assuming the potential profile to be linear, we can reduce equations (3.1.50) to the following set of differential equations:

$$\frac{d^2E_y}{d\xi^2}+k^2\varepsilon_\perp\left(q(\xi),\xi\right)E_y-ik\sin\theta\ \ \frac{\partial E_x}{\partial\xi}=0,$$
$$-ik\sin\theta\ \ \frac{\partial E_y}{\partial\xi}+k^2\left[\varepsilon_\|\left(q(\xi),\xi\right)-\sin^2\theta\right]E_x=0 \tag{3.1.51}$$

The function $\varepsilon_{\|,\perp}(q)$ can be found by introducing the Fourier transform (with respect to the combination of variables ξ - ξ\`, which is contained in the phases $\Phi(\xi, \xi')$ on the right-hand sides of equation (3.1.41). The function $q(\xi,\xi')$ is defined by the dispersion relation

$$(\varepsilon_\| - \sin^2\theta)\varepsilon_\perp - q^2\varepsilon\ = 0$$

For a slight spatial dispersion, we get the following expressions for the quantities specified above:

$$\begin{aligned}
&\operatorname{Im}\varepsilon_{\perp} = \frac{A\sqrt{\pi}}{k^2} 2\operatorname{Re}\left(\frac{1}{\sqrt{iq}} e^{-\frac{k^2}{4iq\beta_{\|}^2}}\right), \\
&\operatorname{Re}\varepsilon_{\perp} = 1 - A\beta_{\|}k^{-3}e^{\xi}, \\
&\varepsilon_{\|}(\xi) = 1 - \frac{\beta_{\|}A}{k^3} e^{-\tilde{\varphi}(\xi)},\ q(\xi) \approx k\sqrt{\cos^2\theta - A\beta_{\|}k^{-3}e^{\xi}}, \\
&E_y(\xi)\big|_{\theta=0} = E_0 K_{2ik}\left(2\exp\left(\frac{\xi}{2} + \frac{1}{2}\ln\frac{A}{\alpha}\right)\right)\frac{2\sinh(2k\pi)}{i\pi}\Gamma(1+2ik).
\end{aligned} \tag{3.1.52}$$

Both approximate methods, (3.1.41) and (3.1.51), are equivalent if the spatial dispersion is weak. In order to pass from (3.1.41) to (3.51.1), $\varepsilon_{\|}(q,\xi)$ must first expanded into series in q and then q should be replaced by the operator $\partial/\partial\xi$. From equations (3.1.41) or (3.1.51), one can find the absorption, which is the ratio of the power absorbed per unit volume, and the Poynting vector of the incident wave. From (3.1.51), we get

$$\eta_a = \frac{k}{2\cos\theta}\int_{-\infty}^{+\infty}\left(\left|\frac{E_y}{E_0}\right|^2 \operatorname{Im}\varepsilon_{\perp} + \left|\frac{E_x}{E_0}\right| \operatorname{Im}\varepsilon_{\|}\right)d\xi = \eta_t + \eta_r, \tag{3.1.53}$$

where E_0 is the incident wave amplitude.

The absorption coefficient contains two terms. The first term η_t varies with $\operatorname{Im}\varepsilon_{\perp}$ and does not vanish at $\theta = 0$ ('transverse' absorption). The integral in (3.1.53) was calculated in [36] for $\theta = 0$ and an exponentially varying plasma density. We have calculated the integral of (3.1.53) approximately, using the saddle point method and found the following expression for η_t:

$$\eta_t\big|_{\theta=0} \approx \sqrt{2}\pi^{3/2}A^{3/4}\left(k/\beta_{\|}\right)^{1/4} e^{-3\frac{(k/\beta_{\|})^{5/4}}{A^{1/4}}} \tag{3.1.54}$$

where $k/\beta_{\|}$ 1.

The second term in expression (3.1.53) describes the familiar 'resonant' absorption corresponding to the case when Langmuir plasma oscillations are excited [36]. In order to find $\eta_r(L)$, we have used the results obtained in [39] to get the following expression for the resonant absorption coefficient of a P-polarized wave:

$$\eta_r = 1 - \left|(\zeta_r - \cos\theta)/(\zeta_r + \cos\theta)\right|^2, \tag{3.1.55}$$

where, for $kL > (\nu_{ei}/\omega)^{3/2}$ and $kL > (n_0/n_{cr})^{-3}$, ζ_r is defined as

$$\zeta_r \approx -1.38\ ix_l^{1/3} \times \left\{1 + x_l^{2/3}\left[0.72 - \sin^2\theta\left(0.21 + 0.48\ln x_l\right)\right]\right\} +$$
$$+\pi x_l \sin^2\theta + 0.38\frac{\nu_{ei}}{\omega}x_l^{-1/3}, \quad x_l = \omega L / c. \tag{3.1.56}$$

We will now consider all the limiting cases and find the dependence $\eta_a(L)$ in the range covering all the scales. In the limiting case of the anomalous skin effect $c/\omega_p < v_T/\omega$ for $0 < L < l_s$, the absorption coefficient is given by the expression for a sharp boundary (3.1.40). For the larger scales, the absorption increases linearly with an increase in L (3.1.46). In this range, η_a can be estimated as $\eta_a = (\omega/c)(L + l_s)$. For the scale of about $L \sim v_{T\|}/\omega$, the absorption coefficient has the first maximum. As L increases, the penetration depth of the electromagnetic field in the plasma increases, and the anomalous skin effect changes to the normal skin effect at L $v_{T\|}/\omega$),. In the latter case, the absorption is described by (3.1.55) and (3.1.56). Finally, the resonant absorption dominates at L comparable with $\sim c/\omega$. In the entire range of scales, the function $\eta_a(L)$ can be approximately described as

$$\eta_a \approx \eta_r + \eta_0 + dx_l^{\gamma}\, e^{-bx_l^{\beta}}, \tag{3.1.57}$$

with $\eta_0 = 0.03$, $x_l = \omega L/c$, $b = 0.3(v_{T\|}/c)^3$, $\gamma = 2.5$, $\beta = 0.5$.

Figure 3.1.16 shows this dependence and demonstrates a good agreement with the results of numerical calculations [76], which are also presented in this figure.

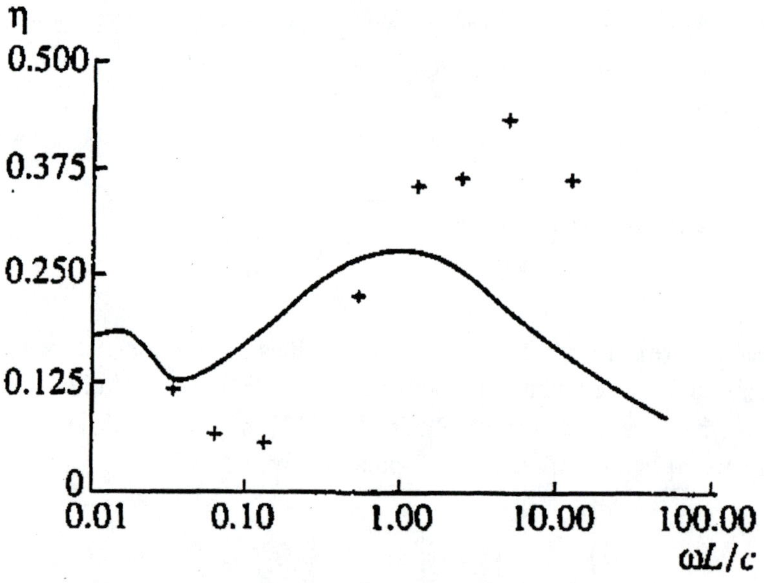

Fig.3.1.16. Dependence of the absorption coefficient η on the plasma inhomogeneity scale L for $\theta = 23.5°$, $v_T/c = 0.1$, and $\omega_p/\omega = 30$. The numerical results are shown by crosses.

Assume $L = 0$ and passing over to the normal skin effect, $c/\omega_p > v_T/\omega$, the first maximum of the curve $\eta_\alpha(L)$ shifts to the right, towards the origin of the coordinates. Its value decreases and the maximum eventually disappears. At $0 < L \quad c/\omega_p$, $\eta_\alpha(L)$ is defined by formula (3.1.40) for a sharp boundary. For the larger scales, $c/\omega_p < L < c/\omega$, $\eta(L)$ is defined by formula (3.1.57), the first maximum is not pronounced, and for $L = c/\omega$, only resonant absorption is important.

3.1.5 Nonlinear Absorption of a Short Intense Laser Pulse in an Over-Dense Plasma

High intensity laser pulse absorbs in over dense plasma by different mechanisms. For example, the fairly strong absorption reported in many experiments is attributed to the mechanism called **J**×**B** heating. This mechanism become acting at high intensities, particularly when electrons become relativistic. It should be pointed out that an increase in the laser radiation intensity reduces the difference between the absorption of the S- and P- polarized light, as found recently on several occasions. This may be attributed to a change in the absorption mechanism and to generation of strong magnetic fields, which alter the resonant properties of the plasma.

In the present section the numerical simulations and analytical solution of non-linear relativistic equations for plasma are considered. There is shown, that at weak non-linearity the resonant absorption arises even at normal incidence of an electromagnetic wave on spatially non-uniform plasma. Absorption coefficient has the rise in comparison with a linear case.

Basic Set of Equations and Numerical Solutions

Let the plane linearly polarized electromagnetic wave comes along the axis X normal to the semi-limited plasma. Plasma temperature is T_e, the density grows from zero to n_e on the distance L and plasma over-dense, max $n_e >> n_{cr} = m\omega^2/4\pi e^2$. The wave with amplitude E_0 and frequency ω_0 is chosen in such a way that the quiver velocity $\upsilon_E = eE_0/m\omega$, is larger than the thermal velocity $(T_e/m)^{1/2} = \upsilon_T$. The reason is that during the laser pulse (< 100*fs*) the movement of ions is negligible and the plasma edge preserves its sharpness. Movement of the plasma electron component is described by the self consistent set of equations, consisting of the collision-less kinetic Boltzman equation ($0<x<l_s$) and Maxwell equations for electromagnetic fields (in covariant form):

$$\frac{p^\mu}{m_e}\frac{\partial f}{\partial x^\mu}+\frac{eF^{\mu\nu}p_\nu}{c}\frac{\partial f}{\partial p^\mu}=0, \qquad \frac{\partial F^{\mu\nu}}{\partial x_\mu}=J^\nu \tag{3.1.58a}$$

$$J_\nu = e\int d^4p\,cp_\nu\theta(p_0)\delta(p_\mu p^\mu - mc^2)f,$$

here $F^{\mu\nu}$- the Maxwell tensor; and of the kinetic Fokker-Plank equation for $x>l_s$:

$$\frac{\partial f}{\partial t}+\upsilon_x\frac{\partial f}{\partial x}+eE_a(x,t)\frac{\partial f}{\partial p_x}=\nu_{ei}\frac{\partial^2 f}{\partial\phi^2} \tag{3.1.58b}$$

Ambipolar field E_a is determined from the zero current along the axis X condition:

$$\int\upsilon_x f(\upsilon_x,\upsilon_y,\upsilon_z)d\upsilon=0 \tag{3.1.58c}$$

or Puasson equation.

Oblique Incidence of Laser Radiation on Plasma Boundary

Oblique incidence is "boosted" into normal incidence by using the method of [83]. Referring to **Fig. 3.1.17**, and denoting the boost (*S*) frame quantities by primes, the inverse Lorentz transformations for the wave frequency and *k* vector are

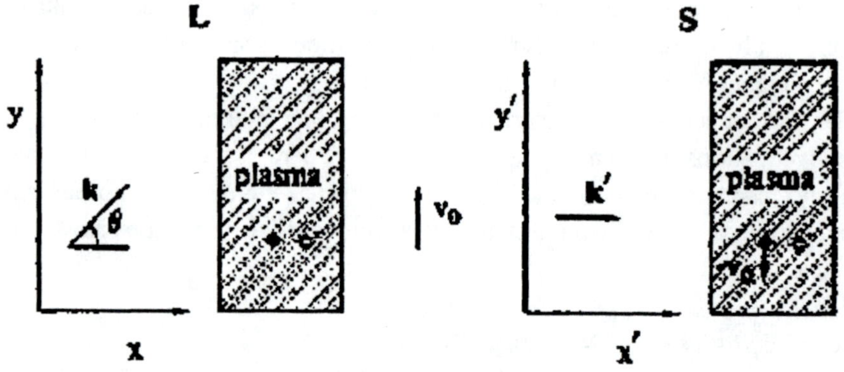

Fig. 3.1.17. Geometry for the transformation from the lab frame (*L*) to the simulation frame (*S*) using a boost along the *y*-axis.

$$\omega'=\gamma_0\left(\omega-\upsilon_0 k_y\right),$$

$$k'_y=\gamma_0\left(k_y-\frac{\upsilon_0}{c^2}\omega\right),$$

$$k'_{x,z}=k_{x,z}.$$

Since $k_y = k\sin\theta = \omega c\sin\theta$ *and* $\upsilon_0 = c\sin\theta$ we have

$$k_y'=0$$
$$\omega'=\omega/\gamma_0 \tag{3.1.59}$$
$$k'=k/\gamma_0$$

where $\gamma_0 = 1/\cos\theta$. For the space and time coordinates, we have (from $S \to L$)

$$t = \gamma_0\left(t' + \frac{\upsilon_0}{c^2}y'\right),$$

$$x = x',\ y = \gamma_0(y' + \upsilon_0 t'),\ z = z' \qquad (3.1.60)$$

Likewise, noting the $B_x' = 0$, the electric and magnetic fields transform as

$$\left.\begin{aligned} E_x &= \gamma_0\left(E_x' - \upsilon_0 B_z'\right) \\ E_y &= E_y' \\ B_z &= \gamma_0\left(B_z' - \frac{\upsilon_0}{c^2}E_x'\right) \end{aligned}\right\} \text{P-polarized light;} \quad \left.\begin{aligned} E_z &= \gamma_0 E_z' \\ B_x &= \gamma_0\upsilon_0 E_z' \\ B_y &= B_y' \end{aligned}\right\} \text{S-polarized light;} \quad (3.1.61)$$

and the density and current for each particle species a (inversely) as

$$\rho_\alpha' = \gamma_0\left(\rho_\alpha - \frac{\upsilon_0}{c^2}j_{\alpha y}\right),$$

$$j_{\alpha y}' = \gamma_0\left(j_{\alpha y} - \upsilon_0\rho_\alpha\right),$$

$$j_{\alpha x,z}' = j_{\alpha x,z}.$$

Initially, $j_{ey} = j_{iy} = 0$, so in the boost frame, $(\rho_0)'_{e,i} = \gamma(\rho_0)_{ei}$ and $(j_{y0})'_{e,i} = -(\rho_0)'_{e,i}c\sin\theta$. In an electromagnetic PIC or Vlasov code, it is convenient to use the standard normalizations:

$$\tilde{t} = \omega t,$$

$$(\tilde{x}, \tilde{y}) = (kx, ky),$$

$$\tilde{\mathbf{E}} = \frac{e\mathbf{E}}{m\omega},$$

$$\tilde{\mathbf{B}} = \frac{e\mathbf{B}}{m\omega}.$$

$$\tilde{n}_{e,i} = n_{e,i}/n_c,$$

$$\upsilon_{e,i} = \upsilon_{e,i}/c.$$

Thus from Eqs. (3.1.59) and (3.1.60), the normalized, time interval $\omega' t' = \omega t$ is invariant, as is the wave phase $\omega t - k \cdot r$. On the other hand, the simulation grid length $k' x_i'$

$= \gamma_0^{-1} kx_l$ shrinks with increasing θ. The scaling of ω' and k' with γ_0 and the fact that $\tilde{B}_x$ = 0 means that the normalized field transformation become

$$\tilde{E}_x = \tilde{E}'_x - \tilde{\upsilon}_0 \tilde{B}'_z,$$
$$\tilde{E}_y = \frac{\tilde{E}'_y}{\gamma_0},$$
$$\tilde{E}_z = \tilde{E}'_z,$$
$$\tilde{B}_x = \tilde{\upsilon}_0 \tilde{E}'_z,$$
$$\tilde{B}_y = \frac{\tilde{B}_y}{\gamma_0}.$$

Note that the critical density in the simulation frame transforms as

$$\frac{n'_c}{n_c} = \frac{\omega'^2}{\omega^2} = \frac{1}{\gamma_0^2}.$$

Hence the initial normalized unperturbed electron density is

$$\tilde{n}_{e0} = \frac{n_e(t=0)'}{n'_c} = \gamma_0^3 \tilde{n}_e.$$

Thus, to initialize the particles in the simulation frame, we give them a charge $q_e' = \gamma_0^2 q_e$, and mass $m_e' = \gamma_0^2 m_e$, which, when combined with the grid length contraction $k'x' = kx_l/\gamma_0$, satisfies $N_e q_e' = - n_e' x_l'$ for the upper density shelf, were N_e is the number of simulation particles. To initialize the particle moment we first specify the thermal distribution in the lab frame, and then perform boosts:

$$\tilde{p}'_{x,z} = \tilde{p}_{x,z},$$
$$\tilde{p}'_y = \gamma_0 \left(\tilde{p}_y - \tilde{\upsilon}_0 \gamma\right), \qquad (3.1.62)$$
$$\gamma' = \gamma_0 \left(\gamma - \tilde{\upsilon}_0 \tilde{p}_y\right).$$

Finally, to launch the EM wave, we must specify its amplitude $a_0 = \upsilon_E/c$ at the left-hand simulation boundary. Since $\upsilon_E/c = eE_0/m\omega c$, we have for a P-polarized wave,

$$a_0' = \frac{eE'_y}{m\omega' c} = \gamma_0 \frac{eE_y}{m\omega c} = \frac{eE_0}{m\omega c} \gamma_0 \cos\theta = a_0. \qquad (3.1.63)$$

One can easily verify this invariance by using Eq. (3.1.61) to recover the lab-frame vacuum fields. According to Eq. (3.63), we let $\tilde{E}'_y(x'=0) = \tilde{B}'(x'=0) = a_0$, so since $E_x' = 0$, we obtain $\tilde{E}_x = -\upsilon_0 a_0 = -a_0 \sin\theta$, $\tilde{E}_y = a_0/\gamma_0 = a_0 \cos\theta$, and $\tilde{B}_z = a_0$, which is just what we expect for a plane wave launched at an angle θ to the density gradient according to Fig. 3.1.15.

It should be stressed that the boost technique cannot always to model oblique incidence interaction. In the general 2-D geometry, all physical quantities and the distribution function depend separately on *t, x, y,* p_x and p_y. The transformation to the system corresponding to normal incidence with one spatial coordinate is only possible if the distribution function and other physical values depend on this set of variables in the following way:

$$f\left(t - \frac{\sin\theta}{c} y; x; p_x; p_y\right). \tag{3.1.64}$$

in this case we effectively consider a narrower class of solutions of the initial system of Vlasov-Maxwell equations.

The boundary conditions in the boosts frame are at first sight paradoxical. To launch the EM wave, we need to specify $E_y'(t' = 0; x')$, which in the lab frame corresponds to defining the held in the semi-infinite space $x > x_0 + (y \sin\theta - ct)\cos\theta$, where $x_0 \tan\theta$ is the point of intersection of the wave front with the plasma-vacuum interface. Specification of all infinite plane wave in the transformed coordinate system turns out to be nonphysical, because in this case we have to know the field in plasma, i.e., for $y < -x_0 \tan\theta$. Therefore, we should consider the pulse to be (transversely) limited in space width the half-width $\Delta Y = x_0\pi/\sin\theta$ and limited in time to the half-width $\Delta t < \Delta Y/c$, interacting with a semi-infinite plasma. In this case, the problem is physically more reasonably posed when passing to the new coordinate system. Another way out of this dilemma would be to introduce the new variable $\eta = t - (\sin\theta/c)y$ in the laboratory system. In this case, there are no difficulties with the initial conditions, because at $t=0$ it is easy to specify the field distribution $E(x,y)$ at the plasma surface.

Numerical Simulation Results

This model was used in our code KINET1D2V [55]. In **Fig. 3.1.18,** we show the distribution function f_e calculated at time $180\omega_0 t$. The calculation was carried out with a laser intensity of $I=10^{18}$ W/cm^2, a plasma temperature of T_e=20keV, and a density of $n_e = 15n_c$. One can see that f_e is significantly distorted in comparison with the starting Maxwell distribution. It means that the laser ponderomotive force push the electrons into the plasma, expelling them out of the interaction region. So, twice during the period of laser oscillation, a flux of fast electrons is produced propagating inside the plasma.

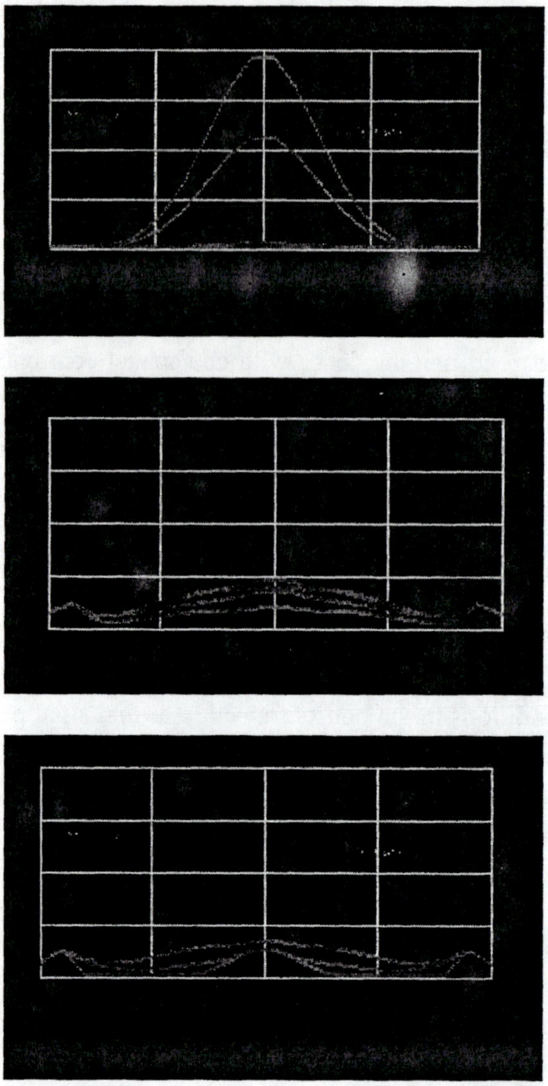

Fig. 3.1.18. Electron distribution function at different time for $v_E/c=1$: (a) $\omega t = 0$ initial Maxwell function, (b) $\omega t = 100$.

Analysis of the calculated distribution function at oblique incidence shows that the constant magnetic field which builds up in the normal incidence frame ("boosted frame") results in an acceleration of the fast electrons in the longitudinal x direction during one half of the laser oscillation period and in a deceleration during the other half of the period.

In Fig. 3.1.3b we have shown the calculated absorption coefficient η_a as a function of laser intensity. We note that the change in the absorption coefficient is small with laser intensity as in [52,56]. Noteworthy, for low intensities the absorption depends on the incidence angle with a characteristic maximum for $\theta \approx \pi/2$. For higher intensities $I > 10^{18}$ W/cm^2 such a dependence with the incidence angle is absent. In this case, the absorption coefficient reaches the same value as for normal incidence, similarly to what was found in [54]. The reason is that high harmonics generation and strong magnetic field strongly perturb the plasma modes. The mode frequencies becoming very different from the laser frequency, this decreases the efficiency of conversion in the plasma waves – the source of the resonant character of the absorption.

Analytical Modeling

Let's consider now the analytical model for absorption coefficient. Solution of equation (3.58) can be written in the general form as:

$$f(x;t;p_z;p_y) = \int \delta(\bar{p} - \bar{p}(x_0;\bar{p}_0;t))\delta(x - x(x_0;\bar{p}_0;t)) f(\bar{p}_0;x_0) d\bar{p}_0 dx_0\,, \quad (3.1.65)$$

where $\bar{p}(x_0;\bar{p}_0;t); x(x_0;\bar{p}_0;t)$ is the phase trajectory of a separate electron with the initial impulse $\bar{p}_0$ and the coordinate x_0; $\bar{p}(x_0;\bar{p}_0;t)$; $x(x_0;\bar{p}_0;t)$: - result of solution of equations of electron motion

$$\bar{p}_\perp = \bar{p}_{0\perp} - e\bar{A}(x,t)/c$$

$$\frac{d}{dt}\dot{x}\sqrt{\frac{m^2 + (\bar{p}_{0\perp}/c - e\bar{A}(x,t)/c^2)^2}{1 - \dot{x}^2/c^2}} = -e\frac{\partial\varphi(x,t)}{\partial x} - \frac{\partial}{\partial x}\sqrt{\frac{m^2c^4 + (\bar{p}_{0\perp}c - e\bar{A}(x,t))^2}{1 - \dot{x}^2/c^2}}\,,$$

These equations define the phase trajectory of electrons with initial momentum $\bar{p}_0$ and coordinate x_0 in self-consistent electromagnetic fields. In (3.1.65) $A(x,t)$ - vector potential of the transverse (div$A$ = 0) electromagnetic fields, $\varphi(x,t)$ – the scalar potential of the longitudinal fields satisfying the Maxwell equations:

$$\left(\frac{\partial^2}{\partial x^2} - \frac{\partial^2}{c^2\partial t^2}\right)\vec{A}(x,t) = -\frac{4\pi e}{c}\int \vec{v} f\left(x;t,p_x;p_y\right) d\bar{p}$$

$$\frac{\partial^2\varphi(x,t)}{\partial x^2} = -4\pi e\left(Zn_i - \int f\left(x;t,p_x;p_y\right) d\bar{p}\right)$$

We assume spatial one-dimensionality of our task and it allows us to use the law of conservation of transverse canonical momentum to find $p_\perp(p_{0\perp}, x_0, t)$.

The initial distribution function $f(\bar{p}_0;x_0)$ in (3.1.65) is the Maxwell – Boltzmann one:

$$f(\bar{p}_0;x_0) = \frac{1}{2\pi mT_e}\exp\left(-\frac{\bar{p}_0^{\,2}}{2mT_e}\right)Zn_{i0}\exp\left(-\frac{e\varphi(x_0)}{T_e}\right),$$

where the potential $\varphi(x_0)$ satisfies the Poisson equation:

$$\frac{\partial^2\varphi}{\partial x_0^2} = -4\pi eZn_{i0}[\exp(-\frac{e\varphi(x_0)}{T_e}) - \frac{n_i(x_0)}{n_{i0}}]. \tag{3.1.66}$$

The profile of ion concentration $n_i(x)$ is assumed specified function with the typical scale L growing from $n_i = 0$ at $x = 0$ to the constant value $n_{i0} = n_i|_{x\to\infty}$.

Analytical Model for ASE

Now we will consider the analytical model for Anomaly Skin Effect regime at the normal incidence of a laser wave. According to the numerical calculation, the electromagnetic field in plasma can be approximated by the following equations:

$$E_y(x;t) = \frac{E_0 \sin(\omega t)}{1 + x^2/l_s^2} \qquad B_z(x;t) = -\frac{2cE_0 x\cos(\omega t)}{\omega l_s^2(1 + x^2/l_s^2)^2} \tag{3.1.67}$$

$$E_x(x) = E_0\left[\Theta(x)\exp(-x/l) - \Theta(-x)\exp(x/l)\right]$$

Here $\Theta(x)$ is the Heavyside step-like function.

The fields are symmetrical with respect to replacement ($x \to -x$); this symmetry is caused by the requirement of mirror-like reflection of the electrons by the plasma edge. Note, that the primary role in the laser radiation absorption is played by the electrons, whose velocity is parallel to the plasma edge. Normal component of their velocity is small, excluding the possibility to overcome the potential barrier (field E_x) on the edge. The lengths l_s and l (i.e., the lengths of the skin layer for the transverse E_y and longitudinal E_x fields) are determined by the comparison with numerical simulation results. Approximate values of these lengths are $l_s = (c^2V_T/\omega\cdot\omega_p^2)^{1/3}$ and $l \approx L$. Non-exponential character of the transverse wave feeding with plasma depth is the result of the anomalous skin effect. Longitudinal field E_x is produced due to the action of pondermotive pressure of laser wave, pressurizing the electrons inside the plasma. Requirement of equilibrium of electrostatic force and this pondermotive pressure makes it possible to evaluate E_x and characteristic scale l:

$$\frac{E_0^2}{4\pi}(1+R) = e\int_0^\infty E_x(x)n_i(x)dx \tag{3.1.68}$$

resulting in characteristic evaluation:

$$E_{x0} \approx E_0\left(\frac{cL}{\omega l^2}\right)\left(\frac{\omega^2}{\omega_p^2}\right)\left(\frac{eE_0}{m\omega c}\right)(1+R) \tag{3.1.69}$$

This evaluation is in agreement in its order of magnitude with the numerical calculation.

Distribution function, numerically calculated in the previous section, can be modeled by evaluation of the electron movement in the fields (3.1.67) and taking Maxwell distribution as the starting one:

$$f(x;p_x;p_y;t)=\frac{n_e(x)}{2\pi mT}\exp\left[\frac{mc^2-(m^2c^4+p_x^2(0)c^2+p_y^2(0)c^2)^{1/2}}{T}\right] \tag{3.1.70}$$

Solution of the relativistic movement equations looks like:

$$\begin{aligned}
&P_y(0)=p_y+e/c\left[A_y(x-V_xt;0)-A_y(x,t)\right]\\
&P_x(0)=p_x+e\int_0^t E(x(\tau);\tau)d\tau+\frac{e(p_y-e/cA_y(x;t)}{mc\gamma}\int_0^t\frac{\partial A_y\left(x(\tau);\tau\right)}{\partial x}\partial\tau+\\
&+\frac{e^2}{2m\gamma c^2}\int_0^t\frac{\partial A_y^2(x(\tau);\tau)}{\partial x}d\tau,\ A(x;t)=\frac{cE_0\cos(\omega t)}{\omega(1+x^2/l_s^2)},\ \gamma=\left(1+\frac{p_x^2+p_y^2}{m^2c^2}\right)^{1/2}.
\end{aligned} \tag{3.1.71}$$

The lengths l_s and l are much shorter than the length of free flight of electron in plasma; hence the trajectory $x(\tau)$ can be approximated by the straight line : $x(\tau)=x+V_x\tau$ $=x-V_x(t-\tau)$. Hence, the equations (3.1.68), (3.1.70) and (3.1.71) make it possible to determine the analytical expression for the distribution function.

In the case of oblique laser pulse incidence on plasma, as we already mentioned it is convenient to calculate the distribution function in the coordinate frame, moving with the speed $V=c\sin(\theta)$ (θ -incidence angle) along the plasma edge. As was explained in previous section in this system the incidence is normal. Longitudinal field E_x recalculation to this set of coordinates results in the additional constant magnetic field and, hence, in the following field configuration:

$$\begin{aligned}
&E_x(x)=E_0\left[\Theta(x)\exp(-x/l)-\Theta(-x)\exp(x/l)\right]/\cos(\theta)\\
&A_y(x;t)=\frac{cE_0\cos(\omega t)}{\omega(1+x^2/l_s^2)}\cos(\theta)-tg(\theta)E_0\left[\Theta(x)\exp(-x/l)-\Theta(-x)\exp(x/l)\right]
\end{aligned} \tag{3.1.72}$$

Solution of this set can be drawn out of the solution of (3.1.71) by replacement of γ in (3.1.71) by $\gamma=\left[1+p_x^2/m^2c^2+(p_y/mc+\sin(\theta))^2\right]^{1/2}$. Noteworthy, that while for low intensities 10^{16} W/cm^2 the absorption depends on the incidence angle with the characteristic maximum for $\theta\approx\pi/2$, for higher intensities such dependence is absent. In this case the absorption coefficient tends its value for the normal incidence $\approx kl$.

Analytical Solution for SIB

We will construct now the expansion of solutions of equation system (3.1.58) in the parameter υ_T/υ_E. In the zero approximation we will obtain a well known set of hydrodynamic equations for cold plasma. Consider now the situation when $\frac{\omega c}{\omega_p v_T} > 1$, i. e., the case of SIB. The conservation law for the transverse canonical impulse permits to reduce the system (3.1.58) to two equations for the vector potential of normally incident electromagnetic wave $\overline{A}(x;t)$ and the longitudinal electric field $E_x(x;t)$ in plasma:

$$(\frac{\partial^2}{\partial \xi^2} - \frac{\omega^2}{\omega_p^2}\frac{\partial^2}{\partial \tau^2})\vec{a} = (\eta_i(\xi) + \frac{\partial E}{\partial \xi})\frac{\vec{a}}{\sqrt{1+a^2}}\sqrt{1-v^2} \; ,$$

$$\frac{\omega}{\omega_p}\frac{\partial}{\partial \tau} v\sqrt{\frac{1+a^2}{1-v^2}} = E - \frac{\partial}{\partial \xi}\sqrt{\frac{1+a^2}{1-v^2}} \; , \qquad (3.1.73)$$

$$v = -\frac{\omega}{\omega_p}\frac{\partial E/\partial \tau}{\eta_i(\xi) + \partial E/\partial \xi} \; .$$

The system (3.1.73) is written in the following dimensionless variables:

$$\xi = \frac{\omega_p}{c}x \; ; \tau = \omega t \; ; \; \vec{a} = \frac{e\vec{A}}{mc^2} \; ; \; E = \frac{eE_x}{mc\omega_p} \; ; \; \eta_i(\xi) = \frac{n_i(\xi)}{n_i(\xi = \infty)}$$

is the dimensionless concentration of plasma ions.

Circular Polarization State of Laser Radiation

Consider the case when (3.1.73) has accurate analytical solutions. One of them is the stationary solution for a wave with circular polarization state ($\overline{A}^2 = const(\tau)$):

$$\vec{a}(\xi;\tau) = a(\xi)(\overline{e}_z \cos\tau + \overline{e}_y \sin\tau) \qquad (3.1.74)$$

In this case, (3.1.73) is reduced to one equation:

$$(\frac{\partial^2}{\partial \xi^2} - \frac{\omega^2}{\omega_p^2})a(\xi) = (\Theta(\xi) + \frac{\partial^2}{\partial \xi^2}\sqrt{1+a^2})\frac{a}{\sqrt{1+a^2}} \qquad (3.1.75)$$

$$E = \frac{\partial}{\partial \xi}\sqrt{1+a^2} \qquad (3.1.76)$$

where the plasma boundary is chosen as $\eta_i(\xi) \equiv \Theta(\xi)$, the step function. Nonlinear equation (3.1.75) can be easily integrated, and its decreasing solution has the following form:

$$a(\xi)=\frac{2\sqrt{1-\frac{\omega^2}{\omega_p^2}}\,\mathrm{ch}[\sqrt{1-\frac{\omega^2}{\omega_p^2}}(\xi+\xi_0)]}{\mathrm{ch}^2[\sqrt{1-\frac{\omega^2}{\omega_p^2}}(\xi+\xi_0)]-1+\frac{\omega^2}{\omega_p^2}},\quad \xi>0, \tag{3.1.77}$$

$$\xi_0=\frac{1}{\sqrt{1-\frac{\omega^2}{\omega_p^2}}}\mathrm{arcch}\frac{a_0\sqrt{1-\frac{\omega^2}{\omega_p^2}}}{\sqrt{1+a_0^2}-1},$$

where a_0 is the value of the field on the plasma surface, $a_0=a(\xi=0)$.

In vacuum, the following incident and reflected waves correspond to solution (3.1.75):

$$\vec{a}(\xi;\tau)=(a_0\cos\frac{\omega}{\omega_p}\xi-\frac{\omega}{\omega_p}\sqrt{1+a_0^2}\sqrt{2(\sqrt{1+a_0^2}-1)-\frac{\omega^2}{\omega_p^2}a_0^2}\sin\frac{\omega}{\omega_p}\xi)\cdot(\vec{e}_x\cos\tau+\vec{e}_y\sin\tau) \tag{3.1.78}$$

At $a_0 << 1$, solution (3.1.77) becomes a simple shielding law $a(\xi)=a_0\exp(-\xi)$. The stationary solution exists not for all values of a_0. When a_0 reaches the value $\frac{3}{2}\frac{\omega^2}{\omega_p^2}$, the electron concentration $(1+\frac{\partial^2}{\partial\xi^2}\sqrt{1+a^2})$ becomes zero at the point $\xi = 0$. At greater a_0, solution (3.1.77) loses its physical meaning, because the concentration becomes negative. Stationary solutions in this case are impossible: plasma does not hold the incident wave, and the field penetrates in it in the form of separate filaments – solitons [84]. The value $a_0=\frac{3}{2}\frac{\omega_p^2}{\omega^2}$ corresponds to the field strength in vacuum of $\frac{3\sqrt{3}}{4}\frac{\omega_p^4}{\omega^4}$, and the value $a_0 = 1$ - the field value of $\frac{\sqrt{2}}{2}\frac{\omega_p}{\omega}$. Thus, we determined the interval of the field strength in plasma, for which the directed relativistic motion of electrons exists. The electron energy in units of mc^2 is $\mathcal{E}=\sqrt{1+a^2}$, its velocity is $v_\perp^2=\frac{a^2}{1+a^2}$, its longitudinal velocity is

$\nu = 0$, because in these conditions the force of the ponderomotive pressure is equal to the ambipolar field $E = \dfrac{\partial}{\partial \xi}\sqrt{1+a^2}$.

Absorption Coefficient for SIB

The equation set (3.1.73) permits to find fields in plasma in the zero approximation in $\upsilon_T^2/\upsilon_E^2$. By means of these fields, one can find the equation for the electron phase trajectory $\bar{p}(x_0;\bar{p}_0;t); x(x_0;\bar{p}_0;t)$ and to construct the distribution function (3.1.63). The results obtained with it will be valid in the first order in $\upsilon_T^2/\upsilon_E^2$. Below, we will calculate thermal corrections only in those cases when they are connected with absorption of energy.

Now consider absorption connected with Landau damping on separate particles. Using distribution function (3.1.65), the dissipated power can be presented as

$$Q = e\int \bar{v}(x_0;\bar{p}_0;t)\bar{E}(t;x(x_0;\bar{p}_0;t))f(\bar{p}_0;x_0)d\bar{p}_0 dx_0 = $$

$$= 2\int v_x \frac{\partial}{\partial x}\sqrt{e^2A^2 + m^2c^4}\Big|_{x(x_0;\bar{p}_0;t)} f(\bar{p}_0;x_0)d\bar{p}_0 dx_0 \qquad (3.1.79)$$

In dimensionless units, the absorption coefficient can be written as

$$\eta_a = 2\frac{v_T}{c}\frac{\omega_p^2}{\omega^2}\frac{1}{A_0^2}\int_{-\infty}^{+\infty}\int_0^{\infty} \left[v\frac{\partial}{\partial \xi}\sqrt{1+a^2}\Big|_{\xi(\xi_0;v_0;\tau)} (\theta(\xi_0) + \right.$$

$$\left. + \frac{\partial^2}{\partial \xi_0^2}\sqrt{1+a^2(\xi_0)})d\xi_0 \frac{\exp(-v_0^2)}{\sqrt{\pi}}dv_0 \right. \qquad (3.1.80)$$

where $\xi(\xi_0;v_0;t)$ and $v(\xi_0;v_0;t)$ determine the law of the longitudinal motion of an electron from the equation:

$$\frac{d\dot{\xi}}{d\tau\sqrt{1-\dot{\xi}^2}} = E\ (\xi(\tau);\tau) - \frac{\partial}{\partial \xi}\sqrt{\frac{1+a^2}{1-\dot{\xi}^2}}\Big|_{\xi=\xi(\tau)} \qquad (3.1.81)$$

For small $\dot{\xi} \sim \upsilon_T/c$, solution (3.1.81) is $\xi(\xi_0;\upsilon_0) = \xi_0 + \dfrac{\upsilon_T}{c}\dfrac{\omega_p}{\omega}\upsilon_0\tau$. In the next orders, small oscillations are superimposed on the uniform motion. Thus, expressions (3.1.78) and (3.1.79) permit to determine the absorption coefficient for electromagnetic wave, connected with the Landau damping on separate electrons. In weak fields ($a << 1$), the absorption coefficient determined from (3.1.78) has the form:

$$\eta_{SIB}^{(0)} = \frac{4}{\sqrt{\pi}} \left(\frac{\upsilon_{T}}{c} \right)^3 \frac{\omega_p^2}{\omega^2} \qquad (3.1.82)$$

This coincides with results of previous section. At $a >> 1$, estimations of the integral (3.1.80) show that $\eta_{SIB} \sim \eta_{SIB}^{(0)} / a$. Thus, collisionless absorption due to the Landau damping on separate particles is small (less than unities of percent in routine conditions of interaction of a short pulse with the matter).

Pondermotive Absorption in a Non-Uniform Plasma

We shall consider now a weak non-linearity ($a^2 < a$), displaying the solution (3.1.73) in a number on degrees a. This approximation limits intensity of an incident radiation to intensity 10^{18} W/cm^2. Result of decomposition (3.1.73) looks like:

$$\begin{aligned} &\frac{\partial^2 a^{(0)}}{\partial \xi^2} - \frac{\omega^2}{\omega_p^2} \frac{\partial^2 a^{(0)}}{\partial \tau^2} = \eta_i(\xi) a^{(0)}; \; E^{(0)} = 0 \\ &\frac{\partial^2 E^{(1)}}{\partial \tau^2} + \frac{\omega_p^2}{\omega^2} \eta_i(\xi) E^{(1)} = \frac{\omega_p^2}{\omega^2} \eta_i(\xi) \frac{\partial}{\partial \xi} \frac{a^{(0)2}}{2} \end{aligned} \qquad (3.1.83)$$

For linear polarisation of a incident wave a pondermotive force depends on time Then the solution of zero approximation of system (3.1.83) has the following form:

$$\begin{aligned} &a^{(0)} = a(\xi) e_y \cos(\tau); \quad \eta_i(\xi) \neq 0; \; \xi \succ 0 \\ &a^{(0)} = a(0) \cos(\omega \xi / \omega_p) + \frac{\omega_p}{\omega} a'(0) \sin(\omega \xi / \omega_p) e_y \cos(\tau); \; \xi \prec 0 \end{aligned} \qquad (3.1.84)$$

Where $a'(0)$ - derivative $a(\xi)$ at $\xi = 0$. In (3.1.84) it means, that η_i is nonzero in area $\xi > 0$. The function $a(\xi)$ depends on a specific structure of concentration. The further calculations will be made for any $a(\xi)$, and the example of absorption coefficient for exponential of growing concentration is resulted in the following paragraph.

The averaged Pointing vector of a laser wave, appropriate to a field (3.1.84) is determined by $a(0)$ and $a'(0)$:

$$<\gamma> = \frac{\omega^2}{32 \pi c} \left(a^2(0) + \omega_p^2 a'^2(0) / \omega^2 \right) \qquad (3.1.85)$$

The solution of the equation of the first approximation for a longitudinal field E at a given right part in view of a resonance looks as follows:

$$E^{(1)}(\xi,\tau)=\frac{1}{4}\frac{\partial}{\partial\xi}a^2(\xi)\left[1+\frac{\omega_p^2\eta_i(\xi)/\omega^2}{\omega_p^2\eta\ (\xi)/\omega^2-4}\cos\left(2\tau-\cos\left(\omega_p\sqrt{\eta_i(\xi)}\tau/\omega\right)\right)\right]$$

(3.1.86)

In a point $\eta_i(\xi) = 4\omega^2/\omega_p^2$ the field $E^{(1)}$ linearly accrues in due course, that corresponds to a resonance between pondermotive force and plasma wave. Energy of a longitudinal field per unit of the area of plasma

$$W=\frac{m^2c^2\omega_p}{8\pi e^2}\int E^{(1)^2}d\xi \tag{3.1.87}$$

Calculated with the help of presentation of δ-function

$$\lim_{\tau\to\infty}\frac{\sin^2\alpha\tau}{\pi\tau\alpha^2}=\delta(\alpha)$$

Also linearly grows in due course:

$$W=\frac{m^2c^3\omega\tau}{2^7e^2}\left[\frac{\partial a^2}{\partial\xi^*}\right]^2\left[\frac{\partial\sqrt{\eta_i(\xi)}}{\partial\xi^*}\right]^{-1} \tag{3.1.88}$$

By dividing absorbed power on Pointing vector of laser wave we shall receive required absorption coefficient η_a

$$\eta_a=\pi a^2(\xi^*)\left[\frac{\partial a^2(\xi^*)}{\partial\xi^*}\right]^2\left[\left(a^2(0)+a'^2(0)\right)\frac{\partial\sqrt{\eta_i(\xi^*)}}{\partial\xi^*}\right]^{-1} \tag{3.1.89}$$

So, the received expression determines absorption coefficient at normal incidence of an electromagnetic wave on non-uniform plasma through the decision of the equation of zero approximation for vector potential. A condition of applicability (3.1.87) is: electron thermal speed less then its hydrodynamic speed, or amplitude of vector potential it imposes a condition $1 >> a^2 >> \upsilon_T\omega_p/c\omega$.

Absorption Coefficient in Plasma with Exponential Growing Density

The function $a(\xi)$ manages to be constructed analytically for a limited set of structures $\eta_i(\xi)$. One of variants is $\eta_i(\xi)=\exp(\alpha\xi)$, where $\alpha = c/L\omega_p$ – un-dimensional scale of plasma inhomogenity. The solution (3.1.83), aspiring to zero at large ξ looks like

$$a(\xi) = -2a_0\left(2\Omega sh(2\pi\Omega/\alpha)/\pi\alpha\right)^{1/2} K_{2i\Omega/\alpha}\left(2\exp(\alpha\xi/2)/\alpha\right) \qquad (3.1.90)$$

In (3.1.90) for Ω is the relation ω/ω_p, $K(\xi)$ – function McDonald, a – amplitude of a laser wave is designated at $\xi \to \infty$. Doing further account according to the above-stated technique, we shall receive the following result for absorption coefficient:

$$\eta_a = \frac{64}{\pi} a_0^2 x_l^3 \, sh^2(2\pi x_l) K_{2i\Omega/\alpha}^{;2}(4x_l); \; x_l = \omega L/c \qquad (3.1.91)$$

The stroke in (3.1.91) designates derivative of function McDonald.

We analyse (3.1.91) to understand the behevior of absorption coefficient. We note, that as well as in case of incident at some angle the resonant absorption does not depend on density of plasma. At small inhomogenity scales

$$\eta_a = 16\pi a_0^2 x_l^3 \ln^2(1/2x_l); \; x_l \prec\prec 0 \qquad (3.1.92)$$

We shall remind, that at oblique incident for small x_l: η $2\alpha x_l \sin 2\theta/\cos\theta$. At large x_l asymptotic (3.1.89) looks as follows:

$$\eta_a = 4\pi a_0^2 x_l \exp\left(-(16-4\pi)x_l\right); \; x_l \succ\succ 1$$

The maximum η_a =0.38 absorption coefficient reaches at x_{lmax}=0.26, as well as in case of oblique incidence. We compare resonant absorption to other channels of absorption. In collision-less plasma this is Landau damping on separate electrons (SIB see previous section):

$$\eta^{SIB} = 8\sqrt{2/\pi}\frac{\upsilon_T^3\omega_p^2}{c^3\omega^2}; \; 0 \prec L \prec c/\omega_p \qquad (3.1.93)$$

At large L ζ^{SIB} exponentially decreases with increase L. Thus, at sufficient intensity of a laser wave $a_0^2 > \upsilon_T\omega_p/c\omega$ in spatially non-uniform plasma ($L > c/\omega_p$) the resonant absorption is dominant, and absorption coefficient reaches tens of percents. We note, that at some angle of incidence the number of the terms proportional a in the equations also results in additional absorption, connected to generation of the second harmonic in reflected radiation and presence of the second resonant point. But at oblique incidence these effects give only amendment to standard linear transformation.

Comparison Between Theory, PIC and Vlasov Simulations

There are different codes for kinetic simulations of laser plasma absorption. As an example, we compare the absorption calculated from theory with that obtained from both

PIC and Vlasov simulations for the following: exponential density profile with $L/\lambda = 0.15$. $n_e/n_c = 2$, and T_e= 10 keV; irradiance $I\lambda^2 = 10^{16}$ W/cm^2, P-polarized.

The results, shown in **Fig. 3.1.19** display good agreement between the theory, PIC [85] and Vlasov simulations, although some discrepancy at intermediate angles is apparent. Better agreement is found if a temperature of 5 keV is taken using the PIC code, but a detailed inspection of the initialization routines in both codes indicate that the discrepancy in absorption is not due to differing definitions of T_e, but more likely a some difference in the initial density profiles. The PIC results [86,87] from both BOPS and EUTERE codes gave identical absorption rates, so only one set of results is displayed. Since a different number of particles was used with the two PIC codes, we can be confident that the results have converged. The absorption values are also in fair agreement with the analytical expression (for the "normal" skin effect with no temperature dependence). The discrepancy here can be attributed to the neglect of "vacuum-transit" electrons (Brunel effect) from the analytical model. Thus, one can conclude that most of the absorption observed in the simulations, in fact, takes place within the skin layer. The error bars attached to the PIC results indicate be typical fluctuation in the absorption about the average value.

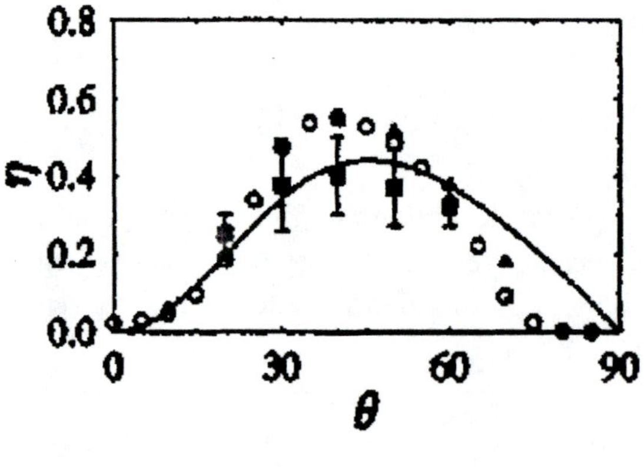

Fig. 3.1.19. The results display the agreement between the theory, PIC and Vlasov simulations.

The second example we consider is a highly over-dense step profile with $L/\lambda = 0.023$, T_e= 10 keV, and $n/n_c = 25$. In this case the density scale length is less than the skin depth: $L/l_s = \omega_p L/c$ 0.7, and υ_T/c ω/ω_p, which means we are between the normal and anomalous skin effect regimes. (For pure, highly over-dense step profiles, the Brunel effect is suppressed, because the electrons are prevented from entering the vacuum by the induced dc fields near the surface.) A comparison of the total absorption calculated from PIC and Vlasov simulations, together with to theoretical value, is summarized in **Fig. 3.1.20.**

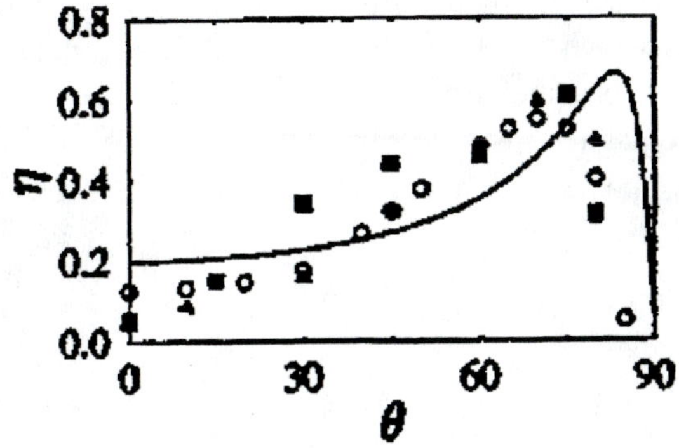

Fig. 3.20. A comparison of the total absorption calculated from PIC and Vlasov simulations, together with to theoretical value.

3.2 Scattering of a Short Laser Pulse in Strongly Inhomogeneous Plasma

Nonlinear Wave Interaction in Plasma

If the wave amplitude rises, the waves begin to interact with one another. This may increase the amplitude of some waves but decrease that of others. This process represents a sort of energy pumping from some oscillations into the others. At the beginning we will omit the quantitative relations that describe the energy exchange between the waves involved in the interaction and restrict ourselves to its qualitative description. The focus will be on the SBS process. Having a low energy threshold, this process plays an important role in the laser energy deposition into the plasma, since the light scattered by ion sound can take some of the laser energy away, reducing the deposited energy. On the other hand, the scattered radiation can provide information important for the diagnostics of some plasma parameters.

In addition to the energy deposition and plasma diagnostic problems, SBS allows the solution of some conventional problems of quantum electronics concerning spatial and time characteristics of a laser pulse in the medium. This requires the account of plasma properties negative to the problems mentioned: absorption, heating, and the non-uniform distribution of its hydrodynamic parameters, which reduce the output laser pulse quality. In spite of this, the plasma may prove to be a perspective medium for nonlinear control, especially, of super high power laser radiation.

The basic features of this non-linear process are as follows. A powerful electromagnetic wave propagating through the plasma interferes with a weak electromagnetic wave arising from its scattering by random thermal density fluctuations. If the interference pattern is in phase with the weak ion sound wave, this wave amplitude begins to grow together with the scattered wave amplitude because of the lower energy of the incident wave. The density modulation is due to the high frequency pressure of the electromagnetic field, which will be discussed below. The generation of ion sound tends to grow because the ion sound wave acts as a three-dimensional diffraction grating, which scatters electromagnetic waves. The scattered waves increase the field amplitude within the interference peaks, increasing the amplitude of the ion sound wave, and so on. For this feedback to take place, certain conditions known as *decay conditions* must be fulfilled:

$$\omega_0 = \omega_1 + \omega_s,$$

$$\boldsymbol{k}_0 = \boldsymbol{k}_1 + \boldsymbol{k}_s, \tag{3.2.1}$$

where ω_0, ω_1, ω_s and $\boldsymbol{k}_0$, $\boldsymbol{k}_1$ and $\boldsymbol{k}_s$ are, respectively, the frequencies and wave vectors of the incident, scattered, and ion sound waves. The first equation in (3.2.1) indicates the amplitude synchronization of the incident and scattered waves with the ion sound wave; the second equation indicates a spatial correlation. In uniform plasma, the correlation may relate to the whole plasma volume, and the scattered wave is amplified in the direction of its propagation, amplifying, simultaneously, the ion sound oscillations. In non-uniform plasma, the decay condition is fulfilled for a given frequency of the scattered wave only in the vicinity of one point. This wave is not amplified at an adjacent point, because a different frequency is required. For this reason, the SBS efficiency in non-uniform plasma is lower than that in a uniform plasma, decreasing with increasing plasma non-uniformity.

The process of stimulated Brillouin scattering is sometimes expressed as

$$t \to t' + s, \tag{3.2.2}$$

where t denotes a transverse electromagnetic wave and s an ion sound wave.

In addition, there are other processes of wave decay and fusion. These are usually represented as

$$t \rightleftarrows t' + l \tag{3.2.3}$$

$$t \rightleftarrows l + l \tag{3.2.4}$$

$$t \rightleftarrows l + s \tag{3.2.5}$$

Here, l stands for a Langmuir wave and the oppositely directed arrows indicate that the process is effective in both directions, i.e., the wave fusion is also possible. The process described by formula (3.2.3) is known as *stimulated Raman scattering,* which represents the scattering of an electromagnetic wave by a Langmuir plasma wave (rather than by ion sound, as in SBS). The reverse process is the fusion of a Langmuir wave and an electromagnetic wave to generate a new electromagnetic oscillation. The processes described by formulas (3.2.2) and (3.2.3) primarily lead to wave backscattering, which means a smaller absorption of laser light by the plasma. When the wave decay produces only Langmuir and ion sound waves, as in (3.2.4) and (3.2.5), the plasma absorbs more laser radiation, because these waves cannot escape the plasma and are eventually damped. The region where the processes (3.2.3) and (3.2.4) are most intensive lies in the vicinity of one quarter critical concentration, $n_e = n_{ec}/4$. In the forward process, the frequency of the produced waves is equal to the half frequency of the incident electromagnetic wave. The reverse process of wave fusion (3.2.3) and (3.2.4) shows interesting physical phenomena, e.g., the generation of incident light harmonics. In particular, the process of resonant absorption or that of (3.2.5) can produce high intensity Langmuir waves with a frequency approximately equal to that of the heating light ω_0. Formulas (3.2.3) and (3.2.4) describe the fusion of these waves with one another or with the incident wave (or the one reflected from the critical concentration point). The generated wave of frequency $2\omega_0$ may escape the plasma. Similarly, an electromagnetic wave of frequency $(3/2)\omega_0$ may be generated by the process (3.2.3) in the vicinity of the point $n_e = n_{cr}/4$. In this case, the reflected wave passing the point $n_e = n_{cr}/4$ in the plasma may be combined with the Langmuir wave of frequency $\omega_0/2$, produced by the processes (3.2.3) and (3.2.4). The resulting wave will have the frequency $\omega_0 + \omega_0/2 = (3/2)\omega_0$.

SBS in a Laser Plasma

Stimulated Brillouin scattering is studied by analysing a set of equations for laser plasma hydrodynamics averaged over a high frequency, together with the electromagnetic field equation. It is usually assumed that a transverse wave is S-polarized and that the plasma has a high temperature, so that one has to allow for the electron thermal conductivity and ion viscosity. At a constant ion temperature, we have the following set of equations to describe the processes of interest [88]:

$$\frac{\partial \eta_e}{\partial \tau} + \nabla(\eta_e \vec{\mu}) = 0$$

$$\frac{\partial \vec{\mu}}{\partial \tau} + \vec{\mu}\nabla\vec{\mu} + G_s\vec{\mu} = -\nabla(\eta_e T)/\eta_e - \nabla|E|^2 \tag{3.2.6}$$

$$\frac{3\partial T}{2\partial \tau} + \nabla(\vec{\mu}T) = \nabla æ_T \nabla T/\eta_e + \nu|E|^2/\eta_e - \delta_i T$$

$$-2i\beta\frac{\partial E}{\partial\tau}+\Delta E+\left(1-\eta_e\right)E=0$$

Here, η_e = n_e/n_{cr}, $\vec{\mu}=\vec{u}/c_s$ and T = $(ZT_e + T_i)/(ZT_{I} + T_{i0})$ are, respectively, the dimensionless plasma density, velocity and temperature varying slowly over the field oscillation period, $2\pi/\omega_0$, when the field strength is dimensionless, $\bar{E}_l$ = Re[$\bar{E}_0\exp(-i\omega_0 t)$], $E = \bar{E}_l(16\pi n_c T_e)^{-1/2}$, and enters the equation of motion via the striction pressure and heating terms in the temperature equation; T_e, T_i are the electron and ion temperatures; Z is the ion charge; $c_s = [(ZT_e + T_i)/m_e]^{1/2}$; ν_{ei} is the frequency of electron-ion collisions; $\nu = 2\nu_{ei}\eta_e/\omega_s$, æ$_T$ is the electron thermal conductivity; $G_s = \nu_s/\omega_s$ is the dimensionless ion viscosity, $k_0\vec{r} = (\xi, \zeta)$, $\tau=\omega_s t$, $\beta=\omega_s/\omega_0$.

As we discussed above, SBS is a decay process involving two electromagnetic waves (incident and scattered) and an ion sound wave. For a long laser pulse duration in a uniform stationary plasma, the scattered wave is amplified and has a logarithmic gain on a distance æ $\cong k_0 L\eta_e|E|^2/2G_s$, while in a non-uniform plasma it has the scale of density non-uniformity L_n with the gain æ $\cong k_0 L_n|E|^2(1-\eta_e)^{1/2}\pi/2$. The SBS process occurs at a frequency $\omega_1 = \omega_0 + 2k_0 c_s[\mu(\cos^2\theta - \eta_e)^{1/2} - (1 - \eta_e)^{1/2}]$ at the points of wave synchronism, where the plasma density is below the critical value. The calculated scattering diagram for a plane wave has a large width and a backscattering maximum. Long time scale experimental SBS studies have been made in laser plasma with a fairly uniform sub-critical density and in strongly non-uniform plasma having a beam reflection surface. It is clear from the foregoing that we will be unable to separate the two reflection components - the mirror reflection and the backscatter - in the normal light incidence onto the target surface, whereas oblique incidence allows their individual analysis. In the latter case, the mirror component will carry information on the absorption processes, since the mirror-reflected light has passed through a region of maximum absorption. The backscatter component can provide information on stimulated scattering occurring most intensively in low density plasma regions.

3.2.1 Stimulated Brillouin Scattering of Short Intense Laser Pulses in a Dense Plasma

From above we see that of the main nonlinear processes with a relatively low threshold and readily detected experimentally is stimulated Brillouin scattering (SBS), which has been investigated quite thoroughly [89] for relatively long pulses. A detailed analysis of the characteristics of this process has been made for low-density plasma interacting with picosecond laser pulses [90,91] and also for pulses shorter than 10 ps [92]. In this section we will analyze SBS of laser pulses of 1 - 10 ps duration in a strongly inhomogeneous short and dense laser-plasma layer formed from solid - targets.

Model Equations

Let us consider a plane layer of totally ionized plasma with the ion density rising to a critical value along the x-axis. An oblique beam of plane waves is incident onto the layer from the region $x < 0$. The set of equations for the complex amplitude of the z-component of the field and for the density perturbations ($\eta_e = \eta_0 + \delta\eta$) in a stationary plasma at constant temperature is

$$2i\beta\frac{\partial E}{\partial\tau}+\Delta E+\left[1-\eta_0\left(1-i\nu_{ei}/\omega_0\right)\right]E=E\delta\eta$$
$$\frac{\partial^2\delta\eta}{\partial\tau^2}+2G_s\frac{\partial\delta\eta}{\partial\tau}-\Delta\delta\eta=0.25\eta_0\left(\xi\right)\Delta\left|E\right|^2 \tag{3.2.7}$$

The solution to this set of equations is sought in a two-dimensional region as the expansion in the plane waves. The boundary conditions for electromagnetic waves are those at the medium boundary for the incident and scattered waves; for the sound wave, this is the free exit from the plasma.

We describe SBS in two dimensions by the system of equations following from (3.2.7):

$$\left(\frac{2ic_s}{c}\frac{\partial}{\partial\tilde t}+\frac{\partial^2}{\partial\tilde x^2}+k_{xn}^2\right)\tilde E_n=\left(1-i\tilde\nu\right)\sum_{m\neq 0}\tilde E_{n-m}\delta\tilde N_m,$$
$$\left(\frac{\partial^2}{\partial\tilde t^2}+2\Gamma_s^{(n)}\frac{\partial}{\partial\tilde t}-\frac{\partial^2}{\partial\tilde x^2}+\left(nk_{11}\right)^2\right)\delta\tilde N_n= \tag{3.2.8}$$
$$=\tilde N_0\left(\frac{\partial^2}{\partial\tilde x^2}-\left(nk_{11}\right)^2\right)\left(\left|\tilde E\right|^2\right)_n/4.$$

where $\tilde E=\upsilon_E/\upsilon_{T_e}=E/\left[4\pi n_c\left(\tilde Z T_e+T_i\right)\right]^{1/2}$ is the dimensionless complex amplitude of an S-polarised wave; $\tilde N=n_e/n_{ic}$; $n_{ic}=n_c/Z$ is the critical density of ions; $n_c = m_e\omega_0^2/4\pi e^2$; $\tilde x,\tilde y=k_0x,k_0y$ are the dimensionless coordinates; x is the coordinate orthogonal to the plasma layer and varying in the interval $[0, D_x]$, where $x = 0$ corresponds to the front boundary of the plasma and $.x = D_x$ corresponds to the region where the plasma density is critical; y is the coordinate parallel to the plasma layer and varying in the interval $(0, D_y)$; $\tilde t=\omega_s\left(k_0\right)t$; $k_0 = \omega_0/c$; $\omega_s(k_0) = c_s k_0$; c_s is the velocity of sound; $\delta N\left(\tilde x,\tilde y,\tilde t\right)=\tilde N\left(\tilde x,\tilde y,\tilde t\right)-\tilde N_0\left(\tilde x\right)=\sum_n\delta\tilde N\left(\tilde x,\tilde t\right)\exp\left(ink_{11}\tilde y\right)$ are plasma perturbations caused by the action of the ponderomotive pressure; c is the velocity of lights:

$$k_{11} = \frac{2\pi}{D_y k_0};\ k_{xn}^2 = \cos^2\theta_n - (1 - i\tilde{\nu})(\tilde{N}_0 + \delta\tilde{N}_0);$$

$$\cos^2\theta_n = 1 - (nk_{11})^2;$$

$$\tilde{\nu} = \nu_{ei}/\omega_0$$

We assume that a harmonic of the field $\widetilde{E}_n(\widetilde{x},\widetilde{t})$ with n = +1 corresponds to a pump field incident at an angle θ_1 relative to the layer normal and to a secularly scattered wave, where as $\widetilde{E}_n(\widetilde{x},\widetilde{t})$ with n = - 1 corresponds to a backscattered field; the harmonics with n = 0, ±2 are included in describing acoustic perturbations and the excitation of these harmonics corresponds to self-interaction of' the incident and reflected radiation and to stimulated scattering. The damping decrement of the ionic sound can be described approximately by

$$\Gamma_s^{(n)} = \Gamma_{s0}\left(nk_{11} - \frac{1}{2k_{x1}}\frac{\partial^2}{\partial\tilde{x}^2}\right) + \Gamma_{si},\ \text{for}\quad \nu_{ii} < \omega_s,$$

$$\Gamma_s^{(n)} = \frac{\gamma_{se}}{\omega_s(k_0)}\left(nk_{11} - \frac{1}{2k_{x1}}\frac{\partial^2}{\partial\tilde{x}^2}\right) + \left(\frac{0.8T_i}{\tilde{Z}T_e}\right)\times$$

$$\times\frac{\omega_s(k_0)}{\nu_{ii}}\left((nk_{11})^2 - \frac{1}{2k_{x1}}\frac{\partial^2}{\partial\tilde{x}^2}\right)\quad \text{for}\ \nu_{ii} > \omega_s$$

where

$$\Gamma_{s0} = [(\gamma_{se} + \gamma_{si})/\omega_s(k_0)];\ \Gamma_{s1} = [\gamma_{sii}/\omega_s(k_0)];\ k_{x1} = (\cos^2\theta_1 - \tilde{N}_0)\quad \text{for } 4k_{x1}^2 > k_{xc}^2$$

$$\text{and}\quad k_{xc} = 2(L_c k_0)^{-1/3}\ \text{for } 4k_{x1}^2 < k_{xc}^2;$$

$$L_c = c_s t_L/2$$

is the inhomogeneity scale of the density at the critical point; t_L is the pulse duration; $\gamma_{se,si}$ are the Landau damping coefficients of electrons and ion; γ_{sii} is the damping coefficient representing ion-ion collisions of frequency ν_{ii}. We shall assume that the plasma has an exponential density profile and that the inhomogeneity scale is L_c along the x axis. The acoustic equation (3.2.8) can be applied to small perturbations of the density if $|\delta\tilde{N}_n(\tilde{x},\tilde{t})| << \tilde{N}_0(\tilde{x})$.

Scattering Processes

In describing plasma perturbations it is necessary to take account of a number of' processes.

Self Interaction of Pump Waves

In this case there are plasma density perturbations caused by the optical pressure,

$$\delta\widetilde{N}_0(\widetilde{x}) = \sum_\sigma \frac{\left|\widetilde{E}_{1\sigma}\right|^2}{4} \quad (\sigma = \pm 1),$$

and the plasma density is modulated by interference between the incident and reflected pump waves:

$$\delta\widetilde{N}_0(\widetilde{x}) \approx \sum_\sigma \exp\left(+2i\sigma k_x^{(0)}\right)\widetilde{E}_{1\sigma}\widetilde{E}^*_{1-\sigma},$$

$$\widetilde{E}_1 = \sum_\sigma \widetilde{E}_{1\sigma} \exp\left(+i\sigma k_x^{(0)}\widetilde{x}\right),$$

where $k_s^{(0)} = \left[\cos^2\theta_1 - \widetilde{N}_0[\widetilde{x}]\right]^{1/2}$; $\widetilde{E}_{11}$, $\widetilde{E}_{1-1}$ are the amplitudes of the incident and reflected pump waves. These plasma density perturbations correspond to self-interaction of the pump waves.

Stimulate Brillouin Backscattering

Now, we may encounter plasma density perturbations as a result of excitation of acoustic waves propagating parallel to the pump waves:

$$\delta\widetilde{N}_2 \approx \exp\left(-i\omega\widetilde{t}\right)b^*(\omega)\sum_\sigma \exp\left(-2i\sigma k_x^{(0)}\widetilde{x}\right)\widetilde{E}_{1-\sigma}\widetilde{E}^*_{-1\sigma},$$

where $\widetilde{E}^*_{-1\sigma}$ are the amplitudes of the scattered waves;

$$\widetilde{E}_{-1} = \sum_\sigma \widetilde{E}_{-1\sigma} \exp\left(+i\sigma k_x^{(0)}\widetilde{x}\right);$$

$$b(\omega) = \omega_{sb}^2\left[2\omega\gamma_{sb}(\omega) - i\left(\omega_{sb}^2 - \omega^2\right)\right]^{-1};$$

$$\omega_{sb} = 2\left[1 - \widetilde{N}_0(\widetilde{x})\right]^{1/2}; \quad \gamma_{sb}(\omega_{sb}) = \Gamma_{s0}\omega_{sb};$$

ω is the dimensionless Stokes frequency.

Double Stimulate Brillouin Scattering

In such scattering, there are plasma density perturbations caused by excitation of acoustic waves travelling parallel to the plasma layer:

$$\delta\widetilde{N}_2 \approx \exp\left(-i\omega\widetilde{t}\right)d^*(\omega)\sum_{\sigma}\widetilde{E}_{1\sigma}\widetilde{E}^*_{-1\sigma};\quad \omega_{sd} = 2\sin\theta_1;$$

$$\gamma_{sd}(\omega_{sd}) = \Gamma_{s0}\omega_{sd};\quad d(\omega) = \omega^2_{sd}\left[2\omega\gamma_{sd}(\omega) - i\left(\omega^2_{sd} - \omega^2\right)\right]^{-1}.$$

This process corresponds to growth of a low-threshold absolute instability [93-95].

Specular Stimulated Brillouin Scattering

There are now plasma density perturbations corresponding to the excitation of acoustic waves travelling at right angles to the plasma layer [93-95]:

$$\delta\widetilde{N}_{01} \approx \exp\left(-i\omega\widetilde{t}\right)c^*(\omega)\sum_{\sigma}\exp\left(=2i\sigma k^{(0)}_x\widetilde{x}\right)\widetilde{E}_{1\sigma}\widetilde{E}^*_{-1-\sigma},$$

where

$$\omega_{sm} = 2\cos\theta_1;\quad c(\omega) = \omega^2_{sm}\left[2\omega\gamma_{sm}(\omega) - i\left(\omega^2_{sm} - \omega^2\right)\right]^{-1};$$
$$\gamma_{sm}(\omega_{sm}) = \Gamma_{s0}\omega_{sm}.$$

Scattering of Picosecond Pulses

We first investigated the scattering of picosecond pulses by plasma created directly by the pulse itself, which has a sufficiently high contrast so that the action of a pre-pulse can be ignored. In numerical simulation of the scattering process we used the improved version of DSMBS code [96]. We also assumed that the total length of the plasma layer is $2L_c$, that its density is $\widetilde{N}_0(\widetilde{x}=0) = 0.135$, the angle of incidence is $\theta_1 = 30^\circ$ and the initial density perturbation is $\left|\delta\widetilde{N}^*_n(\widetilde{x},0)\right| \le 10^{-3}$. The backscattering and specular reflection coefficients are defined as follows:

$$R = \frac{k_{x-1}(0)\left|\widetilde{E}_{-1}(0,\widetilde{t})\right|^2}{8\pi S_{x0}},\quad R_S = \frac{k_{x+1}(0)\left|\widetilde{E}_1(0,\widetilde{t}) - \widetilde{E}^0_1(\widetilde{t})\right|^2}{8\pi S_{x0}}, \tag{3.2.9}$$

where $\widetilde{E}_{10}(\widetilde{t})$ is a given incident field; $S_{x0} = \sum_n k_{xn}(0)\left|\widetilde{E}^0_n\right|/8\pi$ is the radiation flux incident along the x axis; a pump pulse is assumed to have a rectangular time profile and the scattering is then characterized by the coefficient R at the end of the pulse. The range of validity of the adopted plane-layer plasma model is represented by the condition r_f L_c, where r_f is the radius of the pump beam at the focus of a lens [96], known to be satisfied by short pulses.

The pump intensity was represented by the steady-state gain over a distance L_c for sound of frequency ω_{sd},

$$æ_0 = \frac{\omega_{sd}}{16\gamma_{sd}}\left(\frac{\upsilon_E}{\upsilon_{Te}}\right)^2 \frac{\omega_0 L}{c} = 0.785\sin\theta_l \left(\frac{\upsilon_E}{\upsilon_{Te}}\right)^2 \frac{L_c}{\lambda_0 \Gamma_{s0}}, \tag{3.2.10}$$

and the scattering was represented by the parameters L_c/λ_0, Γ_{s0}, $æ_0$, θ_l, and by the pump attenuation $2\pi\nu L_c/\lambda_0$ where λ_0 is the incident radiation wavelength. It follows from hydrodynamic calculations carried out in the model of planar plasma expansion and inverse bremsstrahlung absorption [97], that the electron temperature is $T_e \sim 100$ - 1000 eV for $Z = 3$ - 12 and at pump intensities $I = 0.05 - 4$ PW cm^{-2} and for durations $t_L = 1 - 10$ ps, we have T_i/T_e 10^{-1}, Γ_{s0} 0.1. We also established, using the employed code in that change in L_c during a pulse is much less then λ_0, that the changing of T_e is not too large (~20%) and the value of T_e is distributed homogeneously inside the layer.

Scattering of Pulses of $t_L \geq 1/\omega_s$ Duration

Since $\omega_s t_L = 2L_c k_0$, then in the investigated range of temperatures and for pulse durations 1 ps < t_L < 4ps, we have $L_c k_0 < 1$, $L_c/\lambda_0 < 1$, $L_c/\lambda_0 \sim 0.1 - 0.4$. Such temperatures and the intensities in the range $I = 0.05 - 4$ PW cm^{-2} correspond to $\upsilon_E/\upsilon_{Te} \sim 0.33 - 0.8$, $æ_0 = 0.35 - 2.6$ and, since $2\pi\nu L_c/\lambda_0 < 1$, the absorption over the length of the layer does not exceed $5 10^{-3}$ ($R_S \approx 0.99$).

In this case, a numerical calculation predicts a very low backscattering intensity. For example, if $t_L \sim 1$ ps, then throughout the investigated range of the pump intensities the plasma density perturbations correspond to the initial state, i.e. the scattering is thermal ($R \leq 10^{-5}$). If $t_L \sim 4$ ps, it follows from **Fig. 3.2.la** that $R \leq 10^{-3}$, i.e. this reflection coefficient is considerably larger than for $t_p \sim 1$ ps and the scattering occurs above the threshold; the threshold pump intensity corresponds to $æ_0^{th} \approx 0.3$. Calculations show also a very weak dependence of the scattering parameters on Z for $Z = 3$ - 12, which is typical of $2\pi\nu L_c/\lambda_0 < 1$ when $Z/A \sim 0.5$.

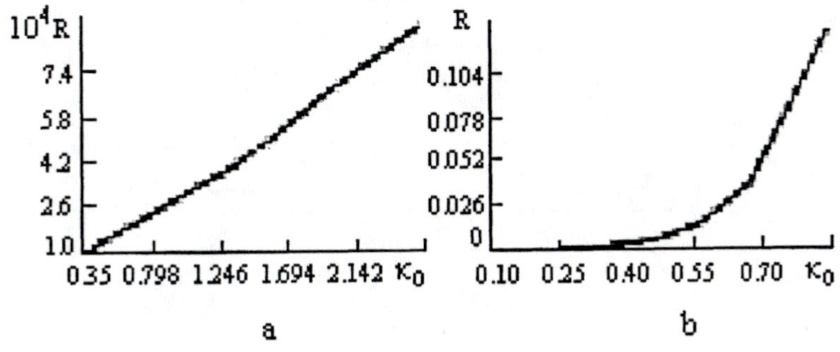

Fig. 3.2.1. Dependence of the back-scattered coefficient R calculated for $\Gamma_{s0} = 0.05$, $L_c/\lambda_0 = 0.4$, $t_L = 4$ ps (a) and $\Gamma_{s0} = 0.1$, $L_c/\lambda_0 = 1.1$, $t_L = 10$ ps (b).

There are two main reasons for such a low backscattering intensity in this case: the small inhomogeneity scale $L_c/\lambda_0 < 1$ and the very transient scattering regime, since $t_L < t_s < t_r$ for $\omega_{spt} \geq 1$, where $t_s = 2\pi/\omega_s$ is the acoustic wave period and t_r is the relaxation time. At sufficiently high pump intensities the amplitude of an acoustic harmonic $\delta\widetilde{N}_0$ is determined entirely by the process described in subsection. The excitation of $\delta\widetilde{N}_0$ is accompanied by the excitation of an acoustic harmonic of amplitude $\delta\widetilde{N}_2$ and focused in the dense-plasma region near the turning point corresponding to $\delta\widetilde{N}_0 = \cos^2\theta_1$. The distance from this turning point to the critical surface is very short if $L_c/\lambda_0 < 1$. The amplitude of $\delta\widetilde{N}_0$ is much larger than that of $\delta\widetilde{N}_2$ and acoustic harmonics of amplitudes $\delta\widetilde{N}_0$ and $\delta\widetilde{N}_2$ are damped out in low density plasma. The nature of excitation of the harmonic with the amplitude $\delta\widetilde{N}_2$ determines entirely the nature of the frequency spectrum of the scattered waves. We can see from **Fig. 3.2.2a** that the spectrum has a red shift $\delta\omega \approx$ -2sinθ_1 = -1.1, typical of the growth of an absolute instability of double stimulated Brillouin scattering [94,95].

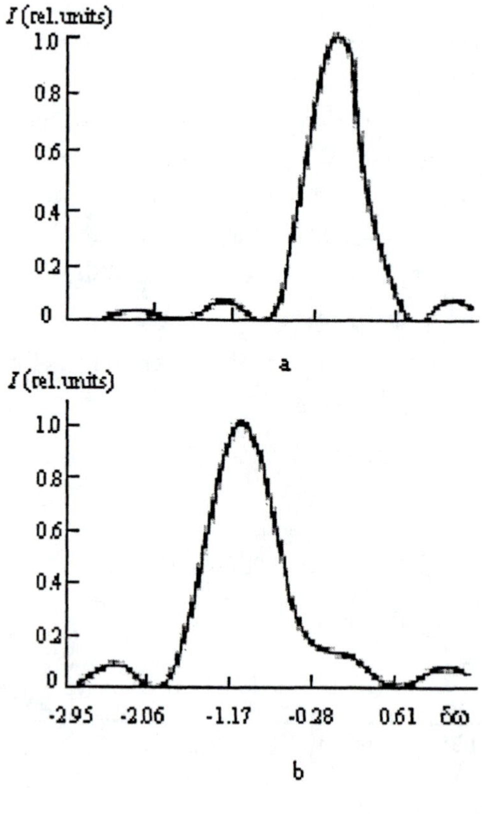

Fig. 3.2.2. Dependence of the intensities of the incident (a) and scattered (b) radiation on the frequency shift $\delta\omega$, calculated for $\kappa_0 = 1.35$, $\upsilon_E/\upsilon_{Te} = 0.65$, $\Gamma_{s0} = 0.05$, $L_c/\lambda_0 = 0.4$, $t_L = 4$ ps.

Scattering of High-Contrast Pulses of $t_L > 1/\omega_s$ Duration

Since $k_o L_c > 1$, $L_c/\lambda_0 \geq 1$, then $t_s/t_r < 1$, which under our conditions corresponds to $5 \leq t_L \leq 15$ ps. We assumed that laser intensity I is in the range 0.5- 2 PW cm^{-2} which in the selected temperature range corresponds to $\upsilon_E/\upsilon_{Tc} = 0.1 - 0.65$, $æ_0 = 0.044 - 1.86$. It can be seen from **Fig. 3.2.3** that for $t_L = 10$ ps the backscattering coefficient is $R \approx 10^{-1}$, i.e. the scattering occurs well above the threshold and $æ_0^{th} \approx 0.05$ is considerably less than in subsection. It can be seen from Fig. 3.2.3 that the excitation of acoustic waves with the amplitudes $\delta\tilde{N}_2$ and $\delta\tilde{N}_0$ corresponds to the condition $\delta\tilde{N}_2 << \delta\tilde{N}_0$ and an acoustic harmonic of amplitude $\delta\tilde{N}_2$ is focused near the turning point, i.e. sound propagates along the plasma layer. When the acoustic damping increment Γ_{s0} increases, the value of R falls: $R = 2.8 \times 10^{-2} - 4 \times 10^{-4}$ for $\Gamma_{s0} = 5 \times 10^{-1} - 2 \times 10^{-1}$ and $I = 0.5$ PW cm^{-2}. We can distinguish four different types of behavior of the spectrum of the scattered waves corresponding to different ranges of intensities when the spectrum of the incident radiation is the same **(Fig. 3.2.4a)**.

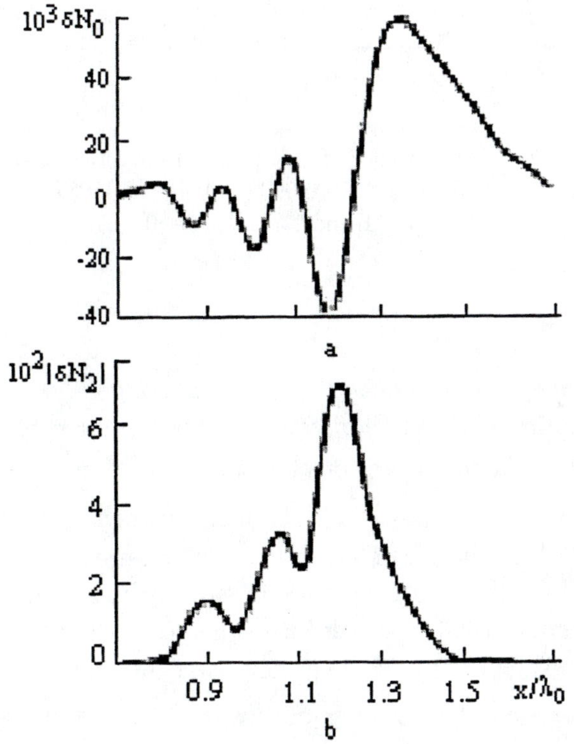

Fig. 3.2.3. Dependence of the amplitude of the density harmonics δN_0 and δN_2 as a function of x/λ_0 at $\upsilon_E/\upsilon_{Te} = 0.33$, $\Gamma_{s0} = 0.1$, $L_c/\lambda_0 = 1.1$, $t_L = 10$ ps.

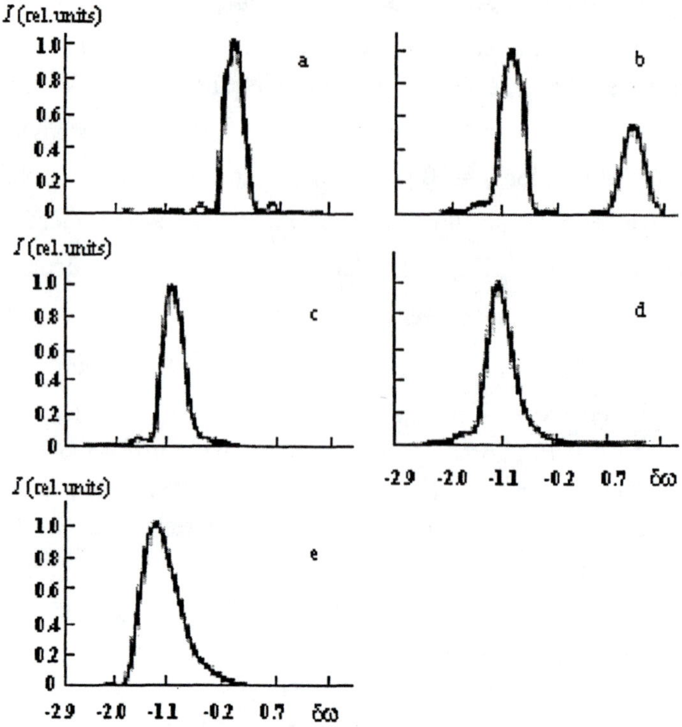

Fig. 3.2.4. Dependence of the intensity of the incident (a) and reflected (b-c) radiation on the frequency shift $\delta\omega$, calculated for $\kappa_0 = 0.099$ (b), 0.48 (a, c), 088 (d) and 1.86 (e) at $\upsilon_E/\upsilon_{Te} = 0.15$ (b), 0.33 (a, c), 0.44 (d) and 0.65 (e); $\Gamma_{s0} = 0.1$, $L_c/\lambda_0 = 1.1$, $t_L = 10$ ps.

Spectra of Scattered Waves

Spectra Governed by Specular Stimulated Brillouin Scattering

It is evident from **Fig. 3.2.4b** that for laser intensities in the range $I = 50$ - 200 TW cm^{-2} and $æ_0 = 0.044$ - 0.22 the scattered-wave spectrum exhibits both the red and the blue shifts. The red shift is $\delta\omega \approx -2\sin\theta_1 \approx -1$ and it is governed by double SBS. The blue shift is $\delta\omega \approx 2\sin\theta_1 \approx 1.1$. The blue shift in the scattered-wave spectrum is related to characteristics of the growth of specular SBS in strongly inhomogeneous plasma (see subsection and also Refs. [93-95]. Specular SBS is the result of interference between the second harmonic of double SBS of frequency $2\omega_{sd}$ and the pump waves, and it corresponds to the excitation of acoustic gratings with $\delta\widetilde{N}_{01}$ and $\omega_{sm} = 2\omega_{sd}$ propagating forward and backward relative to the normal of the plasma layer. Expression $\omega_{sm} = 2\omega_{sd} = 2k_x^{(0)}$ defines the region of localization of the instability, the growth of which leads to a frequency shift of the Stokes satellites $\delta\omega = \pm\sin\theta_1$ (n = 0, 2, 4) in the specular direction and to a frequency shift of the Stokes satellites $\delta\omega = \pm\sin\theta_1$ (m = 1, 3. 5) in the backward direction [93,94]. Since $\delta\widetilde{N}_0 < \delta\widetilde{N}_2$ the influence of specular SBS on R is fairly weak.

Spectra Governed by Double Stimulated Brillouin Scattering

If the intensity is $I = 300 - 700$ TW cm^{-1} and $æ_0 = 0.275 - 0.67$, we can see from **Fig. 3.2.4c** that the scattered- wave spectrum is governed by double SBS and there is only a red shift $\delta\omega = \pm\sin\theta_1$. An increase in $\delta\widetilde{N}_2, \delta\widetilde{N}_0$ with increase in the intensity I increases also R to $10^{-3} - 10^{-2}$. It is interesting that a reduction in the angle of incidence θ_1 from θ_1 ~ 30° and its increase both reduce $\delta\widetilde{N}_2$ and R; the amplitude of the acoustic harmonic $\delta\widetilde{N}_0$ varies as θ_1^{-1}, i.e. the dependence of R on $\delta\widetilde{N}_0$ is weaker than the dependence on $\delta\widetilde{N}_2$. This behavior $\delta\widetilde{N}_0$ and R, observed on reduction in the angle θ_1, is similar to their behavior in the case of a long layer [93-95] when the threshold increment of double SBS obeys $æ_{ih} \propto \theta_1^{-2}$ and we have $R \propto \theta_1^2$, but the changes in $\delta\widetilde{N}_2$ and R with increase in θ_1 are not described by the theory of Refs [93-95], because of the characteristic behavior of double SBS at very high intensities of non-resonant processes in a strongly inhomogeneous short plasma layer.

Spectra Governed by Stimulated Brillouin Backscattering

At intensities $I = 0.8 - 1$ PW cm^{-2} if $æ_0 = 0.74 - 0.93$ we can see from **Fig. 3.2.4d** that the change in the scattered-wave spectrum is governed by the instability of stimulated Brillouin backscattering. The spectral shift of the scattered waves $\delta\omega = -2\left[1 - \widetilde{N}_0(\widetilde{x})\right]^{1/2}$ and the red shift increases with increase in the intensity, which corresponds to a shift of the strongest-scattering region to plasma of lower density [93-95].

Spectra Governed by a Strongly Coupled Instability

It is evident from **Fig. 3.2.4e** that at intensities $I \geq 2$ PW cm^{-2} there is a qualitative change in the scattered-wave spectrum. Not only is there a considerable increase in the red shift compared with the case discussed in subsection but the spectral line width also increases. This change in the spectrum is attributed to the growth of a strongly coupled instability when the frequencies of the waves interacting with the plasma depend on the pump intensity. In the case of a homogeneous layer the time increment G of a strongly coupled instability is greater than the acoustic frequency [88] [($G > \omega_s(k_0)$ where $G = 3/2(\omega_{li}^2 \upsilon_E^2 k_0/2c)^{1/3}/2)$], which is valid if $\omega_{li}^2 \upsilon_E^2/c >> 32k_0^2 c_s^3/3$, where ω_{li}^2 is the Langmuir frequency of ions and the transient gain obeys $æ_n = t_L G > 1$.

These inequalities are obeyed better as the plasma density increases. In our case, i.e. for a strongly inhomogeneous layer when these estimates are incorrect the transition to a strongly coupled instability corresponds to simultaneous excitation of double SBS and stimulated Brillouin backscattering when the pump waves self-interact strongly. An increase in the angle of incidence results in very rapid switching of the scattering to the backward direction caused by double SBS but when the angle of incidence is reduced the spectrum breaks up into separate regions corresponding to the individual contributions of the gratings formed by double SBS and stimulated Brillouin backscattering. Even at these high pump intensities the plasma region dominating the scattering is located fairly close to the turning point and the maximum values of $\delta\widetilde{N}_2$ and $\delta\widetilde{N}_0$ do not exceed 0.15 which corresponds to the acoustic approximation.

Scattering of Picosecond Pulses by a Pre-Pulse Plasma

We shall now consider the scattering of a low-contrast picosecond pulse when the effect of a pre-pulse on a plasma cannot be ignored. When the pre-pulse duration is t_{pr} and the pulse and pre-pulse energies are W_p and W_{pr} the intensity contrast can be expressed in the from $K_I = I_p/I_{pr} = W_p/W_{pr}(t_{pr}/T_p)$. We shall consider the scattering of pump pulses of $t_L \sim 1$ - 4 ps duration and of I_p = 100 TW cm^{-1} which have a pre-pulse of t_{pr} = 10 - 100 ps duration characterized by $W_{pr} = 0.05\ W_p$. In this case the pulse is scattered by a plasma whose inhomogeneity scale is L_c, governed by the pre-pulse ($L_c = c_s t_{pr}/2$, $L_c/\lambda_0 > 1$), and the velocity of sound is determined by the plasma temperature re set in its turn by the pre-pulse intensity: $T_e \propto I_{pr}^{2/3}$ ~1000 eV [96]. These plasma and radiation parameters correspond to χ_0 = 0.1 - 1.0. We can set from **Fig. 3.2.5** that the scattering coefficient $R = 10^{-2} - 10^{-1}$ can be very large compared with the coefficient found in subsection, i.e. in the absence of a pre-pulse and that this coefficient increases with increase in K_I. We can therefore use the process under consideration as the criterion of the influence of the contrast and of a possible presence of a pre-pulse. As in subsection the scattered-wave spectrum is governed by double SBS also when the harmonics $\delta\widetilde{N}_2$ and $\delta\widetilde{N}_0$, $\delta\widetilde{N}_2 << \partial\widetilde{N}_0$ are excited. In general K_I is governed by the quantities t_{pr}/t_L and W_{pr} for given W_p and t_L. The dependence $R(t_{pr}/t_L, W_{pr})$ is single-valued and it represents a surface in three-dimensional space which can be constructed from the results of numerical calculations. If the experimental value of R were to prove to be much larger than in the case considered, then knowing R and W_p, W_{pr}, t_L, we can find t_{pr}/t_L and K_I.

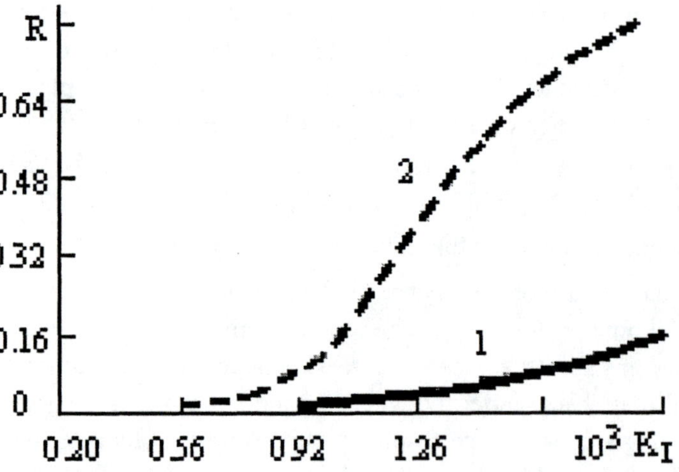

Fig. 3.2.5. Dependence of the back-scatter coefficient R on the contrast K_I, calculated for $\Gamma_{s0} = 0.1$, $I_L = 10^{14}$ W cm^{-2}, $\upsilon_E/\upsilon_{Te} = 0.146$, t_L= 1 ps (1) and t_p = 4 ps (2).

3.2.2 Production of Over-Dense Plasma Cavity by Ultra-Intense Laser Pulse Interaction with Solid Target and Analysis by Scattered Light

In the present section we study the plasma dynamics in strong laser fields by a numerical simulations and analytical modeling. On the basis of the obtained velocity and density the nonlinear plasma oscillations are evaluated. Doppler shift and broadening of the spectrum of scattered radiation are obtained. At considered high laser pulse intensities the spectrum is explained not only by movement of the critical density surface that was obtained earlier [98] but also generation of nonlinear sound oscillations in the plasma.

Model Equations

We find the solution of the self-consistent set of equations (3.1.65 – 3.1.66) by expansion on parameter υ_T/υ_E. In the zero order of approximation, we have instead of collision-less kinetics equation - hydrodynamic equations of motion of plasma electron component. The conservation law of transverse canonical momentum of electrons allows us to reduce a system of hydrodynamic equations of electrons motion and Maxwells equations to two nonlinear partial equations for vector potential $\overline{A}(x;t)$ of electromagnetic wave and longitudinal electric field in plasma [99]. Using the quasi-neutrality approximation, we have one equation for vector potential amplitude ($\overline{A}_z(x;t) = A(x,t) = A(x)e^{i\omega_0 t}$):

$$\left(\frac{\partial^2}{\partial\xi^2}+\frac{\omega_0^2}{\omega_p^2}\right)a(\xi) = \left(\eta_i(\xi)+\frac{\partial^2}{\partial\xi^2}\sqrt{1+a^2}\right)\frac{a}{\sqrt{1+a^2}} \tag{3.2.11}$$

Here $\omega_p=(4\pi Z n_{i0} e^2/m)^{1/2}$- electron plasma frequencies and the following dimensionless variables are used:

$$\xi = \frac{\omega_p}{c}x,\ a = \frac{eA}{mc^2},\ \delta_T = (T_e/mc^2)^{1/2},\ \eta_i(\xi) = \frac{n_i(\xi)}{n_i(\xi=\xi_{max})}$$ - dimensionless profile of ion density.

For the analysis it is convenient to consider (3.1.65) in the following dimensionless variables: instead of velocity v we enter a Mach number $\mu = \mathrm{v}/\sqrt{ZT_{ec}/M}$, ion time τ_i we measure in terms of $\frac{c}{\omega}\sqrt{\frac{M}{ZT_{ec}}}$, i.e. time an ion sound takes to transit a skin layer. Then the dimensionless hydrodynamical equations for ions have the form:

$$\frac{\partial \eta_i}{\partial \tau_i} + \frac{\partial}{\partial \xi}(\eta_i \mu) = 0$$

$$\frac{\partial \mu}{\partial \tau_i} + \mu \frac{\partial}{\partial \xi}\mu + \frac{\partial}{\partial \xi}\ln \eta_i(\xi) - \frac{1}{\eta_i(\xi)}\frac{\partial}{\partial \xi}\tilde{\tau}_i \frac{\partial}{\partial \xi}\mu = \delta_T^{-2}\frac{\partial \sqrt{1+a^2}}{\partial \xi} \qquad (3.2.12)$$

where the dimensionless viscosity $\tilde{\tau}_i$ depends only on temperature of ions:

$$\tilde{\tau}_i = \frac{64}{\Lambda_c}\frac{Z^2 T_e^{\,2}\omega_p}{e^4 n_{0i} c}\left(\frac{T_i}{ZT_e}\right)^{5/2}$$, here Λ_c – Qulomb logarithm

If the laser pulse is shorter than $l_s / (ZT_e/M)^{1/2}$ (where l_s - length of the skin layer) the pressure of a laser field is equivalent to instantaneous shock on the plasma, and the right side of the second equation (3.2.12) is equivalent to the boundary conditions for a homogeneous system (3.2.12). The solution of a problem about such instantaneous shock for a system (3.2.12) is explained by an example in [100].

Under-Critical Plasma Parameters

The analytical solution (3.2.11 – 3.2.12) is difficult, however with the help of the second equation (3.2.12) it is possible to estimate the velocity of shock front. For this purpose we rewrite (3.2.12) as the dimensionless equation for the continuity of momentum flow:

$$\frac{\partial v_i}{\partial \tau} + \frac{\partial}{\partial \xi}\left(\frac{v_i^2}{2} + \frac{c_s^2}{c^2}\ln \eta_i(\xi) - \frac{Zm\sqrt{1+a^2}}{M}\right) = 0 \qquad (3.2.13)$$

Here $v_i = v/c$.

It is possible to omit the term containing plasma pressure, and also we assume that the laser field momentum flow is completely converted to ion momentum flow (by means of fast electrons), then from (3.2.13) we obtain the following ion velocity behind shock waves front

$$v_i = \left(\frac{2Zm}{Am_p}\sqrt{1+a^2}\right)^{0.5} \qquad (3.2.14)$$

We take $Z/A = 0.5$, therefore $v_i \approx 0.02(1 + a^2)^{1/4}$. For example for laser intensity $5{\cdot}10^{18}$ W/cm^2 ($a^2 = 5$) ion velocity is $v_i{\sim}0.03$. We emphasize that the expression (3.2.14) approximates ion velocity near a position $n_e = n_{cr}$, where the ponderomotive pressure is applied. If we want to estimate the velocity inside over-dense plasma, the conservation law of momentum flow produces a factor n_c/n_e inside brackets in (3.2.14). For such a case, the formula (3.2.14) has been represented in [101]. These estimates of ion velocity are confirmed by our PIC simulations (**Fig. 3.2.6**).

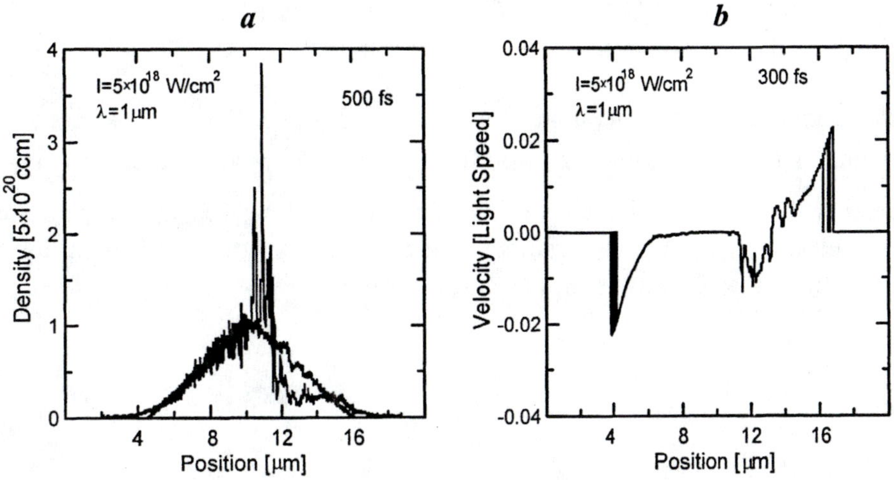

Fig. 3.2.6. (a) Plasma density distribution: black line – before interaction, blue line - at the moment t =500 ps of interaction of laser pulse with duration t_L= 1 ps and intensity $I_L = 5\ 10^{18}$ W/cm^2; (b) Plasma velocity distribution at the moment t = 300 fs and the same laser parameters.

We see in Fig.3.2.6 that numerical simulation gives magnitude of the shock front velocity as $v_i \sim 0.014$. The estimates based on Eqs.(3.2.14) agree well with these numerical results.

Shock Wave of a Small Amplitude

The analytical solution (3.2.12) is possible for laser fields of small intensities ($a < 1$). In this case, the shock wave is feeble, and it is possible to use perturbation theory for solution of the equations (3.2.12). We substitute $n_i = n_{0i} + \delta n_i$ and expand (3.2.12) into a series of δn_i. For the first order perturbation theory, we obtain the sound equation for δn_i. For the second order the density perturbation $\delta\eta$ is described by KdW equation [84]:

$$\frac{\partial \delta\eta}{\partial \tilde{\tau}} + \frac{\partial \delta\eta}{\partial \xi} + \delta\eta \frac{\partial \delta\eta}{\partial \xi} + \beta \frac{\partial^3 \delta\eta}{\partial \xi^3} + \frac{\alpha}{\pi} \mathrm{P} \int_{-\infty}^{+\infty} \frac{\partial \delta\eta}{\partial \xi'} \frac{d\xi'}{(\xi - \xi')} = \frac{c^2}{4 v_{Te}^2} \frac{\partial a^2}{\partial \xi} \tag{3.2.15}$$

where

$$\delta\eta = -\frac{\delta n_i}{n_i}, \tilde{\tau} = \tau \sqrt{\frac{Z T_e}{M c^2}},\ \beta = \frac{v_{Te}^2}{2c^2} \frac{\omega_{pe}}{\omega_0}, \alpha = \sqrt{\frac{\pi m}{8M}} \frac{\omega_{pe}}{\omega_0}$$

The term that contains an integral describes damping of nonlinear ion-sound oscillations, and the term containing parameter β, describes dispersion of plasmons.

The equation (3.2.15) together with the equation for a vector potential a:

$$(\frac{\partial^2}{\partial\xi^2}+\frac{\omega_0^2}{\omega_p^2})a(\xi)=(\eta_0(\xi)(1-a^2/2)+\delta\eta(\xi)+\frac{\partial^2}{\partial\xi^2}\frac{a^2}{2})a \qquad (3.2.16)$$

gives a system describing generation of a weak shock wave. For Al plasma and actual experimental data we substitute $\omega_{pe}/\omega=10$, $v_{Te}/c=0.1$, then β= 0.05 and α = 0.03. The equations (3.2.15), (3.2.16) are valid for $\delta\eta<1$, therefore we can use the expansion in a in the equation (3.2.16). The restriction on amplitude of a field in plasma, for which the perturbation of ion density remains weak, is as follows:

$$\frac{c^2}{4v_{Te}^2}a^2\leq 1.$$

The scale length of ion-sound oscillations is about the length of a skin layer, the velocity is about the velocity of ion sound, so we can conclude that ion-sound frequency is equal to the ratio of these parameters for this case.

For a solution of a system (3.2.15), (3.2.16) and for definition of plasma density profile in a shock wave it is necessary to set boundary conditions along ξ and initial conditions at $\tilde{\tau}$ also.

Solution of the Boundary Problem

To determine the boundary conditions, we integrate (3.2.15) over ξ at the left-hand boundary of plasma:

$$\delta\eta+\frac{(\delta\eta)^2}{2}+\beta\frac{\partial^2\delta\eta}{\partial\xi^2}\bigg|_{\xi=0}=\frac{c^2}{4v_{Te}^2}a^2(\xi=0,\tilde{\tau})$$

At the right boundary we set boundary conditions for function and derivative to zero. The initial conditions also we put as: $\delta\eta|_{\tilde{\tau}=0}=0$. Thus the mathematically problem about excitation of a shock wave has been completely formulated.

First, we qualitatively explore parameters of a shock wave in a homogeneous equation (3.2.15) using the following method. We search for a solution of form $\delta\eta(\xi-\tilde{\tau}-w\tilde{\tau})$, where w - velocity of a shock wave in a system moving with a velocity of an ion sound. Without a dissipative term the equation (3.2.15) becomes the Newton's law equation with an effective potential:

$$\beta\delta\eta''=-\frac{\partial}{\partial\delta\eta}(\delta\eta^3/6-w\delta\eta^2/2)$$

The potential has a maximum at $\delta\eta = 0$ and minimum at $\delta\eta = 2w$. Dissipation reduces the effective driving force from the maximum represented by the potential to a minimum. Thus, the rarefaction wave moves to the right, and the Mach number is connected with density perturbation by a relation which is well known for a weak shock waves: $\mu_s = 1 + w = 1 + \delta\eta/2$. To construct the density profile for a shock wave we solve the equation:

$$\delta\eta'(\delta\eta - w) + \beta\delta\eta''' + \frac{\alpha}{\pi} P \int_{-\infty}^{+\infty} \delta\eta' \frac{du}{(u-u')} = 0 \ . \qquad (3.2.17)$$

Far to the left of the front, where $\delta\eta(\mu) = 2w + \varepsilon(\mu)$, by linearizing (3.2.17), we obtain a linear integral equation for ε:

$$w\varepsilon' + \beta\varepsilon''' + \frac{\alpha}{\pi} P \int_{-\infty}^{+\infty} \varepsilon' \frac{du'}{(u-u')} = 0. \qquad (3.2.18)$$

We search for a solution (3.2.18) as $\varepsilon(\mu) = \exp(i(k - i\gamma)\mu)$. The density oscillations thus have a scale k^{-1} and exponentially rise with an increment γ. Such oscillatory character of a density in a plasma shock wave was noted in [84]. The dispersion equation following from (3.2.18) has a simple form coinciding with the well known dispersion law for an ion sound:

$$\beta(k - i\gamma)^2 - w + i\alpha \frac{\sqrt{k^2 + \gamma^2}}{k - i\gamma} = 0$$

and is solved easily:

$$\gamma = \sqrt{\frac{-w + \sqrt{w^2 + 4\alpha^2}}{8\beta}}$$
$$k = \sqrt{\frac{5w + 3\sqrt{w^2 + 4\alpha^2}}{8\beta}} \qquad (3.2.19)$$

Near to the shock front, the solution (3.2.15) can be found only numerically.

We calculated this numerical solution with the above mentioned boundary conditions, by taking β = 0.05, α = 0, $\frac{Zc^2}{4v_{Te}^2} a^2(\xi = 0, \tilde{\tau}) = 4\tilde{\tau}\exp(-8\tilde{\tau}^2)$. In **Figure 3.2.7**, illustrating the numerical simulation, the spreading shock front and oscillating structure behind are shown (at $\tilde{\tau} = 5$). At a later time the structure described above should be generated.

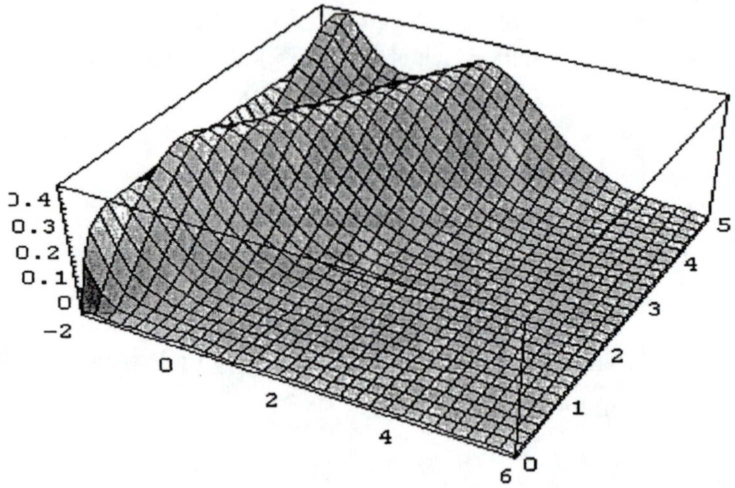

Fig. 3.2.7. Density dependence from ξ and τ. The shock front and oscillated structure behind are shown at $\tau = 5$.

Analysis of Laser Pulse Reflection from Plasma Density Modulated by a Shock Wave

As was shown before, upon action of the laser pulse on plasma, there is a shock front following by the oscillation of density. We can linearize the hydrodynamic equations because laser field, inside over-dense plasma with sharp density gradient, is attenuated in (ω_p/ω_0)>>1 times. Then we consider joint solution of the equation for a:

$$\left(\frac{\partial^2}{\partial \xi^2} + 2i\frac{c_s}{c}\frac{\partial}{\partial \tilde{\tau}} + \frac{\omega_0^2}{\omega_p^2} - \eta_p + \delta\eta(\tilde{\xi},\tilde{\tau})\right)a(\xi,\tilde{\tau}) = 0, \qquad (3.2.20)$$

where $\tilde{\xi} = \xi - \tilde{\tau}$, and equation KdW for $\delta\eta(\tilde{\xi},\tilde{\tau})$:

$$\frac{\partial\delta\eta}{\partial\tilde{\tau}} + \frac{\partial\delta\eta}{\partial\xi} + \delta\eta\frac{\partial\delta\eta}{\partial\xi} + \beta\frac{\partial^3\delta\eta}{\partial\xi^3} = \frac{Zc^2}{4v_{Te}^2}\frac{\partial a^2}{\partial\xi} \qquad (3.2.21)$$

We consider vector potential also as $a_0 + a_1$. We restrict this to terms of the second order, as the equation KdW is written to within the second order. As a result we obtain the next system:

$$\left(\frac{\partial^2}{\partial \xi^2} + 2i\frac{c_s}{c}\frac{\partial}{\partial \tilde{\tau}} + \frac{\omega^2}{\omega_p^2} - \eta_p\right)a_0(\xi,\tilde{\tau}) = 0 \qquad (3.2.22)$$

$$\left(\frac{\partial^2}{\partial \xi^2}+2i\frac{c_s}{c}\frac{\partial}{\partial \tilde{\tau}}+\frac{\omega^2}{\omega_p{}^2}-\eta_p\right)a_1=-\delta\eta(\tilde{\xi},\tilde{\tau}))a_0(\xi,\tilde{\tau})-\delta\eta(\tilde{\xi},\tilde{\tau}))a_1(\xi,\tilde{\tau}) \quad (3.2.23)$$

$$\frac{\partial \delta\eta}{\partial \tilde{\tau}}+\frac{\partial \delta\eta}{\partial \xi}+\delta\eta\frac{\partial \delta\eta}{\partial \xi}+\beta\frac{\partial^3 \delta\eta}{\partial \xi^3}=\frac{Zc^2}{4\mathrm{v}_{Te}^2}\frac{\partial}{\partial \xi}(a_0{}^2+2a_0a_1+a_1{}^2) \quad (3.2.24)$$

Let's select plasma geometry relevant to a figure of the numerical simulations (see **Fig. 3.2.8.**).

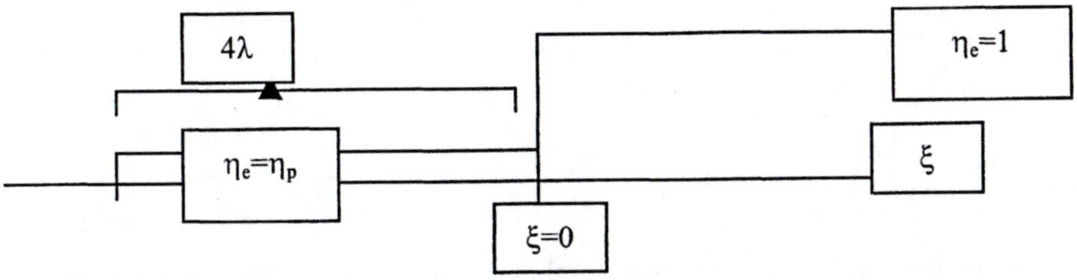

Fig.3.2.8. Laser plasma interaction geometry relevant to the analytical model.

Field a_0 from (3.2.22) take as

$$a_0(\xi,\tilde{\tau})=a^{(0)}\sin(\sqrt{\omega_0^2/\omega_p{}^2-\eta_p}\,\xi) \quad (3.2.25)$$

We show at first, that the system (3.2.23), (3.2.24) in under-dense plasma slab, with η_i=const, describes connected electromagnetic and sound waves and generation of sound harmonics. We search for a solution of the equations (3.2.23), (3.2.24) using a series expansion. A linear approximation gives the following coupled equations:

$$\left(\frac{\partial^2}{\partial \xi^2}+2i\frac{c_s}{c}\frac{\partial}{\partial \tilde{\tau}}+\frac{\omega^2}{\omega_p{}^2}-\eta_p\right)a_1=-\delta\eta(\tilde{\xi},\tilde{\tau}))a_0(\xi,\tilde{\tau}) \quad (3.2.26)$$

$$\frac{\partial \delta\eta}{\partial \tilde{\tau}}+\frac{\partial \delta\eta}{\partial \xi}-\frac{Zc^2}{2\mathrm{v}_{Te}^2}\frac{\partial}{\partial \xi}a_0a_1=\frac{Zc^2}{4\mathrm{v}_{Te}^2}\frac{\partial}{\partial \xi}a_0{}^2 \quad (3.2.27)$$

For solution of homogeneous equations we take:

$$a_1=a_{10}\exp(i(\omega_1-\omega_0)\tilde{\tau}-ik_1\xi)$$
$$\delta\eta=\delta\eta_0\exp(-i\omega_s+ik_s\xi)$$

Then from (3.2.26), (3.2.27) the following conservation laws for the interacting waves are obtained:

$$\begin{aligned} &\omega_0 - \omega_1 = \omega_s \\ &k + k_1 = (\omega_0 + \omega_1)\sqrt{1 - \eta_p \frac{\omega_p^{\ 2}}{\omega_0^2}} = k_s = \frac{\omega_s}{c_s} \end{aligned} \tag{3.2.28}$$

From (3.2.28) the frequency of a sound and scattered waves can be determined:

$$\begin{aligned} &\omega_s = 2\omega \frac{c_s}{c}\sqrt{1 - \eta_p \frac{\omega_p^{\ 2}}{\omega^2}} \\ &\omega_1 = \omega(1 - 2\frac{c_s}{c}\sqrt{1 - \eta_p \frac{\omega_p^{\ 2}}{\omega^2}}) \end{aligned} \tag{3.2.29}$$

Taking finite length $L = 3l_s$ of plasma slab we see that a narrow frequency line is broadened and δ – function transforms to:

$$\delta(\Omega - \omega_1) \to \frac{\sin\left(\dfrac{3(\Omega - \omega_1)}{\omega}\right)}{\pi(\Omega - \omega_1)}, \quad (\Omega - \omega_1) << \omega \tag{3.2.30}$$

Next iteration gives this system:

$$\begin{aligned} &(\frac{\partial^2}{\partial \xi^2} + 2i\frac{c_s}{c}\frac{\partial}{\partial \tilde{\tau}} + \frac{\omega_0^2}{\omega_p^{\ 2}} - \eta_p)a_{12} = -\delta\eta(\tilde{\xi}, \tilde{\tau}))a_1(\xi, \tilde{\tau}) - \delta\eta_2(\tilde{\xi}, \tilde{\tau}))a_0(\xi, \tilde{\tau}) \\ &\frac{\partial \delta\eta_2}{\partial \tilde{\tau}} + \frac{\partial \delta\eta_2}{\partial \xi} - \frac{Zc^2}{2\mathrm{v}_{Te}^2}\frac{\partial}{\partial \xi}a_0 a_{12} = \frac{Zc^2}{4\mathrm{v}_{Te}^2}\frac{\partial}{\partial \xi}a_1^{\ 2} \end{aligned}$$

The conservation laws for fields give the next frequency conditions:

$$\omega_{s2} = \omega_0 - \omega_1 + \omega_s = 2\omega_s$$

It means that a spectral line is shifted on the second harmonics of ion sound. It is broadened as given by Eq.(3.2.30) and has amplitude about the first harmonic ~ $\delta\eta / \eta_p$. Thus in a scatter spectrum there are harmonics of ion sound frequency. Obviously, the solutions of equations (3.2.26) are valid for homogeneous plasma and this rough approximation is good enough. To satisfy initial and boundary conditions it is necessary to use a self-similar solution of the equation KdW (3.2.21) from a self-similar variable

$\delta\eta(\tilde{\xi},\tilde{\tau}) = \delta\eta(\beta\tilde{\xi}/\tilde{\tau}^{1/3})$. As shown in [84], solutions in this case are expressed through Airy functions. Such density profile, containing downward peaks, is qualitatively similar to numerical calculation results.

For the analysis of the spectrum of scattered radiation we use (3.2.26). In under dense plasma $\eta_i(\xi) = \eta_p << 1$ and this plasma is transparent to laser pulse, but deep in plasma $\eta_0(\xi) > 1$ and laser pulse does not propagate inside this region.

Solution (3.2.26) we present through a Green function. Let's take into account, in the Green function the point of reflection $\xi_0(\tau)$ (boundary between over and under dense plasma), where the function equals zero. From the law of motion of a point of reflection we select $\xi_0(\tau) = -v_i\,\tau$, then this function has the form:

$$G(\xi,\xi';\tilde{\tau},\tilde{\tau}') = \sqrt{\frac{c_s}{2\pi ic(\tilde{\tau}-\tilde{\tau}')}}\left(\exp\frac{ic(\tilde{\tau}-\tilde{\tau}')(\omega_0^2/\omega_p{}^2-\eta_p)}{2c_s}\right)\cdot \quad (3.2.31)$$
$$\left(\exp\frac{ic_s(\xi-\xi_0(\tilde{\tau})-\xi'+\xi_0(\tilde{\tau}'))^2}{2c(\tilde{\tau}-\tilde{\tau}')}-\exp\frac{ic_s(\xi+\xi'-\xi_0(\tilde{\tau})-\xi_0(\tilde{\tau}'))^2}{2c(\tilde{\tau}-\tilde{\tau}')}\right)$$

This function is correct for slow dependence of $\xi_0(\tau)$, as from homogeneous equation (3.2.26) it is not satisfied precisely. Further we create a zero order approximation of the solution (3.2.22). Take as this approximation a laser standing wave:

$$a_0(\xi,\tilde{\tau}) = a^{(0)}\sin(\sqrt{\omega_0^2/\omega_p{}^2-\eta_p}\,(\xi-v_i\tilde{\tau}))$$

Thus we neglect the broadening of a spectral distribution of the laser pulse, connected with its duration, considering effects of sound generation more essential. For the correction to a field we get the following expression:

$$a_1(\xi,\tilde{\tau}) = -\int_{-\infty}^{\tilde{\tau}} d\tilde{\tau}' \int_{\xi_0(\tilde{\tau}')}^{\infty} d\tilde{\tau}' G(\xi,\xi';\tilde{\tau},\tilde{\tau}')\delta\eta(\xi',\tilde{\tau}')a_0(\xi',\tilde{\tau}'). \quad (3.2.32)$$

As perturbation of density, we substitute the following solution of the equation KdW – $\delta\eta(\xi,\tilde{\tau}) = \delta\eta_0 \mathrm{Ei}(\beta\frac{\tilde{\tau}+\xi-\xi_0(\tilde{\tau})}{\tilde{\tau}^{1/3}})$, where Ei is Airy function. The ppicture of the scattering looks like the one in **Fig 3.2.9**.

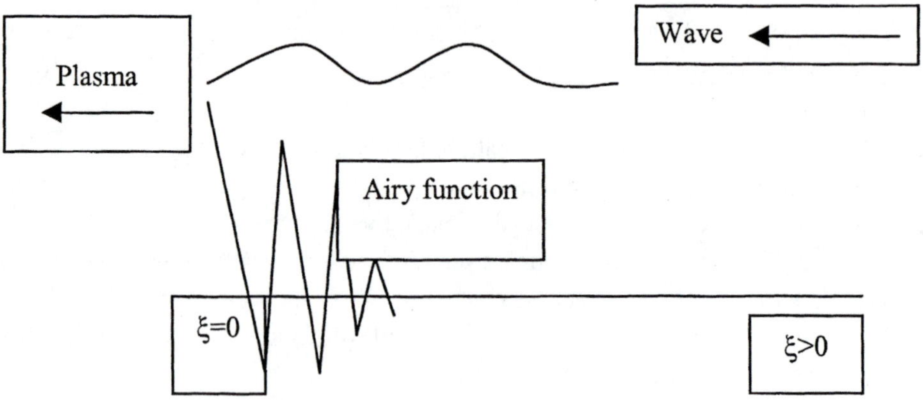

Fig. 3.2.9. The geometry of a laser wave interaction with inhomogeneous plasma.

For the spectral distribution we have the next expression:

$$a_{1\Omega}(\xi)=\frac{1}{2\pi}\int a_1(\xi,\tilde{\tau})\exp(i\Omega\tilde{\tau})d\Omega \tag{3.2.33}$$

We take a limited big interval ξ, where the field has the form of a propagating wave. For further evaluations we insert the new variable to integrate over and interchange the order of the integration. Then for (3.2.32) we obtain the following expression:

$$a_{1\Omega}(\xi)=-a^{(0)}\delta\eta_0\frac{1}{4\pi^{3/2}}\int_0^{\infty}d\xi''\int_{-\infty}^{+\infty}d\tilde{\tau}'\int_0^{\infty}d\tilde{\tau}''\sin\left(\left(\sqrt{\frac{\omega}{\omega_p}}-\eta_p\right)\xi''\right)\exp(i\Omega(\tilde{\tau}'+\tilde{\tau}''))$$

$$\int dkd\Omega_1\exp(i\kappa\xi''-i\Omega_1\tilde{\tau}')\frac{i\Gamma(1/3)\sigma}{\sqrt{3}(2\pi)^2(\sigma^3k^3+k(1-u)+\Omega_1)^{4/3}}$$

$$\sqrt{\frac{c_s}{2\pi ic\tilde{\tau}''}}\left(\exp\frac{ic\tilde{\tau}''(\omega_0^2/\omega_p{}^2-\eta_p)}{2c_s}\right)\left(\exp\frac{ic_s(\xi-\xi''+u\tilde{\tau}'+u\tilde{\tau}'')^2}{2c\tilde{\tau}''}-\exp\frac{ic_s(\xi+\xi''+u\tilde{\tau}''+u\tilde{\tau}')^2}{2c\tilde{\tau}''}\right)$$

The integral on τ' can be easily taken, as it represents the Poisson integral. After that the integrals on τ'' and ξ'' are evaluated. These integrals give δ - functions relevant to the laws of conservation of energy and momentum for generation of sound by a laser pulse:

$$\Omega-\Omega_1=-u\sqrt{\frac{\omega_0^2}{\omega_p{}^2}-\eta_p}$$
$$\kappa=2\sqrt{\frac{\omega_0^2}{\omega_p{}^2}-\eta_p} \tag{3.2.34}$$

With the help δ - functions from (3.2.34) it is possible to fulfil the next integrations on k and ω_1. As result the spectral distribution of the vector potential of scatter radiation for large ξ becomes:

$$a_{1\Omega}(\xi)=a^{(0)}\delta\eta_0\exp(i\kappa_p\xi)\frac{\Gamma(1/3)\sigma}{\sqrt{3}64\pi^{7/2}(\sigma^3\kappa_p{}^3+2\kappa_p(1-u)+\Omega)^{4/3}},\qquad(3.2.35)$$

where $\kappa_p^2=\omega_0^2/\omega_p^2-\eta_p$

From this formula we see that: the correction really is a plane wave propagating from a boundary.and Doppler shift is to the red side and has a large value at negative ω, where ω is related to the optical frequency; the spectrum is almost symmetrical about the maximum. The spectrum is shifted, broadened and its amplitude decreases as ~ $\omega^{-4/3}$. For the shift $\Delta\,\omega=\omega-\omega_0$ we have the following expression:

$$\Delta\omega=\omega_p\frac{c_s}{c}(\sigma^3(\frac{\omega_0^2}{\omega_p{}^2}-\eta_p)^{3/2}+2(\frac{\omega_0^2}{\omega_p{}^2}-\eta_p)^{1/2}(1-u))\qquad(3.2.36)$$

There is frequency ω * at which (3.2.35) converts to infinity. It has taken place because for the evaluation of Fourier components (3.2.32) we used this solution during all time. The introduction of an integral with cut-off at some time removes this limitation. But it is obvious that (3.2.32) is valid while $a_1 << a_0$. Thus, near to maximum of the spectrum our description is inapplicable, but wings of the spectrum we circumscribed correctly. Let's compare the obtained results and our numerical simulation which we have shown in **Fig.3.2.10**.

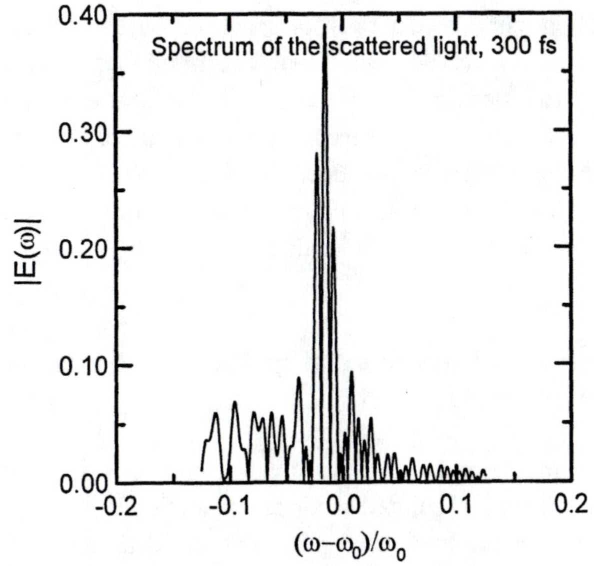

Fig. 3.2.10. Simulated spectrum of the backscattered light at the moment t =300 fs after interaction of the laser pulse t_L = 1 ps with plasma slab.

As we know from Fig.3.2.6 the calculated plasma boundary velocity is- 0.014 c. It gives a shift $\Delta\, \omega/\omega_0$ to the laser spectral line which is also equal to 0.014. It approximately corresponds to the maximum on this graph. From the graph, we see that this peak is broadened and located in the interval of frequencies up to 0.02-0.025. To explain this broadening we should also take into account the generation of sound harmonics in under dense plasma. For this purpose we compare the spectrum from (3.2.35) and the simulation data. In the numerical simulation $\omega_{pe}/\omega_0 = 2,\quad v_{Te}/c = 0.1$ c_s/c = 1.5 10^{-3}, then σ = 10 and from (3.2.36) it follows that $\Delta\omega/\omega_0 =$ 0.05, which is close result to the simulation result.

Finally we discuss the modulation, observed in simulation of the spectrum. Finit time of Fourier transformation is - Δt = 300 fs. It corresponds in Fourier spectrum to dimensionless frequency $\lambda_0/(c\ \Delta t)$ = 0.011. This is approximately the modulation of the spectrum in figure 3.2.10. Thus, by experimentally measuring shift and width of spectrum on experiment we can estimate under dense plasma parameters. This theoretical analysis of short high power laser pulse reflection from a dense plasma resulted in the following conclusion:

Ponderomotive pressure of laser light near critical density produces a shock wave which propagates deep into plasma. This process changes the profile of plasma density and creates two areas: transparent under-dense plasma and above-critical dense plasma. The critical density surface separating these areas moves with supersonic speed deep into the plasma together with the front of the shock wave at ultra high laser intensity. In under- critical and above-critical areas intensive nonlinear ion sound waves generate. In the transparent area the degree of non-linearity is higher, due to higher field amplitude. The indicated processes affect the spectrum of scattered radiation in the following way: Critical surface movement produces Doppler red frequency shift of scattered light.

Scattered radiation has a border peak of spectrum with wide spectral baseline from ion sound oscillations in over dense plasma. The nonlinear ion sound in under-dense plasma produces harmonics in the spectrum of the scattered radiation, widen and modulate it. This process form s a spectrum central peak also. Low frequency modulation of a scattered light spectrum is produced by finite duration of a laser pulse. The measurements of the spectrum of reflected radiation thus allow us to determine the speed of critical surface, and the density in under-dense plasma.

3.2.3 Back Scattering of Ultra Short High Intensity Laser Pulses from Solid Targets at Oblique Incidence

As we already mentioned basic researching methods of laser plasmas include excitation and interpretation of emitted secondary radiation. Various diagnostics covering a range of excited intensities have been proposed and demonstrated experimentally for short pulse interactions. For instance, the mechanisms of back reflection from low density plasmas due to the onset of electronic parametric instabilities have been discussed in detail before. In experiments where laser pulses of relativistic intensity impinge on a solid target at an oblique incidence, back scattering may reach the level of the order of a few percent [102,103].

In this section, a possible mechanism explaining the origin of the back reflection of intense short laser pulse from a plasma surface is suggested and studied theoretically. The basic idea as in 3.2.1 is that a periodic electron density modulation acts as a grating that diffracts the incident laser field into the backward direction in addition to the specular reflection. All the electromagnetic waves together form a ponderomotive force that under proper phase matching conditions drives the initial perturbation. The system presents a parametric instability with a characteristic threshold behaviour, growth rates and saturation at higher pump intensities.

Basic Equations

We consider the interaction between a short laser pulse and a solid target. Ion motion can be neglected during the interaction time when the laser pulse length is of the order of 100 fs or less. Electron motion must be treated relativistically because of the high laser intensity. Maxwell equations together with the fluid equations form the basic set

$$\begin{aligned}
&\frac{d}{dt}\frac{m\vec{\upsilon}}{\sqrt{1-\upsilon^2/c^2}} = -\frac{e\partial\vec{A}}{c\partial t} + e\vec{E} + \frac{e}{c}[\vec{\upsilon}[\vec{\nabla}\vec{A}]] - T\frac{\vec{\nabla}n_e}{n_e} \\
&\frac{\partial n}{\partial t} + \vec{\nabla}(n_e\vec{\upsilon}) = 0 \\
&\vec{\nabla}\vec{E} = 4\pi e(n_e - n_i) \\
&\Delta\vec{A} - \frac{\partial^2\vec{A}}{c^2\partial t^2} = -\frac{4\pi}{c}en_e\vec{\upsilon} - \frac{\partial\vec{E}}{c\partial t}
\end{aligned} \tag{3.2.37}$$

The vector potential $\vec{A}$ describes the transverse electromagnetic field comprising the incident laser beam and the scattered waves; the field $\vec{E}$ is the longitudinal field originating from charge separation in the plasma. The density, velocity, and temperature of electrons is denoted by n, υ, and T, respectively, and n_i stands for the background ion density.

We assume an S-polarized electromagnetic wave with an amplitude $A_l\mathbf{e}_z$ and a wave vector $\mathbf{k}_0$ at oblique incidence on a plasma surface. At reflection both a specular component ($A_r\mathbf{e}_z$, $\mathbf{k}_r$) and a backreflected component ($A_b\mathbf{e}_z$, $\mathbf{k}_b$) are generated (see **Fig. 3.2.11**). For small reflected intensities we can expand Eqs. (3.2.37) in the parameter eA/mc^2 and we obtain the two coupled equations

$$\begin{aligned}
&\frac{\partial^2\vec{E}}{\partial\tau^2} + \frac{\omega_p^2}{\omega^2}\eta_i(\vec{\xi})\vec{E} = \frac{\omega_p^2}{2\omega^2}\eta_i(\vec{\xi})\vec{\nabla}_\xi\vec{a}^2 + \frac{\upsilon_T^2\omega_p^2}{c^2\omega^2}\eta_i(\vec{\xi})\vec{\nabla}_\xi\left[\frac{1}{\eta_i(\vec{\xi})}\vec{\nabla}_\xi\vec{E}\right] \\
&\nabla_\xi^2\vec{a} - \frac{\omega^2}{\omega_p^2}\frac{\partial^2\vec{a}}{\partial\tau^2} = [\eta_i(\vec{\xi})(1-\vec{a}^2/2) + \vec{\nabla}_\xi\vec{E}]\vec{a}
\end{aligned} \tag{3.2.38}$$

The upper equation describes the longitudinal plasma wave driven by the ponderomotive force (~ $\nabla \mathbf{A}^2$) caused by the laser fields. The second equation describes the electromagnetic wave propagation under the influence of plasma oscillations. The fields are written in a normalised form, $\boldsymbol{a}=e\boldsymbol{A}/mc^2$ and $\boldsymbol{E}=e\boldsymbol{E}/mc\omega_p$, the dimensionless time and space coordinates are $\tau = \omega t$ and $\xi=\omega_p r/c$, respectively; v_{th} is the thermal velocity, $\omega_p=[4\pi Z n_i(\infty)e^2/m]^{1/2}$ is the plasma frequency and $\eta(\xi) = n_i(\xi)/n_i(\infty)$ describes the frozen ion density profile normalized to the bulk ion density $n_i(\infty)$. The physics behind the coupling is simple: the laser field excites, by ponderomotive action, density oscillations which, in turn, scatter the laser field. The spatially and temporally varying density grating diffracts transverse electromagnetic waves into new directions [104] and oscillation frequencies. We shall show, for a plasma that is transparent to the pump wave, that this process is unstable and leads to generation of electronic density perturbations at a wave length $(\lambda_0/2)\sin\theta$ where λ_0 is the incident laser wave length and θ is the angle of incidence (see Fig. 3.2.11). Such diffraction grating scatters the incident laser beam not only into the specular but also into the back direction. A completely analogous parametric instability, Double Stimulated Brillouin Scattering (DSBS) (see [3.2.1, e.g., [95]), occurs for longer laser pulses, say one picosecond or more. In DSBS ion sound waves can be excited and consequently the ion density modulations create a corresponding diffraction grating. In the presently considered process, the ions are assumed immobile

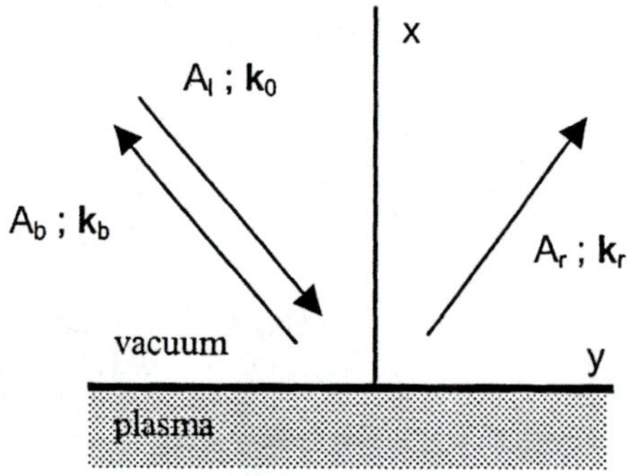

Fig. 3.2.11. Incoming S-polarized (along x-axis) pump wave (A_i, $\mathbf{k}_0$) diffracts from the grating formed on the plasma surface and generates specular (A_r, $\mathbf{k}_r$) and backreflected (A_b, $\mathbf{k}_b$) waves.

Linear Stability Analysis

When the density inhomogeneity scale L of the plasma exceeds the skin depth l_s, i.e., $L > l_s$, Eqs. (3.2.38) simplify to:

$$\frac{\partial^2 N}{\partial \tau^2} - \varepsilon_T^2 n_0 \nabla_\xi^2 N + n_0 \eta_i(\vec{\xi})N = \frac{n_0}{2}\eta_i(\vec{\xi})\nabla_\xi^2 a^2$$
$$\frac{1}{n_0}\frac{\partial^2 \vec{a}}{\partial \tau^2} - \nabla_\xi^2 \vec{a} + \eta_i(\vec{\xi})\vec{a} = N\vec{a} \qquad (3.2.39)$$

Here $n_0 = (\omega_p/\omega_0)^2$ is the relative density given in terms of the critical density and $\varepsilon_T = \upsilon_T/c$ is the normalized thermal velocity. Instead of the longitudinal field **E** we have used the relative electron density perturbation, $N = (n_e - n_i)/n_i = \nabla_\xi \cdot \boldsymbol{E}$, which reveals more transparently the scattering of transverse electromagnetic waves from density perturbations in the inhomogeneous plasma.

In the derivation of (3.3.39) we have neglected the derivatives $\nabla_\xi \eta_i$, because $L > l_s$. Assuming s-polarized waves, $\boldsymbol{a} = (a, 0, 0)$, writing $a = a_0 + a_1$ and $N = N_0 + N_p$ where a_1 and N_p are small perturbations, we obtain from (3.2.39)

$$\frac{\partial^2 N_p}{\partial \tau^2} - \varepsilon_T^2 n_0 \nabla_\xi^2 N_p + n_0[\eta_i(\varsigma) - \nabla_\xi^2 a_0^2/2]N_p = n_0 \eta_i(\varsigma)\nabla_\xi^2(a_0 a_1)$$
$$-\nabla_\xi^2 a_1 + \frac{1}{n_0}\frac{\partial^2 a_1}{\partial \tau^2} + [\eta_i(\varsigma) - N_0]a_1 = N_p a_0 \qquad (3.2.40)$$

where the spatial coordinate $\vec{\xi} = (\xi, \zeta)$ is split into in-plane and perpendicular components ξ and ς components with respect to the the surface. The lowest order electron density N_0 is modified by the intense pump field a_0, and is obtained from

$$\frac{\partial^2 N_0}{\partial \tau^2} - \varepsilon_T^2 n_0 \Delta N_0 + n_0 \eta(\vec{\xi})N_0 = \frac{n_0}{2}\eta(\vec{\xi})\Delta a_0^2 \qquad (3.2.41)$$

Equation (3.2.41) describes the following physical processes: 1) changing of the plasma density profile under the action of the ponderomotive pressure and 2) the generation of surface plasma waves in the planar interaction region. In more detail the generation of a plasma wave by a ponderomotive force has been investigated in Ref. [105] where also Eq. (3.2.41) was first derived.

If the plasma is uniform along the surface, the plasma wave (3.2.41) is independent of the coordinate ξ and, therefore, no back reflection of the pump wave occurs. It is also worth pointing out the problem of boundary conditions related to the plasma spot on the surface: It is artificial to assume reflection of plasma waves occurs on the spot boundaries. In this case, a standing plasma wave that is able to reflect the laser pump wave in the back direction appiers, and no plasma instability is required for reflection to occur. In the present paper we shall not consider the effects of two-dimensional plasma inhomogeneity, in which case the average density, $N_0 = \frac{1}{2}\nabla^2 \langle \boldsymbol{a}_0^2 \rangle$, is taken as a solution for (3.2.41). Then in Eqs. (3.2.40) one can simply introduce a new density profile $\eta_i(\varsigma) - N_0 \rightarrow \eta_i(\varsigma)$ and the model reduces to that described by the original equations (3.2.39).

As discussed above, the instability develops in the transparent plasma region where the fields have their largest amplitudes. To demonstrate the onset of the instability, we shall first consider a uniform, underdense plasma layer which lies in front of a reflecting surface. The solution of Eqs. (3.2.40) can be written as

$$
\begin{aligned}
a_0 &= a_{i0}\exp[i(-\chi_{0\varsigma}\varsigma+\chi_{0\xi}\xi-\ \Omega_0\tau)]+a_{r0}\exp[i(\chi_{0\varsigma}\varsigma+\chi_{0\xi}\xi-\ \Omega_0\tau)]\\
a_1 &= a_{l1}\exp[i(-\chi_{1\varsigma}\varsigma-\chi_{1\xi}\xi-\ \Omega_1\tau)]+a_{r1}\exp[i(\chi_{1\varsigma}\varsigma-\chi_{1\xi}\xi-\Omega_1\tau)]\\
N_p &= n_{pi}\exp[i(-\chi_{p\varsigma}\varsigma+\chi_{p\xi}\xi-\ \Omega_p\tau)]+n_{pr}\exp[i(\chi_{p\varsigma}\varsigma+\chi_{p\xi}\xi-\ \Omega_p\tau)]
\end{aligned}
\tag{3.2.42}
$$

where $\chi_j(\Omega_j) = k_jc/\omega_0$, and $\Omega_j = (\omega_j + i\gamma)/\ \omega_0$ for $j = 0, 1, p$; the projections of the wave vectors are denoted by $\chi_{j\varsigma} = \chi_j \cos\theta$ and $\chi_{j\xi} = \chi_j \sin\theta$. These formulas describe the reflection of the pump wave at $\varsigma = 0$ from the overdense plasma ($\eta_i > 1$ when $\varsigma < 0$). At $\varsigma > 0$ there is an underdense ($\eta_i < 1$) plasma layer with a thickness of a few wave lengths.

Figure 3.2.12 illustrates the wave vectors of the incident and scattered laser fields and the plasma wave described by Eqs. (3.2.42). Momentum and energy conservation, i.e., spatial and temporal phase matching, imply that

$$
\begin{aligned}
k_{0y} &= k_{1y} + k_{py}\\
k_{0x} &= k_{1x} + k_{px}\\
\omega_0 &= \omega_1 + \omega_p
\end{aligned}
\tag{3.2.43}
$$

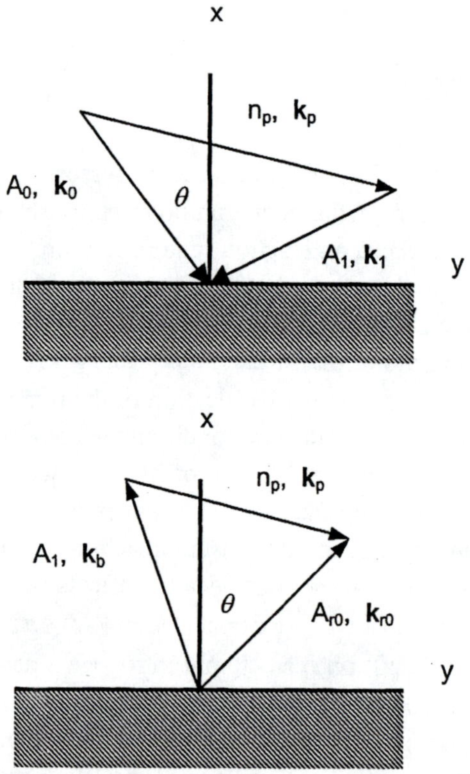

Fig. 3.2.12. In the upper frame the incoming pump wave decays parametrically into a plasma wave and a reflected EM wave; in the lower frame the specularly reflected pump wave experiences a similar decay.

The absolute values of the wave vectors and frequencies are obtained from the dispersion relations:

$$2\sin\theta = \chi + (1/\varepsilon_T)\tan\theta(\delta^2 - n - \chi^2\varepsilon_T^2)^{1/2}$$

$$(\cos^2\theta - n)^{1/2} = [(1-\delta)^2 - n - (\sin\theta - \chi)^2]^{1/2} + 0.5(\delta^2 - n - \chi^2\varepsilon_T^2)^{1/2} \qquad (3.2.44)$$

where $\chi = k_{py} / k_{0y}$, $\delta = \omega_p / \omega_0$, and $n = (\omega_{pe} / \omega_0)^2$, ω_p is the plasma frequency in the transparent underdense region, ω_{pe} is that in the dense bulk plasma.

The growth rate of the instability γ is obtained by inserting (3.2.42) into (3.2.40):

$$\gamma_i = \omega_0(\eta_i/8)(\chi_p^2/\mathrm{Re}\Omega_1\,\mathrm{Re}\Omega_p)|a_0|^2 . \qquad (3.2.45)$$

By calculating χ and δ from (3.2.44), we get an estimate for γ_i. Let us consider the parameter range $\delta < 2$ and $(2\sin\theta\,\varepsilon_T)^2 < n < 4$. Equation (3.2.44) gives $\chi \approx 2\sin\theta\,(1 - 0.5n^{1/2})$ and $\delta \approx n^{1/2}$ leading to an approximation γ_0 for γ_i:

$$\gamma_0 \approx \omega_0(n_e / 4n_c)^{1/4}\sin\theta\,(eA_0/mc^2). \qquad (3.2.46)$$

For a laser intensity I=10^{17} W/cm^2, unperturbed plasma density $n_e = 0.02n_c$, and angle of incidence $\theta = 30^0$, the growth time γ_0^{-1} is approximately 30 fs according to (3.2.46). Thus in a transparent plasma the instability will have enough time to develop during the laser pulse.

The growth rate calculated above using a plasma model consisting of homogeneous slabs is, of course, only indicative, because the actual situation is a spatially inhomogeneous plasma density distribution. In this case we may have either convective amplification or an absolute instability. In the convective case the waves grow exponentially along the surface with a gain coefficient [106]

$$G = \gamma_0^2\, l^2 / \upsilon_{1y}\, \upsilon_{py} \qquad (3.2.47)$$

The inhomogeneity due to density gradients resides in the parameter $l^{-2} = (\partial/\partial y)(k_{0y} - k_{1y} - k_{py})$; $\upsilon_i = \partial\omega_i/\partial k_i$ denotes the group velocity of the wave i. Using the dispersion relations (3.2.44) and approximating the density gradient scale length by L_y, we find the following expression for the convective gain coefficient

$$G \approx k_0 L_y \sin^2\theta\,(eA_0/mc^2)^2 \geq 30 \qquad (3.2.48)$$

for the same parameters as used in connection with Eq. (3.2.46) and for a laser spot size L_y=30 microns. Thus, in an extended laser formed plasma a convective instability generating a back reflected wave can develop. As is known already at $G > 1$ also an absolute instability can be generated. Therefore, we shall continue to analyse the conditions for an absolute instability in inhomogeneous plasmas with an inhomogeneity scale length $l_s << L_z << \lambda$ along the perpendicular direction x.

In the upper frame of Fig.3.2.12 the incoming pump wave decays parametrically into a plasma wave and a reflected EM wave; in the lower frame the specularly reflected pump wave experiences a similar decay. When the plasmon is along the plasma surface the two processes become coupled as shown in **Fig. 3.2.13**.

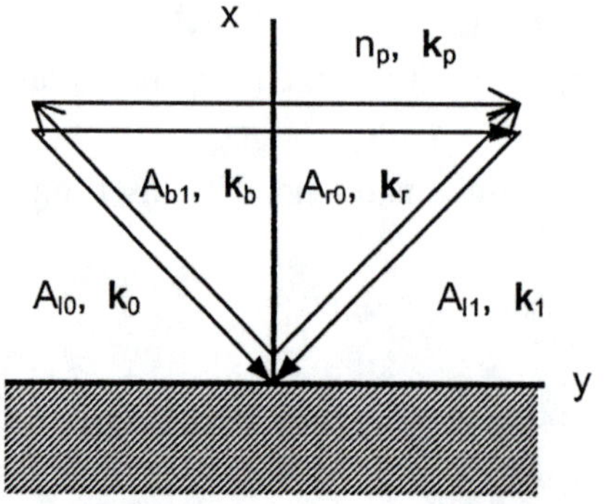

Fig. 3.2.13. An absolute instability scheme when the plasma wave is along the surface.

An absolute instability can be excited when the plasma wave is along the surface (see Fig. 3.2.13). The two decay instabilities $a_{l0} \rightarrow a_{l1} + n_p$ and $a_{r0} \rightarrow a_{b1} + n_p$ become coupled because of the common plasma wave and the specular reflections $a_{l0} \rightarrow a_{r0}$ and $a_{l1} \rightarrow a_{b1}$.

Absolute Instabilities

We shall consider a linear plasma density profile $\eta(\varsigma) = \alpha\varsigma = (c/\omega_p L_x)\varsigma$. An analytical solution of Eqs. (3.2.40) will be given below for that particular case. We assume a surface plasmon, which propagates as shown in Fig. 3.2.13. This configuration enables energy exchange between incident, back and specularly reflected waves. Above a threshold intensity of the pump wave, the wave amplitudes can grow exponentially inside the focal region and an absolute instability develops.

Let us calculate the threshold for this process. For this purpose we Laplace transform Eqs. (3.2.40) with respect to ζ, the normalized x-coordinate along the normal of the surface, and find

$$(-\alpha\frac{\partial}{\partial p} - \varepsilon_T^2\chi_{p\xi}^2 + \tilde{\Omega}_p^2 + \varepsilon_T^2 p^2)\tilde{n}_p = a_0\alpha\frac{\partial}{\partial p}(p^2 - \chi_{1\xi}^2)\tilde{a}_1$$
$$(-\alpha\frac{\partial}{\partial p} - \chi_{1\xi}^2 + \tilde{\Omega}_1^2 + p^2)\tilde{a}_1 = \tilde{n}_p a_0^* \qquad (3.2.49)$$

The Laplace transforms

$a_1(\varsigma) = \frac{1}{2\pi i}\oint \tilde{a}_1(p)\exp(p\varsigma)dp,\ N_p(\varsigma) = \frac{1}{2\pi i}\oint \tilde{n}_p(p)\exp(p\varsigma)dp$ have the same temporal and spatial variation along the surface coordinate ξ as in (3.2.41). When performing the transformation we have neglected the spatial dependence of A_0, although the inhomogeneity scales of a_0 and a_1 are comparable. It is standard approximation, although relative corrections may be of the order of unity [107]. As a result, equations (3.2.49) reduce to an ordinary second order differential equation for $a_1(p)$

$$\alpha^2 \frac{\partial^2 \tilde{a}_1}{\partial p^2} - \alpha b(p)\frac{\partial \tilde{a}_1}{\partial p} + f(p)\tilde{a}_1 = 0,$$
$$b = p^2(1+|a_0|^2) + \hat{\Omega}_p^2 + \hat{\Omega}_1^2 - \chi_{1\xi}^2(1+|a_0|^2) \tag{3.2.50}$$
$$f = \hat{\Omega}_p^2 p^2 - 2\alpha p(1+|a_0|^2) + \hat{\Omega}_p^2(\hat{\Omega}_1^2 - \chi_{1\xi}^2)$$

where $\hat{\Omega}_1^2 = \Omega_1^2 / n_0$; $\hat{\Omega}_p^2 = \Omega_p^2 / n_0$. We have omitted some small terms in (3.2.49) to obtain (3.2.50). To solve Eq. (3.2.50) we shall transform it into a Schrödinger equation with zero energy with the help of the transformation

$$\tilde{a}_1(p) = \tilde{a}(p)\exp(\int \frac{b(p)}{2\alpha} dp) \tag{3.2.51}$$

and obtain

$$\alpha^2 \frac{\partial^2 \tilde{a}(p)}{\partial p^2} - U(p)\tilde{a}(p) = 0 \tag{3.3.52}$$

The effective potential $U(p)$ in the Schrödinger equation (3.2.52) is given by

$$U(p) = \frac{1}{4}(1+|a_0|^2)^2 p^4 + \frac{1}{2}p^2((1+|a_0|^2)(\hat{\Omega}_1^2 - \chi_{1\xi}^2(1+|a_0|^2)) - \hat{\Omega}_p^2(1-|a_0|^2))$$
$$+\alpha p(1+|a_0|^2) + \frac{1}{4}(\hat{\Omega}_p^2 - \hat{\Omega}_1^2 + \chi_{1\xi}^2(1-|a_0|^2))^2 - (\hat{\Omega}_1^2 - \chi_{1\xi}^2)\chi_{1\xi}^2|a_0|^2 \tag{3.2.53}$$

Using the methods described in [107] we obtain the threshold intensity for the absolute instability in the following fashion. The discrete energy eigenvalues Ω_1 can be determined with the aid of semiclassical quantization rules, i.e., by requiring that

$$\int_{p_1}^{p_2} \sqrt{-U(p)}\, dp = \pi(2n+1), \tag{3.2.54}$$

where $p_{1,2}$ are solutions of the equation $U(p)=0$ (turning points). The instability develops only if there are the regions, where $U(p) < 0$. The potential (3.2.53) is a fourth order polynomial in a_0 which can attain negative values. At $a_0=0$ $U(p) = \frac{1}{4}(p^2 + \widehat{\Omega}_1^{\,2} - \widehat{\Omega}_p^{\,2} - \chi_{1\xi}^{\;2})^2 + \alpha p$ which has no positive roots. At normal incidence, when $\chi_{1\xi}=0$, real roots are also missing. The first root occurs, when

$$a_{0th} = (\frac{eA_0}{mc^2})_{th} \approx \left(\frac{\omega_0}{\omega_p}\right)^2 \frac{\lambda_0}{L} \frac{1}{\sin\theta} \tag{3.2.55}$$

This gives an estimate for the instability threshold. At ω_0/ω_p 0.1, L $0.2\lambda_0$, $\theta = 30^\circ$ we obtain a threshold of about 10^{16} W/cm^2 which agrees with experimental data [105].

Analysis of the Nonlinear Behaviour

We shall next consider the saturation properties of the instability. The ponderomotive pressure will steepen the plasma density profile which influences the scattering from the surface and consequently the total electromagnetic field. As the electromagnetic field, on the other hand, is responsible of the ponderomotive force, the feedback loop is thus established. To begin the analysis, we shall assume a given value of the reflection coefficient and from this basis evaluate the electron density profile. A stable operating point is found when the reflection caused by the grating due to the density modulation is consistent with the force sustaining the grating.

Let us study the neighborhood of the plasma surface around the operating point of fully developed instability. From Poisson equation we obtain for the electron density perturbation

$$\frac{\delta n}{n} = \frac{c^2}{\omega_p^2} \Delta(\varphi / mc^2) < 1 . \tag{3.2.56}$$

In a steady state situation, the electrostatic force must compensate the pondermotive pressure

$$\varphi / mc^2 \approx \sqrt{1 + a^2} \tag{3.2.57}$$

So the grating amplitude is approximately given by

$$\frac{\delta n}{n} = \frac{c^2}{\omega_p^2} \Delta\left[\sqrt{1 + a^2}\right] \tag{3.2.58}$$

where the time-averaged potential A^2 includes both the incident and back reflected waves

$$a^2 = a_0^2 e^{-2z/l_s}[1 + R^2 + 2R\cos(2\sin\theta\, y\omega_0 / c)]. \qquad (3.2.59)$$

The diffraction efficiency, and consequently also the reflection coefficient R, of the induced grating can be calculated as a function of the density profile $\delta n/n$. **Figure 3.2.14** shows, as an example, the constant density contours $<\delta n/n> = 0.5$ for two values of back reflection coefficient, R=0.4 (broken line) and R=0.8 (solid line), and for two pump amplitudes $a_0 = 1$ (lower pair of curves) and 7 (upper curves). The ordinate axis denotes the perpendicular direction into the plasma and the abscissa the transverse direction both in units of $l_{s0}=c/\omega_p$. In the unperturbed situation the plasma-vacuum boundary is at x = 0. We have assumed an angle of incidence $\theta = 45°$, and plasma density such that $l_{s0}/\lambda_0 = 0.2$ where $l_{s0}=c/\omega_p$ (note that the effective skin-layer depth, $l_s=l_{s0}(1+a_0^2)^{1/4}$, depends on the pump field amplitude a_0). From Fig. (3.2.14) we can conclude that:

1) The plasma deformation clearly increases with a_0.
2) At small pump amplitude values a_0 and reflection coefficients R, the induced density profile modification is rather insensitive to R, and its shape is nearly harmonic. *a* good approximation is obtained by expanding Eq. (3.2.58) in power series of a_0.
3) At large pump amplitudes the profile shape is more sensitive to the level of back reflection R and it exhibits highly aharmonic features.

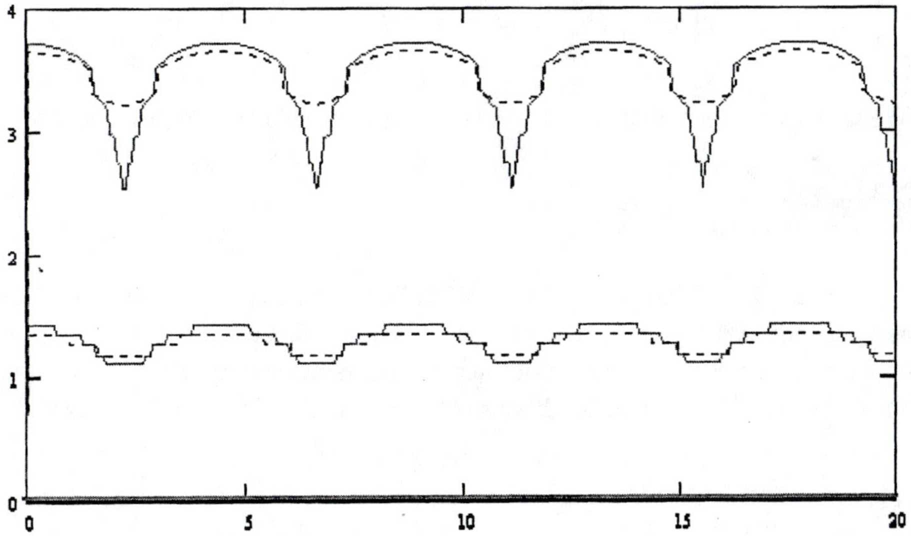

Fig. 3.2.14. Constant density contours $<\delta n/n>=0.5$ for R=0.4 (broken lines) and R=0.8 (solid lines); in the upper pair of the curves the pump amplitude is a_0=7 and in the lower ones a_0=1. The ordinate is the distance into the plasma (down the x–axis) and the abscissa the spatial coordinate along the input surface; both are given in units of $l_{s0}=c/\omega_p$. The unperturbed plasma vacuum surface is at x=0.

The introduction of higher harmonics into the density profile modification makes the reflections from the grating more diffuse and can thus lead to a saturation of the instability. We shall estimate the value of the reflection coefficient under saturation conditions taking into account the incident and both of the reflected waves (in the back and specular directions). Such approximation corresponds to the two-wave theory of diffraction, i.e., to a model where the incoming wave diffracts from a purely sinusoidal perturbation, sin(qy), with k_y=k_{y0}±q. This model is valid for lowest order expansion in a. At a>>1, higher harmonics sin(Nqy) appear, introducing new diffraction orders with k_y=k_{y0}±Nq . The simple two-wave model is a reasonable approximation even close to the saturation level where a is of the order of one. Instead of the vector potential A_x, it is more convenient to use the wave electric field E_x. In the vacuum region, the total electric field is of the form

$$\begin{aligned} & \mathrm{E}_z = E_i \exp(ik_{0y}y + ik_x x) + \sum_q E_q \exp(i(k_{0y} - q)y + \Gamma_q x) + c.c. \\ & \Gamma_q^{\,2} = -\omega_0^2 / c^2 + (k_{0y} - q)^2 \end{aligned} \quad , \quad (3.2.60)$$

The first term stands for the incident field and the sum term contains the reflected components. Within the plasma region we have

$$\begin{aligned} & E_z^{\,p} = \sum_q E'_q \exp(i(k_{0y} - q)y - \gamma_q x) + c.c. \\ & \gamma_q^{\,2} = -\omega^2 \varepsilon(\omega) / c^2 + (k_{0y} - q)^2 \end{aligned} \quad (3.2.61)$$

We shall assume a high dielectric permeability of the plasma, so that

$$\gamma_q^2 \cong -\omega^2 \varepsilon(\omega) / c^2 = \gamma^2 \quad (3.2.62)$$

The amplitudes of the spatial harmonics of the fields are determined by the usual boundary conditions on the plasma-vacuum surface. Since the depth of the surface modulation, as also the scale length of the plasma inhomogeneity, are less than the wave length, we actually apply these conditions at the critical surface inside the plasma:

$$\begin{aligned} & E_z = E_z^{\,p} \Big|_{z=f(y)} \\ & \frac{\partial E_z}{\partial x} = \frac{\partial E_z^{\,p}}{\partial x} \Bigg|_{x=f(y)} \end{aligned} \quad (3.2.63)$$

The equation of the plasma surface (rippled by the instability) $x = f(y)$ is expanded in spatial harmonics

$$f(y) = \sum_q \xi_q \exp(-iqy) + c.c.. \tag{3.2.64}$$

The amplitudes ξ_q are assumed smaller than the relevant vacuum wavelengths, but they can be comparable to the width of the plasma skin-layer. An estimate of the depth of modulation ξ is obtained by equating the ponderomotive and ambipolar forces, i.e., $\ln\xi = \frac{\partial}{\partial x}\sqrt{1+a^2}$, from which we find

$$\chi_x \xi \le (\omega_0 / \omega_p) a_0^2 \le 1 \tag{3.2.65}$$

We have assumed, above that the width of the skin layer is of the order of c/ω_p.

The boundary conditions (3.2.63) involve an infinite set of equations for the amplitudes E_q. We shall truncate this system to contain only the wave numbers $q = 0$ and $q = 2k_y$ which is consistent with the two-wave approximation of the diffraction theory. The equation of the surface is thus of the form

$$f(y) = \xi_0 + \xi_2 \cos(2k_{0y}y - \varphi_2) \tag{3.2.66}$$

where $\xi_{0,2}$ are the modulation amplitudes and φ_2 the relative phase. When substituting (3.2.60) and (3.2.61) into (3.2.63) we shall check that the depth of modulation of the surface is less than one wavelength (in agreement with (3.2.65)), although it can exceed the skin-layer depth, i.e., $\chi_x\xi \le 1$, $\chi_x\gamma_q \le 1$ which also follows from (3.2.65) at $a_0 \sim 1$.

In the vacuum region, the electric field, which is consistent with (3.2.66) is given by

$$\begin{aligned} E_z = E_i[&\cos(k_{0x}x + k_{0y}y - \omega_0 t) - \cos(-k_{0x}x + k_{0y}y - \omega_0 t) + \\ &+2\chi_x(\xi_0 + 1/\gamma)\sin(-k_{0x}x + k_{0y}y - \omega_0 t) + 2\chi_x\xi_2 \sin\varphi_2 \cos(k_{0x}x + k_{0y}y + \omega_0 t) - \\ &-2\chi_x\xi_2 \cos\varphi_2 \sin(k_{0x}x + k_{0y}y + \omega_0 t)] \end{aligned} \tag{3.2.67}$$

Accordingly, the amplitude back reflection coefficient equals $2\chi_x\xi_2$. For its determination we need the fields in the plasma. Their general form within the two-wave approximation is as follows:

$$E_z^{\,p} = \exp(-\gamma_q\xi)[E_0' \cos(\mathrm{k}_{0y}y - \omega_0 t + \alpha_0') + E_2' \cos(\mathrm{k}_{0y}y + \omega_0 t + \alpha_2')] \tag{3.2.68}$$

Substituting (3.2.67) into (3.2.62), we find:

$$E_0' \exp(i\alpha_0') = 2E_i \exp(\gamma_q \xi_0 + 2i\phi - i\pi/2)\frac{(\chi_x/\gamma_q)[I_0 \exp(-i\phi) + 2\xi_0 I_0 \sin(\phi) + \xi_2 I_1 \sin(\phi)]}{I_0^2 - I_1^2}$$
$$E_2' \exp(i\alpha_2') = 2E_i \exp(\gamma_q \xi_0 + 2i\phi + i\varphi_2 - i\pi/2)\frac{(\chi_x/\gamma_q)[I_1 \exp(-i\phi) + 2\xi_0 I_1 \sin(\phi) + \xi_2 I_0 \sin(\phi)]}{I_0^2 - I_1^2} \quad (3.2.69)$$
$$\phi = arctg(\chi_x/\gamma_q) << 1$$

where $I_0 = I_0(\gamma_q \xi_2); I_1 = I_1(\gamma_q \xi_2)$ are modified Bessel functions of order 0 and 1, respectively.

It is worth emphasizing that the system depends in an essentially nonlinear manner on the amplitudes ξ. If the second wave amplitude ξ_2 is zero, the surface is planar, and we obtain the standard Fresnel formulas:

$$E_0' = 2(\chi_x/\gamma_q)E_i$$
$$E_2' = 0 \quad (3.2.70)$$
$$\alpha_0 = 2\phi - \pi/2$$

The nonlinearity starts to appear, when the surface oscillation amplitude becomes of the order of skin layer depth, $\gamma_q \xi_2 \sim 1$, that is when the arguments of the Bessel functions reach the value of about one. From this observation it is possible to deduce the amplitude of the back reflection coefficient (which is equal to $2\chi_x \xi_2$) to be of the order of χ_x/γ. Also it is obvious that for large modulation amplitudes ξ_2 ($\gamma_q \xi_2 > 1$) the electric field amplitudes decrease. The back reflected wave disappears both when $\xi_2 \to 0$ and when $\xi_2 >> 1/\gamma_q$, reaching a maxima at about $\xi_2 \sim 1/\gamma_q$.

The electron density in the plasma is determined by the average plasma wave intensity:

$$n(x,y) = Zn_i[\Theta(x) + \frac{1}{\gamma_q^2}\frac{\partial^2 \sqrt{1 + < E_x^p(x,y,t) >^2}}{\partial x^2}] \quad (3.2.71)$$

where Zn_i is the background density of electrons and $\Theta(x)$ is the Heaviside step function. From (3.2.71) we get the constant density contours. In particular, the reflecting surface $x = f(y)$ is determined by the critical density contour $n(x,y) = n_c$. Inserting the plasma field from (3.2.68), we find

$$f(y,\xi_0,\xi_2)=\frac{1}{2\gamma}\ln(\frac{E_0'^2+E_2'^2+2E_2'E_0'\cos(2k_{0y}y+\alpha_0'+\alpha_2')}{n_{cr}/Zn_i-1}) \quad (3.2.72)$$

We again decompose $f(y)$ into spatial harmonics and keep only the first two terms of the ensuing series. Thereafter we compare the result to the initial assumption made for the surface profile:

$$f(y,\xi_0,\xi_2)\approx\frac{1}{2\gamma_q}\left(\ln(\frac{E_0'^2}{n_{cr}/n-1})+2\frac{E_2'}{E_0'}\cos(2k_{0y}y+\alpha_0+\alpha_2)\right)= \quad (3.2.73)$$

$$=\xi_0+\xi_2\cos(2k_{0y}y-\varphi_2)$$

As a result we find for $\xi_{0,2}$ and for φ_2 the expressions:

$$\frac{1}{2\gamma_q}\ln(\frac{E_0'^2}{n_{cr}/n-1})=\xi_0$$

$$\frac{E_2'}{\gamma_q E_0'}=\xi_2 \quad (3.2.74)$$

$$\alpha_0'+\alpha_2'=-\varphi_2$$

As was pointed out before, the amplitude back reflection coefficient $R = 2k_{0x}\xi_2$ is determined by ξ_2. By inserting the second equation of (3.2.74) into (3.2.69) we obtain

$$1+\frac{\chi_x\xi_2}{2}\sin(2\phi)\frac{(I_0^2(\gamma_q\xi_2)-I_1^2(\gamma_q\xi_2))}{I_1(\gamma_q\xi_2)I_0(\gamma_q\xi_2)}=\frac{\gamma\xi_2 I_0(\gamma_q\xi_2)}{2I_1(\gamma_q\xi_2)}$$

$$\phi=arctg\frac{\chi_x}{\gamma_q} \quad (3.2.75)$$

which can be used for solving ξ_2. The amplitude of the incident field does not enter in (3.2.75) (although it appears in (3.2.74)). This is connected to the expansion in E_0 used in the derivation of Eqs. (3.2.74): The fields are proportional to the incident amplitude (as we have linear boundary conditions), and as a result the ratio E_2'/E_0' is independent of E_0.

The dependence on E_0 enters only in higher order terms. Equation (3.2.75) has a solution

$$\gamma_q \xi_2 = 0.4; R_{0r} = 2.5(\chi_x / \gamma_q)^2 = 0.025 \cos^2 \theta \tag{3.2.76}$$

at ω_0/ω_p=0.1. Physically it is clear that in the diffraction grating a nonlinear aharmonic profile develops, when the amplitude of perturbations becomes comparable with those caused by E_0, i.e., when the modulation amplitude ξ_2 becomes of the order of the depth of the skin layer. In that case the profile of the surface looses its harmonic structure (field is inhomogeneous along x), and higher harmonics at the expense of ξ_2. Note that the amplitude ξ_0 describes a global shift of the surface into the plasma and is not connected with a saturation from aharmonic behavior. Thus the back reflection coefficient approaches zero both at small pump intensities, when the surface modulation is small, and at large intensities, $a_0 > 1$, when the surface modulation acquires higher harmonics resulting in a diffuse scattering of a radiation on them. The maximum backreflection is reached at $a_0 \leq 1$. We shall further elaborate on this by writing an approximate analytical expression for the reflection coefficient R as a function of the pump intensity: $R(I)$. According to Eqs. (3.2.47) and (3.2.48), for small values of the pump amplitude a_0, the amplitude of the back reflected wave increases exponentially the gain factor being dependent on the pump intensity. From (3.2.75) it follows that saturation at a level R_r occurs when the pump intensity is further increased. Beyond $I > 1$, the magnitude R_r should decrease, because according to (3.2.69) the amplitude E_2 decreases exponentially, when $\gamma\xi_2 > 1$. Thus a reasonably good approximate expression for R (I), in the range $I > I_{th}$, can be written as:

$$R(I) = \frac{\exp(G(I)) - \exp(G(I_{th}))}{1 + \dfrac{1}{R_r(I)} \exp(G(I))} \tag{3.2.77}$$

where G (I) is given by (3.2.48) and the threshold value I_{thr} by Eq. (3.2.54). From (3.2.69) one obtains for $R_r(I) = R_{0r}\exp(-2I_{18})$ where the reference intensity I_{18} corresponds to the value a_0=1. A plot of the amplitude reflection coefficient R(I) is given in **Fig. 3.2.15** for a numerical values G, I_{th}, and R_{0r}, taken from (3.2.48), (3.2.54), and (3.2.76), respectively.

In Fig. 3.2.15 the approximate reflection coefficient $R(I)$, according to the fit (3.2.77),.versus laser intensity I in units of 10^{18} W/cm^2. The constructed profile of the plasma surface becomes essentially aharmonic at relativistic laser intensity and the reflection coefficient in back direction becomes saturated. Reflection coefficient of back scattered light can be several percent and it correlates with experimental data.

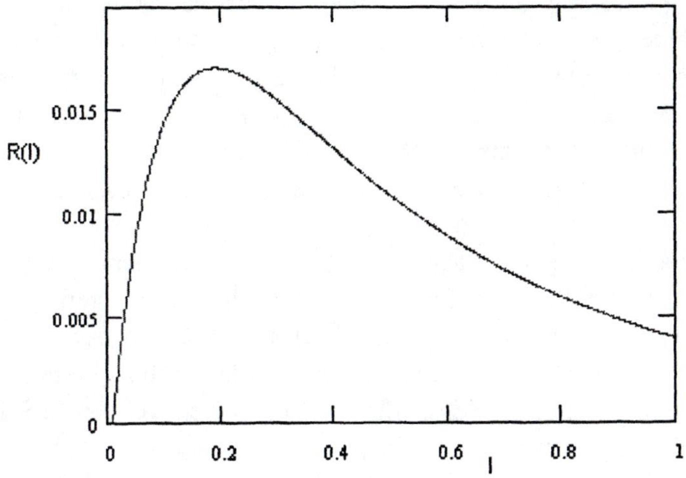

Fig. 3.2.15. The approximated reflection coefficient $R(I)$, versus laser intensity I in units of 10^{18} W/cm^2

3.2.4 Second Harmonic Emission from Solid Target Irradiated by Short Laser Pulse

Second harmonics (SH) emission is a very important diagnostics tool, which can provide a detailed information about particularly important region of critical surface neighbourhood. The parametric instabilities and resonance absorptions are the well known sources of SH emission in laser-produced plasmas. While parametric instabilities are unimportant in short-pulse interactions with solid targets due to submicron density scale length of plasma, resonance absorption of obliquely incident p-polarized laser radiation is often a dominant mechanism of laser absorption. Both energy conversion to SH and SH spectrum has been measured in short pulse experiments [61,108].

In this section we investigate the interaction of ultra-short laser pulses with a planar aluminium target in the frame of 1D hydrodynamics model. We focus our study on the dependence of critical surface motion on laser intensity and we compare the calculated SH intensity and spectra with results from experiments [103].

Analytical Model

The dynamics of plasma is described via one fluid two temperature Lagrangian hydro-code SKIN with electron and ion thermal conductivity, both natural and artificial ion viscosity and ponderomotive force impact on plasma motion. The ionic populations in plasma are calculated via set of atomic rate equations for populations of charge states. The energy loss by bremsstrahlung and recombination radiation is taken into account.

Laser absorption and electromagnetic fields are calculated for P-polarized laser radiation by numerical solving Maxwell equations in hot plasma, taking into account spatial dispersion. The model includes collision absorption, Landau damping and wave-breaking of the plasma wave. The acceleration of electrons by plasma wave damping is treated in each time step via stationary electron diffusion in the velocity space.

The energetic electrons travelling to the plasma vacuum boundary are reflected there in a double layer and a part of their energy is transferred to a group of energetic ions. The transport of electrons into the target is then described in a simple way. The fast electrons penetrate deep into the target, where they preheat the solid density target material as a precursor to the thermal wave.

Here we have calculated the intensity and the phase of SH emission by solving of the Maxwell's equations for SH field. Spatial dispersion is not included for SH field, as it does not influence the intensity and phase of SH emission. We have also neglected the influence of SH fields back on the basic harmonics, which limits the range of validity of the model to medium laser intensities, where the energy conversion to SH is small. The equations for SH fields are written, as follows

$$\frac{\partial}{\partial x}E_{2y} - 2i\frac{\omega_0}{c\varepsilon_2}\left(\varepsilon_2 - \sin^2\theta\right)B_2 = \frac{4\pi\sin\theta}{c\varepsilon_2}j_{2z}$$
$$\frac{\partial}{\partial x}B_2 - 2i\frac{\omega_0\varepsilon_2}{c}E_{2y} = -\frac{4\pi}{c}j_{2y}, \qquad (3.2.78), (3.2.79)$$

where B_2, E_{2y} are magnetic and transverse electric field of SH emission, ω_o and θ denote laser frequency and angle of incidence, ε_2 is plasma dielectric constant for SH emission. The source of SH emission is the non-linear current

$$\vec{j}_2 = -i\frac{e}{4\pi m\omega_0}\left(\vec{E}_1\mathrm{div}\vec{E}_1 + \frac{\omega_p^2}{4\omega_0^2}\nabla\vec{E}_1^2\right) \qquad (3.2.80)$$

where E_1 is electric field at laser frequency.

The source of SH emission is maximum in the neighbourhood of the critical surface, where also resonance condition for the sum of laser and plasma wave with SH electromagnetic wave are fulfilled. The phase of SH emission is controlled mainly by the motion of source of SH emission which is near to critical surface and thus the shift of SH spectra near to the Doppler shift of radiation reflected from the moving critical surface.

When the density scale length L is small $L < \lambda_0$, the laser magnetic field B_r at resonance surface ($\mathrm{Re}(\varepsilon) = 0$) is approximately equal to field at the plasma-vacuum boundary $B_r \cong 2\ B_o$, where B_o is the field of incident laser wave in vacuum. Thus the terms with the maximum powers of ε, ε_2 in denominator dominate the SH source (the right side of eq.(1)) and the equation for SH magnetic field B_2 is transformed to the following form

$$\tilde{B}_2'' - \frac{\varepsilon_2'}{\varepsilon_2}\tilde{B}_2' + \left(\varepsilon_2 - \sin^2\theta\right)\tilde{B}_2 = \sin\theta\frac{\tilde{B}\tilde{B}'}{\varepsilon^2}\left(2\frac{\varepsilon'^2}{\varepsilon_2^2} + \frac{\varepsilon'\varepsilon_2'}{\varepsilon\,\varepsilon_2}\right) \tag{3.2.81}$$

where (′) denotes $d/d\xi$, $\xi = k_o x = \omega_0/cx$ and $\tilde{B} = eB/m\omega c$. The solution of Eq.(3.2.81) that meets the following boundary conditions

$$\tilde{B}_2 \to 0\,(\xi \to 0),$$
$$\tilde{B}_2 \to R_2 \exp(-2i\xi)(\xi \to -\infty)$$

can be easily found using Green function of Eq. (3.2.81). The amplitude R_2 of the emitted SH radiation can be expressed, as follows

$$|R_2| \approx \pi(\sin^3\theta/\cos\theta)|eB_0/mc\omega|/\gamma_e^2 \tag{3.2.82}$$

Here a linear profile of plasma density was assumed in the vicinity of the critical surface, and parameter $\gamma_e = v_{eff}/\omega_0$ describes the damping rate of resonance peak of longitudinal electric field, which is dominated either by collisions or by propagation of plasma waves or by wave-breaking. The expression (3.2.82) differs only by a coefficient of order 1 from similar expression derived by [107]. Thus a simple formula of the efficiency of laser energy conversion to SH emission reads

$$\eta_2 = I_{2\omega}/I_0 \approx 4I_{18}(\omega_0 L/c)^{4/3}\Phi^2(q)$$

where laser intensity $I_{18} = I/10^{18}$ W/cm^2, $\Phi(q)$ is the Ginzburg function and $q = (k_o L)^{2/3}\sin^2\theta$. This expression differs by the multiplier 2.5 $(k_o L)^{2/3}$ from the classical formula, derived in [60], for long plasmas $k_o L >> 1$.

It is important to note that in case of high intensities investigated here, the dominant mechanism of limitation of the longitudinal laser field is the wave-breaking mechanism and the density scale length L depends on laser intensity. When the formula from paper [103] is used for $L(I)$, the conversion efficiency is expressed, as follows

$$\eta_2 \approx 2.10^{-4} T_{keV}^4 I_{18}^{-1} \sin^2\theta \tag{3.2.83}$$

As the electron temperature rises with laser intensity $T \sim I^\alpha$, where $\alpha > 0.3$, the transformation efficiency η_2 still grows with laser intensity, but the increase is slower than linear.

When the phase characteristics of second harmonics emission are investigated, it can be shown that the phase of the term in the Green function integral (product of the source and of the Greeen function) is negligible. Thus the frequency shift in the SH emission is given only by the motion of the critical surface and thus the influence of the plasma density profile may be neglected. Now, we shall consider generation of sidebands of SH emission due to the decay of plasma wave into plasma wave and ion sound wave, proposed in paper [109]. However, as the propagation of ion sound normal to the critical surface, investigated in [108,109], is here inhibited due to sharp density profile, we assume surface sound waves propagating along the critical surface. Role the phase condition of the parametric decay $k_l = k_{l'} + k_S$ and $\omega_l = \omega_{l'} + \omega_S$, the frequency of the sound waves is given by

$$\omega_s \approx 2\frac{Z m_e}{M_i}\omega_l \tag{3.2.84}$$

As the amplitude of the plasma wave, generated by the decay, may be high, it can decay again into plasma wave and ion sound and a series of sidebands, shifted by $j\omega_s$, where j=0, 1, 2,... may be formed.

Numerical Results

Presented results have been obtained for the conditions of experiment [103], where 1.5 ps FWHM Gaussian pulse of Nd-laser was incident onto a flat Al target. The angle of incidence was 45°, the intensity contrast was 10^6 to 10^7. The peak laser intensity I_o was varied in range 10^{16} - 10^{17} W/cm^2.

The overall efficiency of energy transformation to SH emission is plotted in **Fig.3.2.16** versus laser peak intensity. The experimental results are plotted together with the results of numerical simulations. When plasma parameters and the density profile do not depend on laser intensity and linear resonance absorption is assumed, the transformation efficiency grows linearly with the laser intensity $\eta_2 = I_2/I_0 \sim I_0$. This scaling is plausible for the experimental data. However, the computed transformation efficiencies are generally somewhat lower than the experimental values and they indicate a weaker dependence of transformation efficiency on laser intensity, especially for higher laser intensities $I > 5\ 10^{16}$ W/cm^2. For these intensities the simulation is very sensitive to the detailed conditions (e.g. the shape of laser pulse) due to enormous value of the ponderomotive force. Also many non-linear effects (wave-breaking, acceleration of fast electrons) are described only phenomenologically in the frame of hydro-code, and thus they may become a source of inaccuracy.

While the blue shift $\Delta\omega$ = -1.0 ± 0.2 nm of the SH spectrum was observed in experiment for laser intensity 10^{16}W/cm^2, the situation changed qualitatively for the higher intensity 10^{17} W/cm^2 and red shift $\Delta\omega$ = 0.8±0.2 nm has been measured. It indicates a qualitative change in the motion of the critical surface. Our simulations really indicate possibility of such change in plasma dynamics in the studied laser intensity range. It is presented in **Fig. 3.2.17**, where plasma density and velocity profiles are plotted at maximum of laser puke for the respective intensities. Fig.3.2.17b shows that at the higher intensity a period exists when the plasma motion in the vicinity of the critical surface is directed inwards the target. Then the critical surface moves also inwards and plasma density scale length is shortened. In our simulations this period is relatively short when absence of laser pre-pulse is assumed.

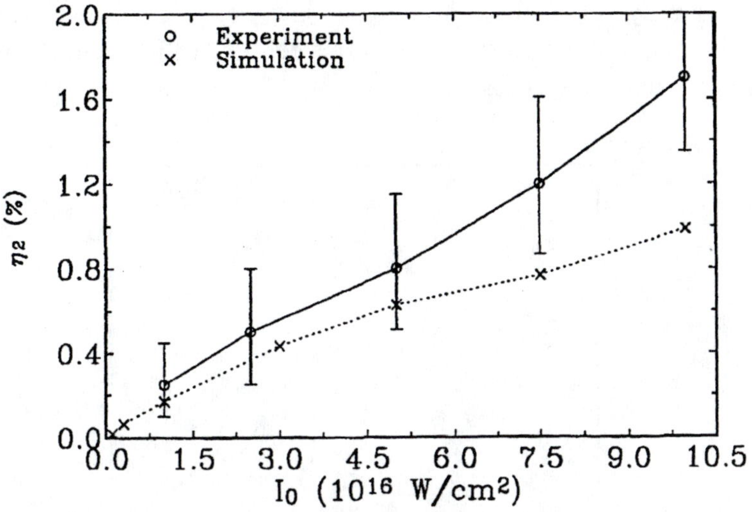

Figure 3.2.16. Efficiency $\eta_2 = I_2/I_1$ of laser energy transformation to second harmonic emission from plasma versus peak laser intensity. P-polarized Gaussian 1.5 ps FWHM pulse of Nd-laser is assumed to be incident at 45° onto a solid aluminium target.

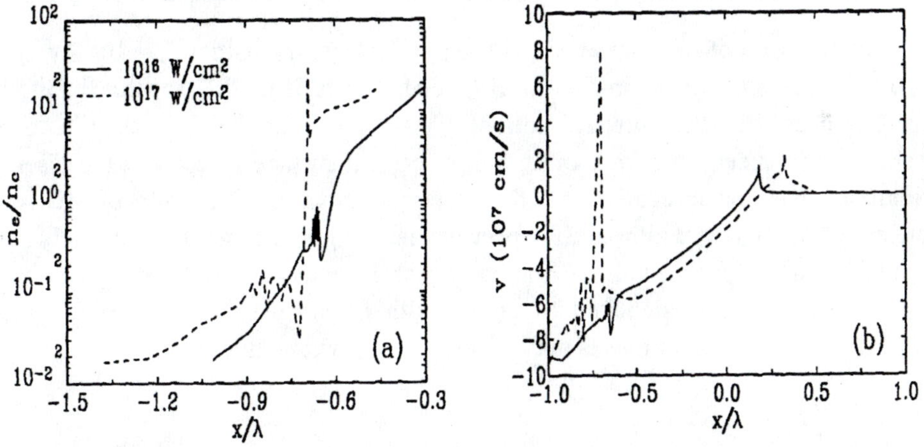

Figure 3.2.17. Spatial profiles of (a) electron density n_e (normalized on critical density n_c) and of (b) plasma velocity υ at the maximum of Nd-laser pulse and laser peak intensities $I_L = 10^{16}$ W/cm^2 and $I_L = 10^{17}$ W/cm^2 P-polarized Gaussian 1.5 ps FWHM pulse of Nd-laser is assumed to be incident at 45° onto a solid aluminium target.

The sensitivity of SH emission on the plasma dynamics is demonstrated in **Fig. 3.2.18**, where the evolution of SH emission intensity is plotted for incident laser intensities $I_0 = 10^{16}$ W/cm^2 and $I_0 = 10^{17}$ W/cm^2, respectively. For the lower intensity the maximum of SH emission is approximately at the maximum $t = 2.9$ ps of the laser pulse. The smooth shape of the SH emission pulse is near to the shape of I_0^2. The SH emission pulse is qualitatively different for the higher intensity when two peaks are formed separated with a period when a very sharp plasma density profile is unfavourable for laser energy transformation to SH emission.

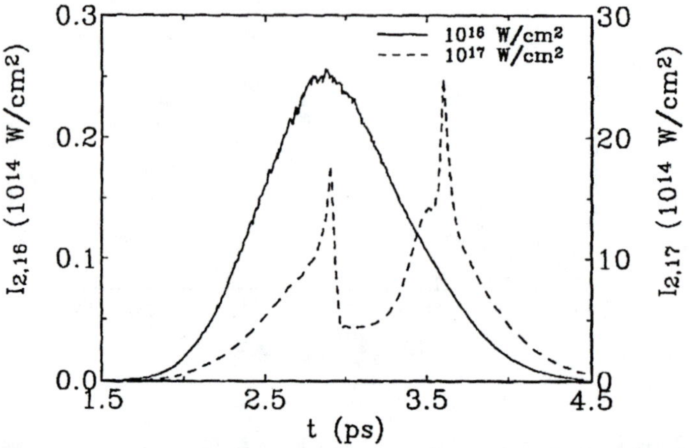

Figure 3.2.18. Evolution of intensity of second harmonic emission for laser peak intensities $I_0 = 10^{16}$ W/cm^2 and $I_L = 10^{17}$ W/cm^2 (left and right y axis are used for the lower and higher laser intensities, respectively). P-polarized Gaussian 1.5 ps FWHM pulse of Nd-laser is assumed to be incident at 45° onto a solid aluminium target.

The comparison of the computed SH spectra for laser intensities $I_0 = 10^{16}$ W/cm^2 and $I_0 = 10^{17}$ W/cm^2 is presented in **Fig. 3.2.19.** A blue shift of the SH spectra for the lower intensity is caused by the outward motion of the critical surface. The magnitude of the blue shift of SH spectrum corresponds well to critical surface velocity at the laser pulse maximum is in a good agreement with the experimental value. The model of SH emission underestimates SH line width as SH spectrum is computed assuming infinitely narrow laser line. As the experimental SH line width corresponds to broadening given by laser spectral width and the calculated SH line width is narrower than in experiment for $I_0 = 10^{16}$ W/cm^2, the simulations is not in contradiction with experiment.

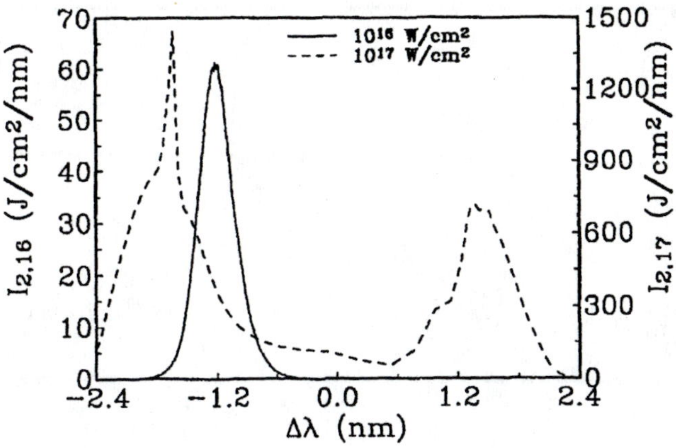

Figure 3.2.19. Computed time integral spectra of second harmonic emission for the same conditions as in previous figure.

The situation is more complicated for the higher intensity. At the laser pulse maximum an inward motion of the critical surface is observed (Fig. 3.2.17a) and thus SH emission during the first peak of SH intensity (Fig. 3.2.18) is red shifted. However, the inward motion of the critical surface lasts only part of the laser pulse duration and at the time of the second maximum of SH emission the critical surface moves in the outward direction. Thus blue shift of SH emission is calculated at the end of laser pulse and two peaks are observed in SH spectrum. Spectra have been occasionally observed in experiment [103]. However, the typical experimental spectrum that is measured for laser intensity 10^{17} W/cm^2, is plotted in **Fig. 3.2.20**. This spectrum contains only red shifted component, the magnitude of the red shift in experiment is less but comparable to the calculated value. A rather small broadening of SH spectrum compared to that correspond to the width of laser spectrum indicates a rather stable motion of the critical surface inwards with velocity $\upsilon \sim 4\ 10^7$ cm/s during the most intense part of SH emission. For such a motion the density scale lengths $L \sim 0.3$ μm, obtained in simulations without laser pre-pulse, are insufficient. The discrepancy in plasma dynamics may be caused by non-linear character of laser-plasma interaction, which is difficult to describe correctly in the frame of hydrodynamics model. Also the dynamics of the critical surface may be caused by two dimensional character of interaction, as the motion of the critical surface inwards may be stabilized by radial ponderomotive force pushing plasma radially out of the focus. This idea is supported by the fact that the spectrum presented in Fig. 3.2.20 is emitted by the central part of the laser focus, while the spectrum of SH emission from the edge of the focus is blue shifted. Also the discrepancy may be explained by the presence of laser pre-pulse. According to experimental data the pre-pulse intensities are likely to be in range 10^{10} - 10^{11} W/cm^2, which can lead to a significant evaporation of the target material.

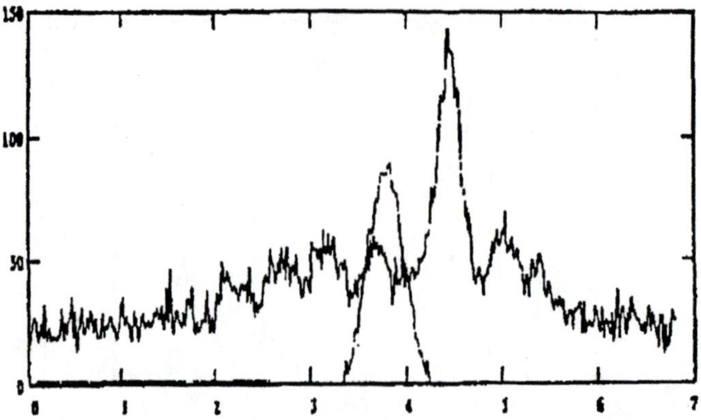

Figure 3.2.20. SH spectrum, measured at the axis of laser spot, when P-polarized Gaussian 1.5 ps FWHM pulse of Nd-laser with peak intensity $I_L = 10^{17}$ W/cm^2 is incident at 45° onto a solid aluminium target. The shift of SH wavelength is measured from the plotted laser pulse (wavelength divided by 2), the wavelength axis scale is 1nm.

The motion of the critical surface cannot explain long sidebands in SH spectrum, observed in Fig. 3.2.20. We have presented here a proposal of the sidebands generation due to a decay instability, physically similar to paper [109], but with a different geometry. The shift of the j-th sideband is according to Eq. (3.2.84) $|\Delta\lambda_j/\lambda_0| \cong 2jZm_e/M_i \cong 5\ 10^{-4} j$, which is $\Delta\lambda_j \cong 0.5\ j$ nm. This is near to the distance of the secondary maxima, observed in Fig.3.2.20.

3.3 X-Ray Emission from Laser Plasma

Compact short-pulse high-power x-ray sources can be produced by focusing on a solid target ultra-short laser pulses of low or moderate energy (from several to ten or a hundred millijouls) but of high intensity ($\geq 10^{15}$ W/cm^2). The absorbed laser energy can be converted to x-radiation for different applications, including the investigation of biological objects and lithograthy. For example, subpicosecond laser pulses can produce X-ray pulses of high power and short duration (in the picosecond range) [110,111]. With respect to keV and soft x-ray emission, such laboratory x-ray sources offer the following advantages over conventional X-ray sources such as synchrotrons: a greater compactness, an easier access, a lower cost, a shorter pulse length and a smaller source size. Short x-ray pulses are also necessary for the investigation of some chemical and biological processes (see, e.g., [112]). To give a specific example, soft x-ray microscopy with picosecond or subpicosecond "water window" X-ray pulses (i.e., with emission between 23.3 and 43.6 Å) has the advantage of preventing the sample heating or movement during the imaging [113].

It is of great importance to control carefully the laser-plasma interaction by choosing an appropriate laser wavelength, pulse duration, intensity, etc., because for example, short wavelengths allow the laser energy to be deposited at high electron density, which increases the emission. Laser pulses of less than 1 ps duration can also provide short high-power x-ray pulses; longer pulses, e.g., of the order of 100 ps, are only preferable if the x-ray energy is to be optimized. It is noteworthy that the laser light polarization and incidence are also of importance because they affect the dominant interaction mechanism and the generation of hot electrons and, hence, the x-ray generation.

3.3.1 Analytical Model for Continuum X-Ray Radiation Yield from Laser Plasma

To understand the main processes for father analysis of X-ray radiation, let's consider at first Bremstrahlung and recombination emission from over-dense plasma created by short laser pulse. We have used the following assumptions to calculate the x-ray yield of picosecond laser plasma [114,115].

1. The ion charge does not change during the laser pulse, t_L: $Z(t) = \text{const}$. For light elements ($Z < 20$) $Z \approx 16T_e^{1/3}$ at $Z < Z_{nu}$ of the nuclear charge and $Z = Z_{nu}$ at $Z \geq Z_{nu}$. For heavy elements, ($Z \geq 20$) $Z \approx 76T_e^{5/6}$, where T_e is the plasma electron temperature in keV.
2. The characteristic size of the hydrodynamic plasma expansion region $l_h = c_s t_L < \lambda_0$ is less than the laser wavelength.
3. The linear absorption coefficient η is considered which is valid at $\upsilon_E < \upsilon_T$. Here, υ_E is the thermal velocity of electrons. This condition restricts the laser radiation intensity I to $10^{17} W/cm^2$.
4. For classical thermal conductivity with the thermal wave path l_T, we have $T_e = 2t_L^{2/9}(\eta I)^{4/9}$, $l_T = 350Z^{-1}(\eta I)^{5/9} t_L^{7/9}$, $l_h = 0.44(Z/A)^{1/2}(\eta I)^{2/9} t_L^{10/9}$. The following units of measure were used here:

$$I[10^{18}\,W/cm^2],\quad l_{T,h}[10^{-7}cm],\quad T_e[1keV],\quad t_L[1fs],\quad \lambda_0[1\mu m]$$

5. We first take into account only the bremsstrahlung component of plasma emission during the laser pulse duration, which means that we evaluate the hardest component of x-ray emission, corresponding to the highest plasma temperature. Then the coefficient of the laser radiation conversion to x-ray emission by an optically transparent plasma will be

$$k_{x,b} = \sigma T_e^4 (l_T/\lambda_R)/I \approx (e^2/hmc^3)(Zn_i T_e^2 v_{ei} l_T/I),$$

where $\lambda_R = c(\omega_x^2/\omega_p^2)\nu_{ei}^{-1}$ is the path length of an x-ray quantum with the frequency $\omega_x = T_e/\hbar$ and $n_i \approx 6\cdot 10^{22} cm^{-3}$ is the ion concentration for most materials in the solid state.

The conditions typical of most experiments with $I \le 0.1, t_L \ge 100$ correspond to the normal skin effect with a smooth density profile, when the expressions for the plasma temperature and conversion coefficient are written at $l_h < 0.1\lambda$ as

$$T_e = 0.36 Z^{0.22} A^{0.04} I^{0.29} t_L^{\ 0.06}$$
$$k_{xb} = 5.6\cdot 10^{-9} Z^{2.38} A^{0.08} I^{-0.5} t_L^{\ 0.6}$$

for complete ionization ($Z = Z_{nu}$) and as

$$T_e = A^{0.05} I^{0.4} t_L^{\ 0.08}$$
$$k_{xb} = 3\cdot 10^{-4} I^{0.2} t_L^{\ 0.75}$$

for incomplete ionization of heavy elements.

For short wavelength lasers with $l_h > 0.1\lambda$, the conversion coefficient is higher and may reach the values of $k_{xb} \approx 2\cdot 10^{-11} \lambda_0^{-41/12} I^{17/24} t_L^{\ 49/16}$, $T_e \approx 0.1 A^{-1/4} \lambda_0^{-1} I^{1/2} t_L^{\ 3/4}$.

The conversion of bremsstrahlung emission in the definite spectral range $\Delta\omega_x = \omega_{2x} - \omega_{1x}$ can be evaluated as

$$k_{xb}^{\Delta\omega} = k_{xb}[\exp(-\frac{\hbar\omega_{1x}}{T_e}) - \exp(-\frac{\hbar\omega_{2x}}{T_e})]$$

At $\hbar\omega_{1,2x} < T_e$, we will have for heavy elements

$$k_{xb}^{\Delta\omega} \approx \lambda^{-29/12} A^{1/4} I^{5/24} t_L^{\ 37/16} \Delta E, \quad \Delta E = \hbar(\omega_{2x} - \omega_{1x})$$

This approach provides the lower limit for the evaluation of x-ray emission by the laser plasma. Additional absorption mechanisms, non-classical transfer, recombination and linear radiation will increase this yield. In particular, for recombination radiation (integrated over the whole spectrum), we have the following expression for the conversion coefficient in a complete ionization:

$$k_{xr} \approx 10^{-10} Z^{62/13} A^{-3/26} \lambda_0^{-6/13} I^{-10/13} t_L^{\ 11/13}$$

For heavy elements, the coefficient is

$$k_{xr} \approx 10^{-13} A^{-1} \lambda_0^{-4} I^{1/3} t_L{}^4$$

Like in the case of backscatter bremsstrahlung radiation, we have

$$k_{xr}^{\Delta\omega} = k_{xr}[\exp(-\frac{\hbar\omega_{1x}}{T_e}) - \exp(-\frac{\hbar\omega_{2x}}{T_e})]$$

It is known that non-linear mechanisms such as anomal skin effect, Bruinell effect and some others do not practically change the dependence of fast electron temperature on ($I\lambda^2$), so the conversion coefficient does not change either. We will assume that the plasma is scattered randomly after a laser pulse of duration t_L is applied. The temperature is then supposed to decrease adiabatically, $T_e \approx T_{\max}(\frac{x}{x_m})^{-2}$, where $x = c_s t$ is the distance to the target surface. Since the x-ray emission intensity is $I_x \sim T_e^{5/2}$, we have $I_x \approx I_m(\frac{c_s t}{x_m})^{-5}, x_m = c_s t_L$. If the x-ray emission is assumed to cease at $I_x \approx 10^{-2} I_m$, its duration can be evaluated as $t_x \approx 3t_L$.

Let us find the target emission intensity of the K_α-line due to a fast electron flux produced by resonance absorption. Similarly to [42], the concentration of fast electrons is taken to be $n_e^{(h)} \approx 3 \cdot 10^{-2} n_c$, n_c is the electron concentration at the critical point. For the concentration $n_i^{(+1)}$ of ions with the charge + 1, we will have the expression: $n_i^{(+1)} = n_i[1 - \exp(-n_e^{(h)} v_i^{(z)} t_L)]$, $v_i^{(z)} = 10^{-8}(Ry/I_z)A_i(\gamma_i^{-1/2} + \chi_i \gamma_i^{1/2})^{-1}\exp(-1/\gamma_i)$ as the rate of ionization of an atom from its ground state, where $\gamma_i = T_h/J_z, J_z$ is the ionization potential, $A_i = 60, \chi_i = 0.7, \quad n_i = 6 \cdot 10^{22}\, cm^{-3}$.

The K_α-line intensity will be

$I_{k_\alpha} \cong \Delta E_{k_\alpha} n_i^{(+1)} A_{k_\alpha} l_{Th} \sim \Delta E_{k_\alpha} Z_{nu}^5 J_z^{-1/2} \lambda_0^{-2} T_h^{3/2} t_L$, $A_{k_\alpha} \approx 5 \cdot 10^7 Z_{nu}^4$ [s] is the radiation decay probability and l_{Th} is the path length of a fast electron. At $T_h \sim (I\lambda_0^2)^{2/3}$, we get $I_{k_\alpha} \cong \Delta E_{k_\alpha} J_z^{-1/2} Z_{nu}^5 \frac{E_L}{S}$.

The x-ray yield for an aluminum target, calculated within the present model for the parameters of the work [116], is found to be $I_{k_\alpha} = 7 \cdot 10^{12}\, W/cm^2$, in good agreement with experimental data.

3.3.2 Numerical Code and Laser-Plasma Coupling

To calculate more precisely *x*-ray emission parameters, we must know the plasma temperature and density at every moment of the laser pulse interaction and following it. This problem can be solved using numerical method. So we first describe the simulation code and then additional simple model of X-ray emission to interpret the experimental results.

Laser Pulse Absorption and Plasma Hydrodynamics

The plasma dynamics can be described by the one-fluid two-temperature Lagrangian hydro-code SKIN with the electron and ion thermal conductivities, both the natural and artificial ion viscosities, as well as the ponderomotive force impact on the plasma motion. For a detailed description of the hydro-code, the reader is refereed to Section 3.1.

A one-dimensional planar description is usually sufficient for short pulse interactions with a solid target, as the laser focus diameter is much larger than the scale length of the plasma expansion by the laser pulse. A simple model built in terms of atomic physics is included in the hydrodynamics model in order to calculate the mean ion charge Z and the average squared ion charge $\overline{Z}^2$. The populations of the ion charge states are calculated from a set of atomic rate equations. The rates of tunnel ionization by laser radiation, collision ionization, radiative and three-body recombination include the suppression of the ionization potential in dense plasma. For a solid target, the tunnel ionization is important only in a thin surface layer of a few Å. The plasma recombination is important when a laser pre-pulse is present. Also, it has a significant impact on the time integrated x-ray emission yield from the target. The energy loss by radiation consists of the bremsstrahlung and recombination emission multiplied by the escape factor, so that only the emission that reaches the plasma-vacuum boundary is included in the energy conservation.

X-Ray Emission Model for Simulations

The resulting ion distribution in the laser plasma was used to calculate the power density of x-ray emission P_x [W/cm^3] as a function of time t and spatial coordinate x along the target normal. The local power density of x-ray emission P_x [W/cm^3] was obtained as a function of time t and the spatial coordinate x. The integration of P_x over the distance yields the x-ray intensity I_x [W/cm^2], and further integration over t provides the energy per unit surface area of the source (further referred to as energy density) of the bremsstrahlung X_{brems}, recombination radiation X_{rec} and the total continuum emission X_{cont}=X_{brems}+X_{rec} [J/cm^2]. The energy density emitted by both Ly-α and He-α lines is X_{lines}=X_{Ly}+X_{He}. The power and energy density of the Ly-α line P_{Ly} and X_{Ly} were calculated in a similar way.

Here we are interested mainly in transforming the laser energy to a very narrow spectral region, e.g., to a single line or a series of consecutive lines. The interpretation of emission spectra of laboratory plasmas and astrophysical sources is a fairly difficult problem. The assumption of local thermodynamic equilibrium is often inapplicable either because of the insufficiently large thickness of the source and/or to the transient nature of laboratory plasmas. On the other hand, the plasma thickness is usually large enough so that the impact of the reabsorption of the line emission on the ion population cannot be neglected. One cannot, therefore, ignore the relation between the radiation transfer and the ion population in the plasma.

In astrophysics, this problem is fairly well understood [117], but laboratory plasma may differ in many aspects from astrophysical plasma. First, not all spectral lines are optically thick in laboratory plasmas, and continuous radiation (bremsstrahlung and recombination emission) is usually optically thin. Second, laboratory plasmas are often so dense that many atomic levels are collision dominated and their populations are practically independent of radiation processes. Finally, laboratory plasmas are often so short-lived that the assumption of statistical equilibrium becomes invalid.

Some of the approaches used in astrophysics have been modified for model studies of homogeneous and static laboratory plasmas [117]. For example, a radiation transfer model based on the escape factors was developed for laboratory plasmas in [118]. The main drawback of this approach is, however, that the macroscopic Doppler shift is ignored when the escape factors are derived. We have implemented here the 'localized' Newton-Raphson method proposed in the work [119]. When the rate equations are linearized, the spectral intensity dependence on the populations is limited to one Lagrangian cell in each iteration step, and the radiation transfer through the entire grid is solved between the iteration steps. A reasonably fast convergence of this scheme has been confirmed.

The radiation transport in the x-ray lines is solved as a postprocessor to the 1D hydro-code in planar geometry. The plasma parameters (density, velocity, electron and ion temperature) are discretized into a grid of Lagrangian cells in the hydro-code output. The radiative field is calculated only for the selected bound-bound transitions, which can be both optically thick, and the radiative transition may influence the populations significantly. The spectral radiation intensity I_ν for these lines is discretized into a frequency grid, and the radiation transfer equation is solved in planar geometry

$$\frac{1}{c}\frac{\partial I_{x\nu}}{\partial t}+\mu\frac{\partial I_{x\nu}}{\partial x}=j_\nu-k_\nu I_{x\nu} \tag{3.3.1}$$

where $\mu=\cos\theta$, θ is the radiation propagation angle with respect to the target normal, j_ν and k_ν are the spectral emission and absorption coefficient, respectively. Since the transit time of the radiation through the plasma is negligible, the radiative transfer is solved in the approximation of infinite light velocity.

The atomic physics model is designed for K-shell spectroscopy and it includes a detailed set of resonance levels for Li-, He- and H-like states, as well as several di-electronic states important for x-ray diagnostics, so that the emission in satellite lines to Ly-α a and He-α lines can be calculated. The number of states included is limited either by the atomic database or by the lowering of the ionization potential in dense plasma. The atomic physics database is implemented for different ions. It includes all collision and radiation processes but its present version implies a Maxwell electron distribution. The rate equation governing the population of the k- level is written as

$$\frac{\mathrm{d}\ N_k}{\mathrm{d}\ t} = \sum_l \left[-N_k A_{kl} + (B_{lk} N_l - B_{kl} N_k)\ \frac{4\pi}{c}\ \bar{J}_{kl} \right] - \sum_n C_{kn}(n_e, T_e)\ N_k + \\ + \sum_m \left[N_m A_{mk} - (B_{km} N_k - B_{mk} N_m) \frac{4\pi}{c} \bar{J}_{mk} \right] + \sum_n C_{nk}(n_e, T_e)\ N_n \tag{3.3.2}$$

where N_k is the population density of the k- level; A_{kl}, B_{lk} and B_{kl} are Einstein's coefficients for spontaneous emission, absorption and stimulated emission; the first term represents radiative bound-bound transitions between the k-level and the lower l-level, while the third term stands for radiative bound-bound transitions between the k-level and the upper m-level. The sum of all possible collision transitions from and to the k-level is described by the second and fourth terms. The rate coefficients C_{nk} depend generally on the electron density n_e and temperature T_e and include collision excitation and de-excitation, radiative and three-body recombination, auto-ionization and di-electronic recombination. Photo-ionization is not included, as the plasma is assumed to be optically thin for continuous radiation.

The absorption and stimulated emission of the spectral lines are the only non-linear terms in the rate equations, their non-linearity arising from the dependence of the mean integrated intensity $\bar{J}_{kl}$ on the population densities. The integration is performed over the angle and frequency

$$\bar{J}_{kl} = \frac{1}{2} \int_{-1}^{1} \mathrm{d}\mu \int_0^\infty I_{x\nu}(x, \mu)\ \Phi_\nu^{kl}(x, \mu)\ \mathrm{d}\nu \tag{3.3.3}$$

where an identical emission and absorption line shape Φ_ν^{kl} is assumed. The spontaneous emission coefficient j_ν and the coefficient k_ν of absorption and stimulated emission for the transition between the upper k-level and the lower l-level are specified by the following expressions:

$$j_\nu^{kl}(x, \mu) = \frac{h\ \nu_{kl}}{4\ \pi}\ A_{kl}\ N_k(x)\ \Phi_\nu^{kl}(x, \mu) \\ k_\nu^{kl}(x, \mu) = \frac{c^2}{8\ \pi\ \nu_{kl}^2}\ \frac{g_k}{g_l} \left(N_l(x) - \frac{g_l}{g_k}\ N_k(x) \right) \Phi_\nu^{kl}(x, \mu) \tag{3.3.4}$$

where ν_{kl} is the kl transition frequency and g_k is the k-level degeneracy.

In the implemented model for the simulation of non-equilibrium line transfer, the line transfer is computed within the core saturation approximation [64]. This concept was used in order to accelerate the convergence and to describe the radiative intensity variation within each hydrodynamic cell. We assume the Voigt profile of the emission and absorption lines

$$\Phi_{\nu}^{kl}(x,\mu) = \frac{\Gamma^{kl}(x)}{4\ \pi^{5/2}\ln 2}\int_{-\infty}^{\infty}\frac{e^{-y^2}\ \mathrm{d}y}{\frac{\Gamma^{kl}(x)^2}{16\pi^2}+\left[\nu-\nu_{kl}(x,\mu)-\frac{\nu_D^{kl}(x)}{2\sqrt{\ln 2}}y\right]^2} \tag{3.3.5}$$

where ν_D is the Doppler width and Γ is the Lorentz width with the account of the natural, lifetime and electron impact Stark broadening. The macroscopic Doppler shift due to the plasma velocity is included in the transition frequency $\nu_{kl}(x,\mu)$ depending on the coordinate and the radiation propagation direction in the laboratory frame. The radiative transfer is solved together with the level populations only for potentially optically thick lines.

The postprocessor implies planar geometry even when a laser pre-pulse is present. This model is substantiated by the fact that only a relatively thin layer (typically 10 – 20 μm for 1 ns delay) behind the critical surface is ionized to a He-like state and that this material provides most of the K-shell line emission. During the x-ray emission, this layer is typically located at a distance of only 0–30μm from the original target surface for a 1 ns delay. This is comparable with the focal radius of the main pulse, which is several times smaller than the ordinary pre-pulse radius, so the plasma expansion is very close to planar expansion here. Moreover, the bulk Doppler shift due to the plasma motion suppresses the reabsorption of the spectral lines far from the emission region. Thus, planar geometry is a good approximation in the computation of x-ray emission lines. The postprocessor code computes in the first phase atomic level populations with the account of the radiation transfer effect on the populations. Then, the computed populations are used to find the spectra emitted in any direction, and the finite transverse plasma dimension can be accounted for at this step. In order to find the time-integrated emission in the direction of experimental observation, we take the integral over the spectral line frequency and time

$$E_x^{kl} = \int \mathrm{d}t \int_{\nu_{kl}} I_{x\nu}(x=-\infty,\mu)\ \mathrm{d}\nu \tag{3.3.6}$$

This value can be compared with the experimental energy emission divided by the main pulse focal spot area.

3.3.3 Prospects of 'Water Window' X-Ray Emission from Laser Plasma

We consider now the important example of laser plasma application a generation of x-ray emission in "water window" region. Absolute measurements of "water window" $[2\,\text{nm} \le \lambda_x \le 4\,\text{nm}]$ X-ray emission have shown that high brightness water window emission is achievable even with small scale femtosecond laser system (**Fig.3.1.1**). For instance, a 18 mJ, 248 nm, 700 fs P-polarized laser pulse focused at an incidence of 30° to an intensity of 10^{16} W/cm^2 onto a flat polished carbon target can produce over 10^{19} photons per second, mm^2 source area, mrad^2 in 0.1% relative bandwidth per shot at 3.4nm (the carbon Ly-α line). This corresponds to 0.36% conversion efficiency in this wavelength region, or to 5% efficiency for the whole region between 1.5 and 5nm. Further emission increase can be easily achieved by increasing the laser intensity and by other optimizations.

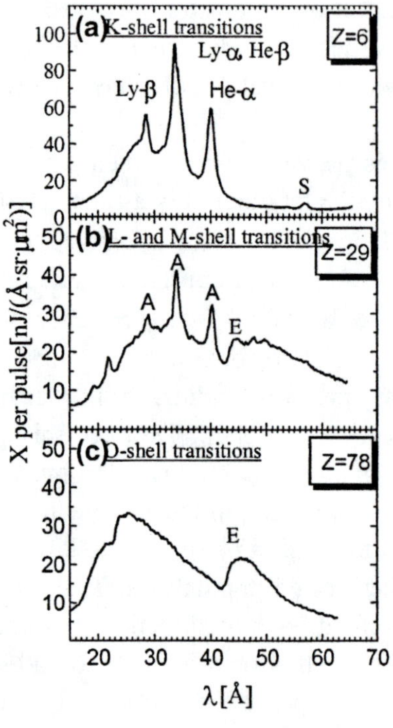

Fig.3.3.1. Spectra of massive targets consisting of (a) carbon, (b) copper and (c) platinum obtained for t_L=700fs, λ_0=0.248μm, θ=35°, P-polarized, I_0=10^{16} W/cm^2.

Here, we consider possibilities for optimal source with respect to "water window" X-radiation for a picosecond laser pulse and a high X-ray power. The results are then compared to the experiments carried out under various experimental conditions [110,111].

Carbon was used as the target material of interest. Fig. 3.3.1 [120] shows that this element provides the largest X-ray yield in the water window, as compared to elements with a higher atomic number Z. Other elements, e.g. aluminium, were also measured but their X-ray energy yield in this spectral range is smaller. A carbon target also gives minimum bandwidth emission (e.g., a short X-ray line width). So we will concentrate on the results obtained from different carbon targets.

X-Ray Emission versus the Incidence Angle

We first carried out a numerical study of the dependence of X-ray emission on the laser pulse incident angle θ. We also used experimental data of the emission from carbon targets irradiated by subpicosecond 0.248 μm pulses [121]. The data showed a clear difference in the emission for S- and P-polarized laser light as well as for line and continuous emission. The numerical results we compared with the measurements.

Fig. 3.3.2 presents the time, space, and wavelength integrated energy in the continuous spectrum and in the emission lines as a function of θ. The circles indicate the experimental data [121] and the crosses are the calculations. X_{cont} denotes the energy density of continuous emission, X_{lines} is the total x-ray energy over all x-ray lines in the "water window", and X_{tot} is the sum of both. The absolute emission value was found from the comparison of experimental spectra measured at laser intensity $I_0=10^{16}$ W/cm^2 [121] with those from [120] recorded under identical experimental conditions. So, the left-hand ordinates show the x-ray energy emitted per 1 μm^2 source area per laser shot while the right-hand ordinates correspond to the x-ray efficiency with respect to the focused laser energy, assuming isotropic emission into 4πsr.

One can see a qualitative agreement between the experiment and simulation for X_{cont} (Figs. 3.3.2a,d). In particular, the maximum yield is obtained for P-polarized laser light at an angle $\theta = 45°$. This means that the optimal conditions for continuous emission in the experiment and simulation were similar. The resonance absorption plays a major role at $\theta = 45°$, leading to a much hotter plasma. The contribution of the Ly-α and Ly-α lines becomes more important, so that the simulation does not differ very much from the measured Ly-α emission at intermediate incidence.

The difference in the simulated and observed emission at large angles (e.g., $\theta > 55°$) results from the opacity and experimental conditions. Although the emission is nearly isotropic at small and intermediate observation angles, opacity makes an essential contribution at larger θ. This was the case at $\theta > 55°$ due to the experimental arrangement in [121]. Therefore, the calculations made for large values of θ probably used a lower re-absorption than that observed experimentally, so the calculated emission is larger than the observed emission. Another reason for the difference is that the laser spot focused at these angles may have had two-dimensional inhomogenities that could not be taken into account by simulations in 1D geometry.

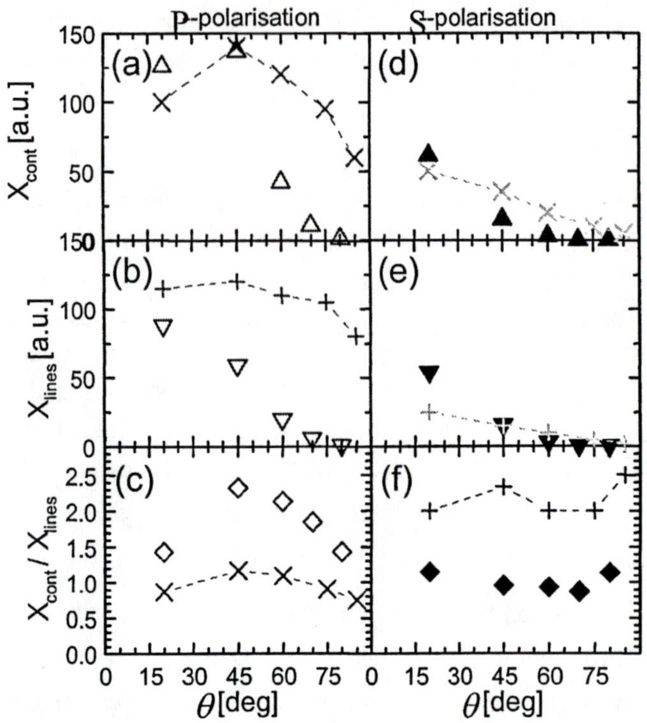

Fig. 3.3.2. Space, time, and wavelength integrated (10-60 Å) x-ray emission from a carbon plasma as a function of the angle of incidence (for fixed laser intensity $I_0 \approx 10^{16}$ W/cm^2 on the target surface): simulation (crosses) and experimental results (circles): x-ray continuum X_{cont} ((a), (d)), entire line emission X_{lines} ((b), (e)), and total emission X_{tot} ((c), (f)) for P- (a)-(c) and S-polarized (d)-(f) pump pulses respectively.

Thus, most of the differences between the simulation and experiment are due to the deviation of the experimental conditions from perfect ones, on which most of the calculations were based. To illustrate, the non flat-top intensity distribution in the focal laser spot in the experiment [121] leads to conditions that may vary with the position in the focal spot, which is, in particular, true for fast electron generation. However, the optimum conditions for water window X-radiation as a function of the incidence angle are well reproduced, namely 20 -45° and P-polarized laser light (Fig. 3.3.2c).

Intensity Dependence of X-Ray Emission

The dependence of X-ray emission on the laser intensity has been studied numerically. It was shown experimentally [121] that the emission of carbon plasmas irradiated by subpicosecond 0.248 μm laser light strongly increases with the light intensity on the target surface I_0.

Fig. 3.3.3 shows the simulated and measured x-ray energy densities at $\theta = 45°$. It can be seen that the numerical dependencies for the continuous and line emission, in particular, for the Ly-α line emission do agree well with the experimental results. The scaling of the x-ray emission follows the power law, i.e. $X \sim (I_0)^\gamma$ [121], and the simulations are seen to have nearly the same exponent ($\gamma_{cont} \approx 1.55$ for the continuous emission, $\gamma_{lines} \approx 1.1$ for the line emission and $\gamma_{tot} \approx 1.4$ for the total emission in the water window. It was mentioned above, however , that there is a slight disagreement in X_{lines} due to the lack of the full contribution of the He-α line emission. In addition, the X_{cont}/X_{lines} ratio in Fig. 3.3.3b shows that the continuous emission increases much faster than the line emission. Again, there is a small disagreement with respect to the total line emission, X_{lines}, which we explained above. In the other respects, the simulation reproduces the experimental data fairly well.

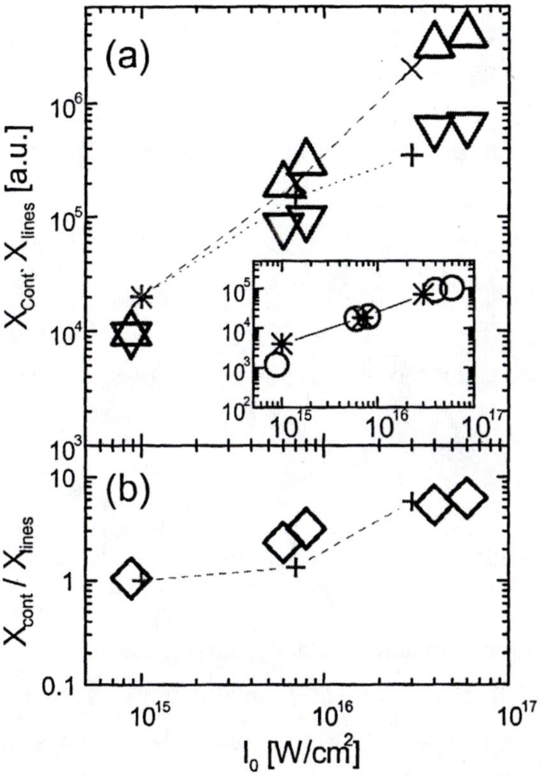

Fig. 3.3.3. (a) Space, time, and wavelength integrated (10-60 Å) x-ray emission from a carbon plasma as a function of the laser intensity on the target surface: experimental values for X_{lines} (down-triangles), X_{cont} (up-triangles), X_{tot} (circles), and X_{Ly} (open squares); simulated values for X_{lines} („+“), X_{cont} („x“) and X_{Ly} (solid squares). The angle of incidence was 45° and the 700 fs pump pulses were P-polarized. (b) Ratio X_{cont}/X_{lines}. The data points in (b) are calculated from the data shown in (a) (diamonds: from experiment, „+“: from simulation). The typical error of the data points in (a) and (b) is given by the symbol size.

Analytical Scaling and X-Ray Emission Variation with the Laser Wavelength

To make the above picture complete, we will describe briefly the analytical scaling laws for x-ray emission and discuss its variation with the laser wavelength and pulse duration, as observed in another experiment [110].

It will be shown that optimal conditions correspond to increased laser intensity and temperature, although there is a limit to the plasma temperature imposed by the requirement of sufficiently ionised but not over-ionised plasma. For a wavelength of λ_0= 0.248 μm, this limit is about 10^{18} W/cm^2.

Another condition for optimal x-ray line emission from a bulk target is that the x-ray line absorption length $l_{x\text{-}abs}$ exceeds the extension of the hot plasma l_{pl} (the plasma volume containing the majority of H-like ions): $l_{x-abs} > l_{pl}$

The x-ray line absorption length can be estimated as [122]

$$l_{x-abs} = 1500 Z^* a_B \frac{c}{v_T} \qquad (3.3.7),$$

where a_B is the Bohr radius and the electron temperature can be estimated from:

$$\frac{T_e}{keV} \approx 2\left(\frac{I_{abs}}{10^8 W \times cm^{-2}}\right)^{4/9}\left(\frac{t_L}{fs}\right) \qquad (3.3.8)$$

If the laser intensity is not too high, l_{pl} is mainly given by the propagation length of the thermal wave

$$l_T = v_T\left(\frac{t_L}{v_{ei}}\right)^{1/2}, \quad \frac{l_T}{nm} \approx \frac{150}{Z^*}\left(\frac{I_{abs}}{10^8 W \times cm^{-2}}\right)^{5/9}\left(\frac{t_L}{fs}\right)^{7/9} \qquad (3.3.9)$$

here v_{ei} is the electron-ion collision frequency, Z^* is plasma average charge state.

Thus, one can find from (3.3.7) that optimal conditions for x-ray line emission are present for intensities up to 10^{18} W/cm^2. Due to the strong scaling of collision-less absorption processes with laser intensity, fast electrons become increasingly important, in particular, at intensities above 10^{17} W/cm^2. At such intensities, (at 10^{18}W/cm^2 this is the case even for λ_0= 0.248 μm radiation), the target heating depth l_{pl} is given by

$$\frac{l_h}{nm} \approx 200\left(\frac{T_h}{keV}\right)^2\left(\frac{\rho}{g \times cm^{-3}}\right) \qquad (3.3.10),$$

where the hot electron temperature can be evaluated in the model from

$$\frac{T_h}{keV} \approx 300\left(\frac{I_{abs}}{10^{18} W \times cm^{-2}}\right)^{2/3}\left(\frac{\lambda_0}{\mu m}\right)^{4/3} \qquad (3.3.11)$$

It can be easily found that l_{pl} then becomes very large, exceeding 100 μm. Therefore, the plasma volume heated by a laser pulse V_h (~ illuminated spot size x l_h) also becomes very large, such that the x-ray emission may no longer increase with the laser intensity. The more general characteristic temperature of the fast electrons T_h scales with $(I_0\lambda_0^2)^{\zeta}$ (at $1/2 < \zeta < 2/3$) [42] and the corresponding mean free path l_h scales from (3.3.11) as

$$l_h \propto T_h^2 \propto I_0^{2\zeta} \qquad (3.3.12).$$

From this, the target mass M_h heated by these electrons is approximately proportional to ρ x V_h , so we get $M_h \propto I_0^{2\zeta}$. (Due to the small wavelength of 0.248 μm, the plasma density is approximately equal to the solid density). The *plasma* temperature due to fast electron heating (it is not the temperature of fast electrons!) can then be evaluated from

$$T_e^{(h)} \propto \frac{E_{abs}}{M_h} \propto I_0^{1-2\zeta} \qquad (3.3.13).$$

With $\zeta > ½$, this scaling leads to a reduction of the plasma temperature due to the larger volume to be heated. This means that, as $T_e^{(h)}$ decreases, a smaller number of H-like ions will be produced reducing the Ly-α emission down to $T_e^{(h)} < I_H$ (where I_H is the ionization potential for the production of H-like ions) when the Ly-α line emission ceases.

As long as the laser intensity is not too high (i.e. well below 10^{18} W/cm^2), the plasma electron temperature T_e, the length l_T of the plasma heated by a non-linear thermal wave (or, the heat wave propagation length), and the average ionisation stage Z^* are given by direct plasma heating and non-linear heat wave propagation. These quantities can be derived from the following scaling laws [114]:

$$\begin{aligned} T_e &\propto I_0^{4/7}\lambda_0^{-4/7}t_L^{6/7} \\ L_T &\propto I_0^{11/12}\lambda_0^{-11/12}t_L^{9/7} \\ Z^* &\propto I_0^{4/21} \end{aligned} \qquad (3.3.14)$$

The energy density of bremsstrahlung radiation, X_{brems}, and that of recombination radiation, X_{rec} , can be found from following scaling formulas:

$$X_{brems} \propto I_0^{10/7}, X_{rec} \propto I_0^{33/21} \qquad (3.3.15).$$

From Eqs. (3.315), the total continuous emission $X_{cont} = X_{brems} + X_{rec}$ is approximately proportional to $I_0^{1.5}$. Similarly, the line emission is given by

$$X_{lines} \propto I_0^{20/18} = I_0^{1.1} \tag{3.3.16}$$

This scaling shows the same dependence of X_{cont} and X_{lines} on the laser intensity I_0 as that observed in the experiment (see Fig. 3.3.3).

Laser Wavelength Dependence of X-Ray Emission

For wavelengths longer than UV laser light, e.g., $\lambda_0 = 0.8$ μm, fast electron generation is even more important, so fast electrons dominate even at lower intensities. If the plasma extension length is then given by the mean free path of fast electrons l_h, i.e. $l_{pl} \sim l_h \sim I_0^{4/3}$, even a stronger dependence on the laser intensity will be observed. This was found in a recent experiment [111]: $X_{tot} \sim (I_0)^{1.4}$ for $\lambda_0 = 0.248$ μm and $X_{tot} \sim (I_0)^{2.2}$ for $\lambda_0 = 0.8$ μm, both for carbon plasmas produced by ~700 fs laser pulses at intensities in the range from 10^{14} to 10^{17} W/cm^2. Besides, due to the larger heated volume (see above) and the lower electron density where the laser pulse absorption occurs [111], the x-ray energy emitted at $\lambda_0 = 0.8$ μm is lower than at $\lambda_0 = 0.248$ μm, provided that the only difference in the experimental conditions is the laser wavelength.

In our model, the line emission increases with shorter laser wavelengths λ and weakly depends on longer pulse durations. The laser wavelength must have such a magnitude that at the respective electron temperature the laser plasma would consist of H-like ions: $T_e \leq I_H$, where I_H is the H-like ion ionisation potential. This condition holds for $\lambda_0 > 0.1$ μm. It can be concluded from this that it is necessary to decrease the wavelength to get a higher line emission, so $\lambda_0 = 0.248$ is preferable at real experimental parameters when the pulse duration is about several hundreds of femtosecons.

To conclude, the x-ray energy is greater for UV laser plasmas with a pulse duration of several hundreds of femtoseconds than in the case of a longer laser wavelength and a shorter pulse duration (even for constant fluency, as indicated by experiments [111]).

X-Ray Emission of Foil Targets Illuminated by Oblique Laser Pulses

We have shown above that hot plasma extension l_{pl} is of great importance, and it will be demonstrated below that the target thickness optimal for maximum Ly-α line emission is defined by this length. The simulations with foil targets at varying thickness d were performed under conditions similar to those used in the previous simulations and experiments described in [121], with the only difference that now massive targets were replaced by thin foils. The simulation parameters were as follows: $t_L = 700$fs, $\lambda_0 = 0.248$ μm, P-polarized laser light, $\theta=40°$, $I_0 = 10^{16}$ W/cm^2. Besides, there was no pre-pulse and, hence, no steep plasma-vacuum boundary. The simulation results can be seen in Figs.

3.3.4 3.3.6 which show the electron density and electron temperature profiles as a function of the spatial coordinate x for three time values, together with the positions of emission volumes of the Ly-α line, the He-α line and the continuum.

Emission Profiles of Massive Targets

Fig. 3.3.4 shows that a thick, or massive, target is heated up to the regions far behind its original surface (x = 0 μm). This is responsible for the ionisation deep into the target bulk, although the temperature is not high enough to reach the highest ionization conditions. In contrast, H-like ions are produced only in the vicinity of the original target surface, where the electron density and temperature are sufficiently high ($ne \sim 10^{24}$ cm^{-3}~ $55n_c$ and $T_e \sim 200$ eV; the high electron density is due to the shock wave).

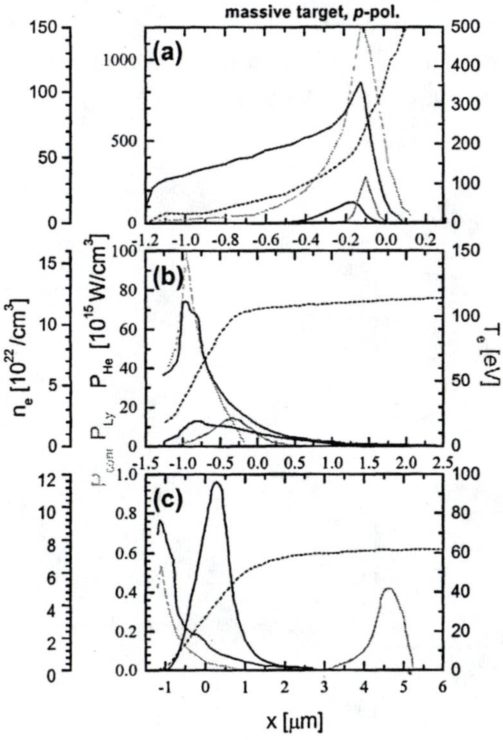

Fig. 3.3.4. Results from simulations for a massive carbon target at 3 different times, i.e. maximum of the laser pulse (at $t = 700$ fs) (a), 5 ps later (b), and 10 ps later (c): n_e (solid line; most left axis), T_e (dashed line; right axis). The power densities of the x-ray Ly-α emission P_{Ly} (solid line and area; second axis on left hand side), the He-α emission P_{He} (dotted line; same axis), and the continuum emission P_{cont} (dotted-dashed line; same axis), respectively, are shown as a function of the spatial coordinate The $t_L = 700$ fs, $\lambda_0 =$ 0.248 μm, $I_0 = 10^{16}$ W/cm^2 laser pulse was p-polarized. No pre-pulse was present. The laser was incident from the right hand side and focused at an angle of incidence of $\theta =$ 40°. The original target surface was located at $x = 0$ μm.

As a result, the intense Ly-α emission is concentrated in definite regions, whereas the emission by He-like ions is distributed over a larger volume and comes from cooler areas (Fig. 3.3.4a). At later times, both emission volumes begin to expand decreasing the Ly-α emission (Fig. 3.3.4b). At still later times, e.g. 10 ps after the laser pulse maximum, the Ly-α emission decreases further so that the line emission is dominated by the emission by He-like ions (Fig. 3.3.4c). This can be better seen in Fig. 3.3.8 where the calculated space-integrated emission in the carbon Ly-α and He-α lines is plotted as a function of time.

For comparison, the continuous emission volumes are given in Figs. 3.3.4 and 3.3.6As long as the electron temperature is high enough, the spatial emission profiles are not much different from the electron density profiles. In contrast to p-polarized laser pulses, where target heating by hot electrons may be very important, these are hardly present in a plasma produced by S-polarized light. Thus, the heated region is not given by l_h but by the propagation length of the thermal wave l_T, which is much smaller than l_h. Besides, the laser pulse absorption is much weaker, leading to a much smaller region to be heated and to a much lower x-radiation (**Fig. 3.3.5**).

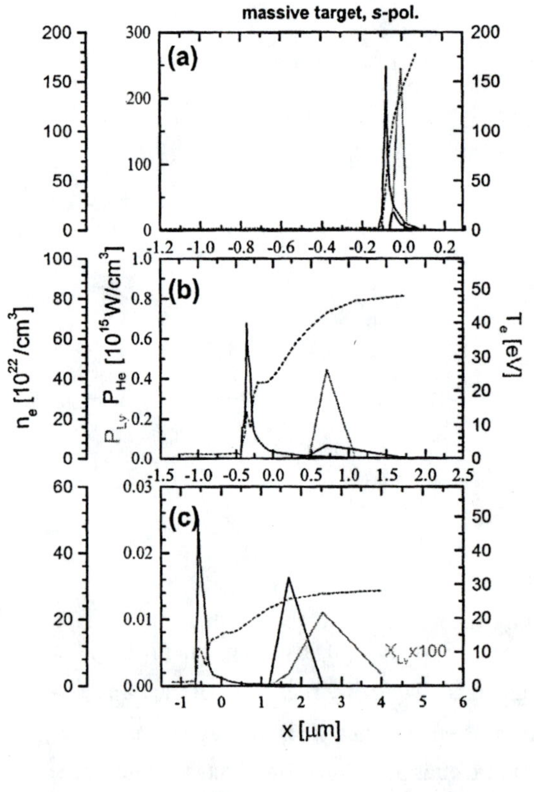

Fig. 3.3.5. Same as in Fig. 3.3.4, but for S-polarized laser light.

Emission Profiles of Thin Foil Targets

X-ray emission by massive targets may also be compared to that of thin foil targets, both illuminated with p-polarized laser light (**Figs. 3.3.6, 3.3.7**).

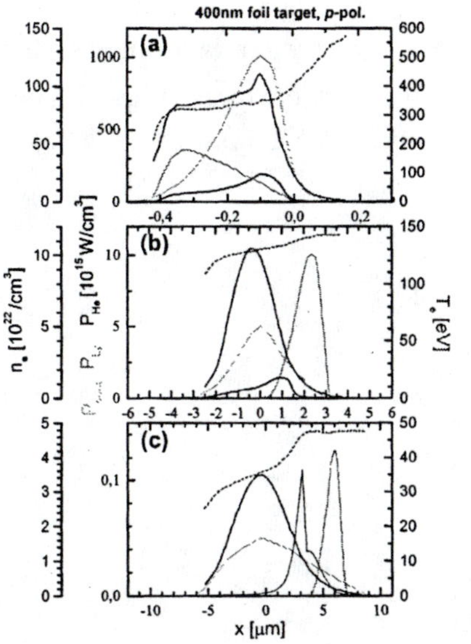

Fig. 3.3.6. Same as in Fig. 3.3.4 but for a foil target with $d = 400$ nm. The original target surface was located at $x = 0\mu m$ (front side) and at $x = -0.4\mu m$ (backside).

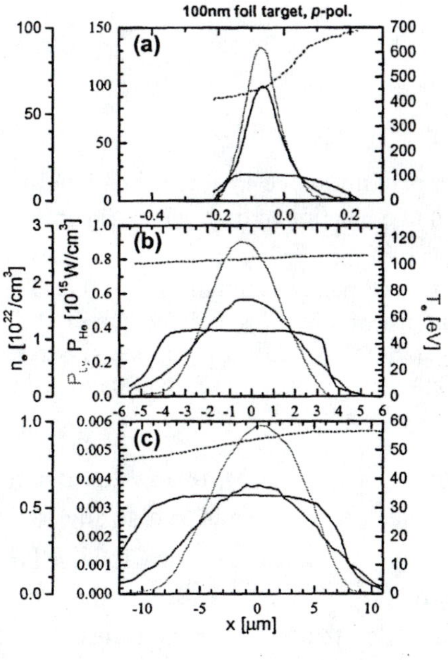

Fig. 3.3.7. Same as in Fig. 3.3.4 but for a foil target with $d = 100$ nm. The original target surface was located at $x = 0$ μm (front side) and at $x = -0.4\mu m$ (backside).

In particular, at $l_{pl} > d$, the absorbed energy is distributed throughout the target thickness d, leading to a fast target heating and, as a consequence, to a rapid expansion of the target material (compare the extensions of the heated regions in Fig. 3.3.4 and Fig. 3.3.6). The thermal wave which arises later than the jet of fast electrons only present during the laser pulse then propagates through a preheated target, contributing to further heating, particularly at the target rear where the shock wave is toppled (see Fig. 3.3.6a). As a result, the Ly-α line intensity is high only at earlier times (Figs. 3.3.6a), becoming with time lower than that of a massive target (Figs. 3.3.6b, **3.3.8**).

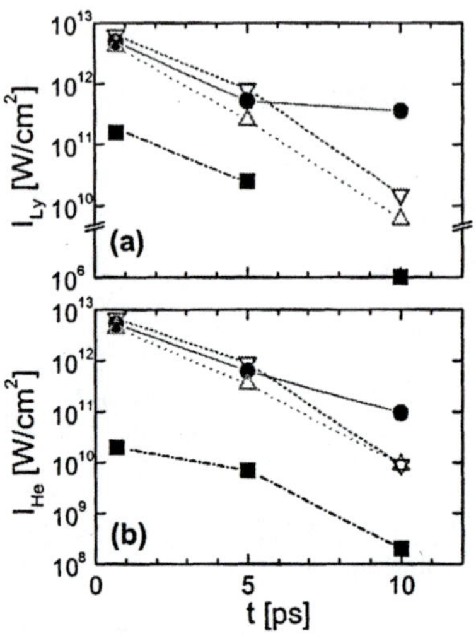

Fig.3.3.8. X-ray intensity (X-ray power per unit source area) in the carbon Ly-α line I_{Ly} (a) and the carbon He-α line I_{He} (b) as a function of time. The x-ray intensity is calculated from Figs. 3.3.4-3.3.5 and is the space integrated (in x-direction) power density of the X-ray power density. The laser pulse maximum is at 700 fs. The data points correspond to the following conditions: solid circles: massive target, P-polarized laser light; solid squares: massive target, S-polarized laser light; open up-triangles: 100 nm foil target, P-polarized laser light; open down-triangles: 400 nm foil target, P-polarized laser light.

When the target is very thin, say, 100 nm (Fig. 3.3.7), this behaviour is even more pronounced so the whole target is strongly heated. This leads to a strong ionisation and a fast bilateral expansion of target material. The optimal conditions for the Ly-α emission are present nearly in the whole foil centre (Fig. 3.3.7). However, the total number of atoms available for ionisation is much smaller than in a thicker target. (This number correlates with the number of line emitters and, hence, with the x-ray yield). Thus, for thin foils, additional energy may only lead to a higher temperature and to a faster expansion, so that the foil becomes 'over ionised'. Due to the high temperature and lack of the target bulk, He-like ion density is, at first, low and there is only a weak He-α emission (Fig. 3.3.7a). A more intense He-like emission occurs only later, after the recombination of H-like ions (Figs. 3.3.7b,c).

Comparison of Thick and Thin Targets

Because very thin targets have the disadvantage of having an insufficient number of atoms and of becoming over-ionised, thicker targets are obviously preferable. This can be seen in **Fig. 3.3.9** where the calculated Ly-α line conversion efficiency η_{Ly} is plotted as a function of d.

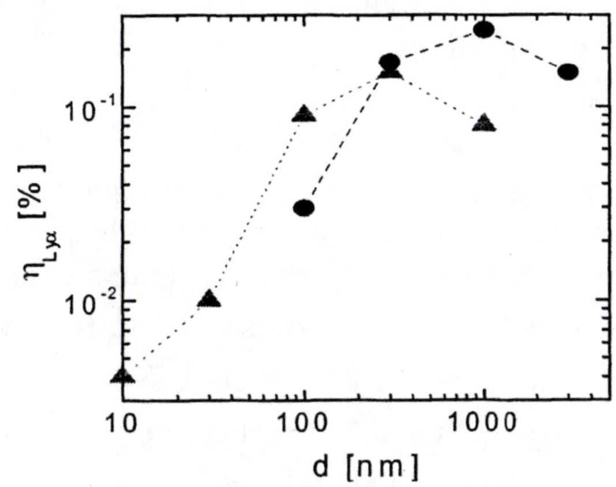

Fig. 3.3.9: Simulation of the Ly-α line conversion efficiency η_{Ly} as function of target thickness d for a 248 nm, 700 fs, 10^{16} W/cm^2 pulse incident on a steep plasma vacuum boundary (dots) and a 308 nm, 500 fs, 10^{16} W/cm^2 pulse incident on a smooth plasma vacuum boundary (triangles).

On the other hand, very thick, or massive, targets have a relatively small region with the optimum conditions for H-like ion production. In this case, a large fraction of the absorbed laser energy is spent on inefficient heating of bulk material (in Fig. 3.3.4, the region of Ly-α emission is much smaller than, e.g., in Fig.3.3.6). Fig. 3.3.9 shows that for the present conditions the optimum target thickness is approximately $d \sim 1$ μm (circles), or the length which is heated, $d \sim l_{pl}$. In contrast to the case of a thin target, the plasma density is then high enough, the target bulk is hot ($Z^* \sim 5$) but not overheated because of the greater mass involved. . Moreover, the spectral shape of x-ray line emission is also influenced by the target thickness. It was shown above and can be seen from Figs. 3.3.4, 3.3.6, 3.3.7 that the density of expanded plasmas is lower in thin foils than in thick targets. Due to the less pronounced collision de-excitation and the longer lifetime of the upper transition states, one should expect smaller line widths of the observed resonance lines. Furthermore, because of the high temperature and moderate density, higher excitation states may be present so one may observe the respective x-ray spectral lines. The plasma produced on a foil target has a smaller emission continuum than on a massive one.

Qualitative Effect of Laser Parameters on X-Ray Emission

In addition to target thickness, other experimental conditions can also influence the x-ray emission of a foil target. For instance, if λ_L, t_0 or the plasma electron density scale length L is different from the value chosen before, there will be an optimal value for the foil thickness but at a different value of *d*, which may be smaller (Fig. 3.3.9) or larger (e.g., for λ_0>1 μm). Fig.3.3.9 illustrates the efficiency η_{Ly} calculated for a carbon foil irradiated by a t_L = 500 fs, λ_0 = 0.308 μm, I_0 = 10^{16} W/cm^2 laser pulse at normal incidence. Two density profiles were simulated. One was a smooth initial density profile with L = 0.3 μm, or the same one as in the work [116], and the other was an initially abrupt target-vacuum boundary (L = 0). In the first case when the profile was due to a pre-pulse produced, say, by amplified spontaneous emission, the results of the present work (top triangles in Fig. 3.3.9) show a good agreement with those of [116] (bottom triangles). Even in the case of a negligible pre-pulse, η_{Ly} was not much different in spite of the greater pulse absorption to be expected for a smooth boundary.

When the target is very thin, it is heated uniformly throughout its entire thickness and expands into two opposite directions. TAs a result, the electron temperature is very high ($T_e > I_H$). The plasma becomes overheated and consists of nearly fully ionised atoms (Z^* ≈ 6). Due to the fast foil expansion, the ion and electron densities decrease. The low density of H-like ions $n_i^{(H)}$ then leads to only a negligible Ly-α line emission because of $X_{ly} \sim n_i^{(H)} d$ (see the emission yield for d = 10, 30 nm in Fig. 3.3.9).

For the conditions considered in this section, namely, the normal incidence and a smooth density profile, the absorption is dominated by inverse bremsstrahlung, producing a propagation length of l_T ~ 200-300 nm. So, for thicker targets with $d \sim l_T$ and l_T given by Eq. (3.3.9), η_{Ly} reaches its maximum. The shock wave very reaches the opposite side of the target very fast, the electron density decreases and the thermal wave efficiently heats the entire target. A relatively dense, but not overheated (Z^* ~ 5) plasma, is produced and it means that the conditions for H-like ion line emission are optimal (see Fig. 3.3.9).

As the target thickness increases further, the line intensity falls gradually to a certain value, until the line yield becomes independent of the thickness. In such a case ($d > l_T$), the target may be regarded as being massive: the shock wave does not reach the opposite side of the target where, otherwise, it would be toppled producing a rarefaction wave.

The electron density of the plasma covered by the shock wave is then high (approximately 10-20 times the critical density n_c), and the thermal wave is unable to heat this region efficiently, so that the target material becomes ionised only by a factor of 2 to 4. Thus, the plasma mass in the region with optimal emission conditions (including the region of ionisation up to H-like ions) is reduced and concentrates in a lower density region where $n_e \sim n_c$.

Remarks on Foil Targets Affected by Very Intense Laser Light

If a massive or thin foil target is illuminated by laser light exceeding the intensity of 10^{18} W/cm^2, the ponderomotive pressure onto the plasma is very large even at short laser wavelength. The planar foil geometry, used in the present 1D simulations, becomes distorted producing a curvature of the target surface during the interaction, which increases the laser pulse absorption and x-ray emission.

Although such 2D effects in thin foil targets illuminated by intense laser light are beyond the scope of the present work, we have made, by way of example, a 1D simulation for a 10^{18} W/cm^2 laser pulse obliquely incident onto a 400 nm foil. X-ray emission of approximately 1 GW was obtained for a source area of 10 μm diameter, which corresponds to the x-ray intensity $I_x \sim 10^{15}$ W/cm^2 or the peak brilliance of approximately 10^{22} photons per s, mrad^2 and mm^2 in a 0.1 % relative bandwidth [123]. This is consistent with the analytical scaling. It should be emphasized that a laboratory subpicosecond laser-plasma soft x-ray source has the same order of magnitude for the peak brilliance in the water window as advanced synchrotrons, such as BESSY II, but this was done with a much smaller effort. The next physical advantage essential for applications could be the ultra-short duration of x-ray pulses from a subpicosecond laser-plasma (see **Fig. 3.3.10**).

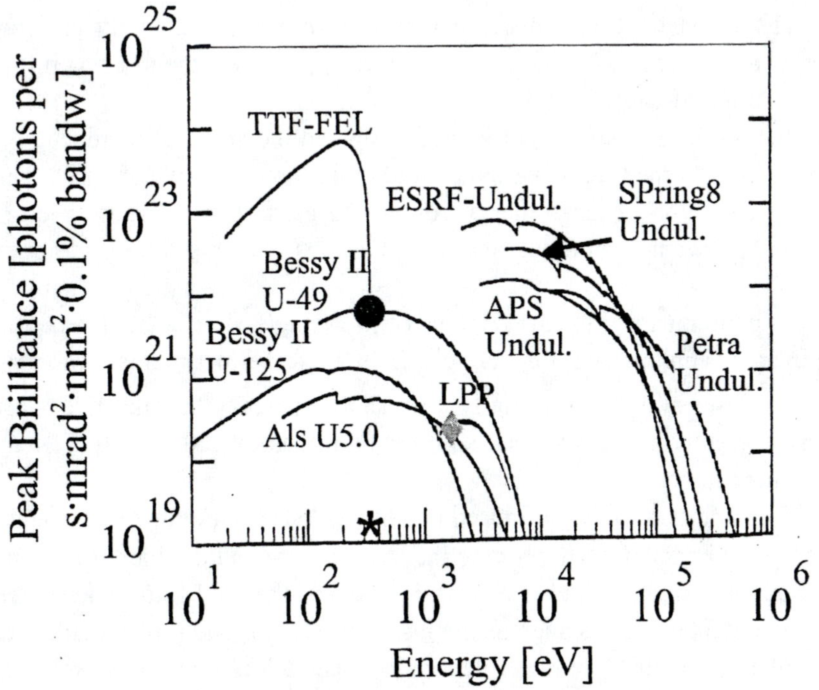

Fig. 3.3.10. Peak brilliance of various X-ray sources (synchrotrons, free electron lasers and incoherent K-α emission LPP). The asterisk corresponds to the N-Ly-α line emission of a typical pulsed discharge source (e.g., plasma focus), as well as to the measured C-Ly-α emission of a plasma produced by fs-laser system. The solid circle is the expected peak brilliance for the optimal conditions with foil target.

Our theoretical studies and scaling have made to find optimal laser parameters and target conditions to achieve short x-ray pulses and high x-ray power. We can make the following conclusions from this analysis.

First, it has been shown that carbon is the most suitable target material with respect to both overall "water window" x-ray energy and intense line emission. In particular, two strong resonance lines, the Ly-α and He-α lines, are in the desired wavelength interval.

Second, the laser wavelength is to be small, for instance 0.248 μm. This allows the laser energy to be deposited at a high electron density. If the laser intensity is high enough even at this wavelength, a fraction of absorbed laser energy will be transformed to fast electrons. Due to the small wavelength, however, their energy and propagation length are not very high, which allows a controlled heating of target material

Third, the laser intensity is be high so that sufficient laser energy should be available for conversion to x-ray energy; however, the intensity should not exceed 10^{18} W/cm^2 (for 0.248 μm laser light) to protect the target from overheating and to reduce the ponderomotive pressure on the plasma. For longer wavelengths, the optimal laser intensity is to be lower due to the stronger coupling to fast electrons. This has the disadvantage of having less laser energy available for conversion to x-ray energy.

Fourth, the laser pulse duration is to be less than 1 ps to achieve short pulse duration for x-ray Ly-α line emission (approximately 5 ps or less) and continuous emission (of the order of a picosecond or less) resulting in a high X-ray intensity. As has been discussed elsewhere [124], longer laser pulses, of the order of 100 ps, are only preferable if the x-ray energy is to be optimised. However the x-ray power will then be much lower due to the longer duration of x-ray emission.

Fifth, the P-polarized laser light is preferable and the angle of incidence of the laser pulse with respect to the target normal should be in the range of 20° - 40°. Under these conditions, the fraction of laser energy coupled to the target is largest, the target mass heated by fast electrons is optimal, and the intensities of X-ray lines and continuum are highest.

Sixth, in addition to high x-ray intensity in a single shot, a high repetition rate is desirable in many applications, for example, in biological experiments. This requires high repetition rate laser systems, especially with short wavelengths, like KrF*-lasers. Such systems are presently being designed so repetition rates of 200 to 400 Hz may become available in near future.

Seventh, in addition to the optimised conditions listed in points one through five, foil targets with a thickness of 1 μm provide the largest X-ray yield which is approximately a factor of 2 larger than with thicker or massive targets. Optimal foil thickness varies with other experimental conditions such as the laser wavelength and pulse incidence. Another advantage of foil targets is reduced debris, because debris may be a problem in some applications.

To conclude, the optimised conditions described above can provide high-power picosecond laser-plasma X-ray sources with x-ray pulse duration of about 5ps or less, emitted in the spectral range of 1 Å. For a source diameter of 10 mkm, this corresponds to soft x-ray power of about 10 GW in a spectral window of 1 Å. These values can be improved by the proper choice of experimental conditions [125].

3.3.4 Enhancement of X-Ray Line Emission by Shaping Short Intense Laser Pulses

Here, we consider the possibility of enhancing X-ray emission from laser plasma, using the pre-pulse method, in which the main short pulse and the pre-pulse divided by a time interval interact with a solid target. In many experiments, the delay time of the main pulse is fairly large, so that the plasma expansion is no longer planar and the expansion is self-similar in the transverse direction. Let us consider the situation when, the transverse radius of the focal spot during the laser pre-pulse is much larger than the plasma expansion radius so that the interaction is independent of geometry. During the pause between the pre-pulse and the main pulse, the transverse plasma expansion should not be neglected, so we assume the spherical geometry of the expansion. The focal length of the main pulse is taken to be larger than the scale length of the plasma density, and planar geometry is assumed for the interaction of the main pulse, starting from the profiles of the plasma parameters (density, velocity and temperatures) taken at the end of spherical expansion.

The laser radiation has normal incidence on the target in the experiment [126]. While the absorption efficiency of P-polarized oblique radiation is generally high, the absorption efficiency of a short intense pulse at normal incidence on a perfect solid target is generally rather low, ~15%, in the absence of a laser pre-pulse. The laser pre- pulse produces plasma, which expands and cools down before the main pulse arises. The absorption efficiency of normally incident radiation depends on the plasma extension and density when the main pulse is incident onto the target. Here we present the dependence of the absorption efficiency of the main normally incident pulse on the delay time of the main pulse relative to the pre-pulse, under the experimental conditions of [127].

It is shown that the absorption efficiency for rather short pulse delay $\Delta\tau \leq 100$ ps is even lower than in the absence of a pre-pulse. A thin plasma layer suppresses laser penetration into the dense plasma due to the skin effect, so the absorption of a normally incident wave is reduced. The absorption efficiency rises significantly for a longer pulse delay when a dense plasma region is formed where collision absorption is efficient (see **Fig.3.3.11**). The laser absorption gradually decreases at a pulse delay longer than 1 ns, at least in the idealized model, since the density of the plasma produced by the pre-pulse is insufficient to maintain the extensive absorption efficiency.

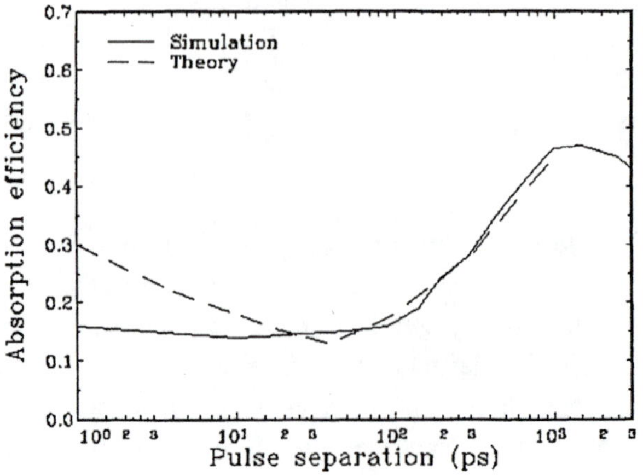

Fig. 3.3.11. The absorption efficiency of laser radiation incident normally onto an Al target versus the pulse delay time $\Delta\tau$. Gaussian laser pulses with t_{FWHM}=130 fs and wavelength λ_0 =800 nm are assumed. The main pulse intensity is I_L=2.10^{16} W/cm^2, while the pre-pulse intensity is I_{pL}=10^{15} W/cm^2.

After the main pulse is absorbed, a relatively thin layer of dense plasma is formed on the target. While multi-photon ionization is unable to ionize Al to a Li-like state at the laser intensity of 10^{16}W/cm^2, the dominant ionization mechanism is collision ionization. It is mostly efficient in the dense plasma, so the highest ionization is produced in the over dense plasma under the critical surface. Collision ionization is highly non-stationary and non-equilibrium in the case of a 130 fs laser pulse. The highest emission intensity of He- and H-like ions is achieved, approximately 1 ps after the main pulse maximum. The profile of the hydrodynamic parameters at this time is presented in **Fig. 3.3.12.**

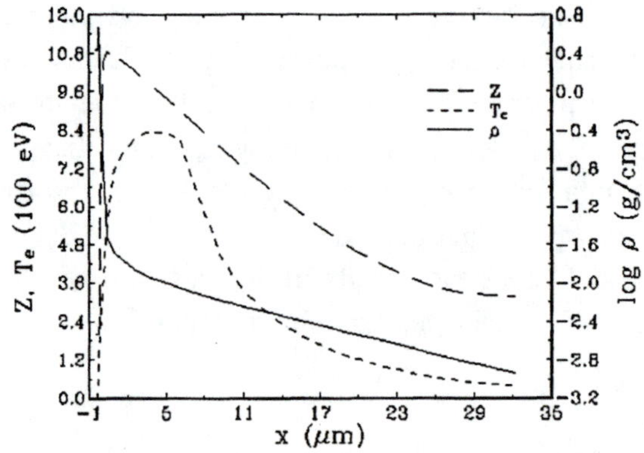

FIG. 3.3.12. The profiles of the mean ion charge Z, electron temperature T_e and plasma density ρ in the moment of maximum emission of He-α line (1 ps after the main pulse maximum). The laser pulse separation is $\Delta\tau$=890 ps, I_L=2.10^{16} W/cm^2, I_{pL}=10^{15} W/cm^2.

X-Ray Emission from the Plasma Formed by a Picosecond Pulse Train

In the experiment [128], Nd laser generated picosecond pulses with energy up to 0.2 J. A pulse consisted of three peaks of 3 ps duration separated by 50 ps. The energy was shared by the peaks in such a way that the first carried 1/16, the second 2/16, and the third (the main pulse) 13/16 of total energy. These pulses were injected into a vacuum chamber and focused on an aluminium target by an aspherical lens of 30 mm diameter with the focal length of 120 mm. The pulses were directed at 45° to the target surface. The focal spot diameter was about 20 μm, and the maximum laser radiation intensity on the target was (6-8) 10^{17} W/cm^2.

The number of photons predicted by the numerical calculations was in satisfactory agreement with the experimental results. This agreement served as the basis for a more detailed numerical calculation of the x-ray yield dependence on pre-pulse intensity. We studied this dependence in detail for the x-ray yield of the plasma formed by two pulses of 3 ps duration with the wavelength of 1.06 μm. The total incident laser energy was 0.1 J, and the energy carried by the first pulse was an order of magnitude less than that carried by the second pulse. We calculated the intensity of the bremsstrahlung and recombination radiation over the whole spectral range, in addition to the intensity of the He-α line with the transition energy of 1.6 keV.

Our calculations and experimental data show that the bremsstrahlung and recombination radiation intensities increase with the delay time between the pulses and that the intensity of the He, line had a maximum at 30 ps delay (**Fig. 3.3.13**).

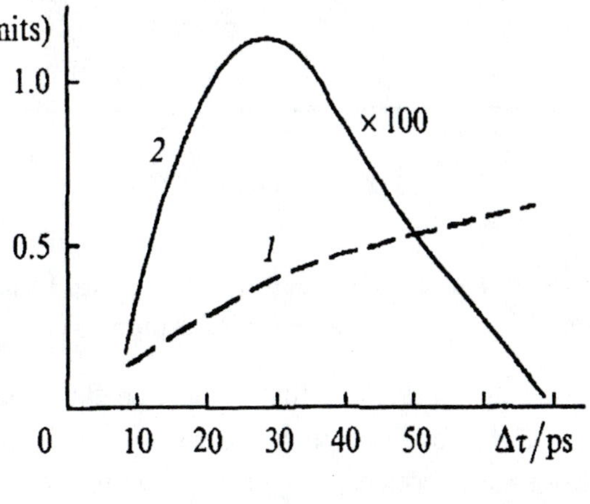

Fig. 3.3.13. Dependence of the X-ray yield I_R. from a plasma on the delay time Δτ between two pulses, representing the integrated bremsstrahlung and recombination radiation (dashed curve), and the radiation from the He-α line with the transition energy of 1.6 keV (solid curve).

Simulation Results

The computer simulations provide information about temporal form of x-ray emission pulse that is difficult to measure. However, the length of x-ray emission pulse may be very important for certain applications of line x-ray source. The calculated temporal profiles of emitted He-α line intensities are plotted in **Fig. 3.3.14**. The shapes of and Ly-α pulses are displayed for pulse separation $\Delta\tau$=0 ns in the inserted small figure. Generally, when no laser pre-pulse ($\Delta\tau$=0 ns) is present, the pulses of H-like resonance lines are subpicosecond, however their energy is rather small for intensities assumed here. Pulses of He-like lines are generally longer, here a subpicosecond pulse of He-α emission is accompanied with an energetic tail approximately 3 ps long. When laser pre-pulse is used, the energy conversion into is considerably increased, but lengthening of pulses of x-ray resonance lines is inevitable. Here He-α line pulse with FWHM length of about 5 ps is supplemented with ~30 ps long tail.

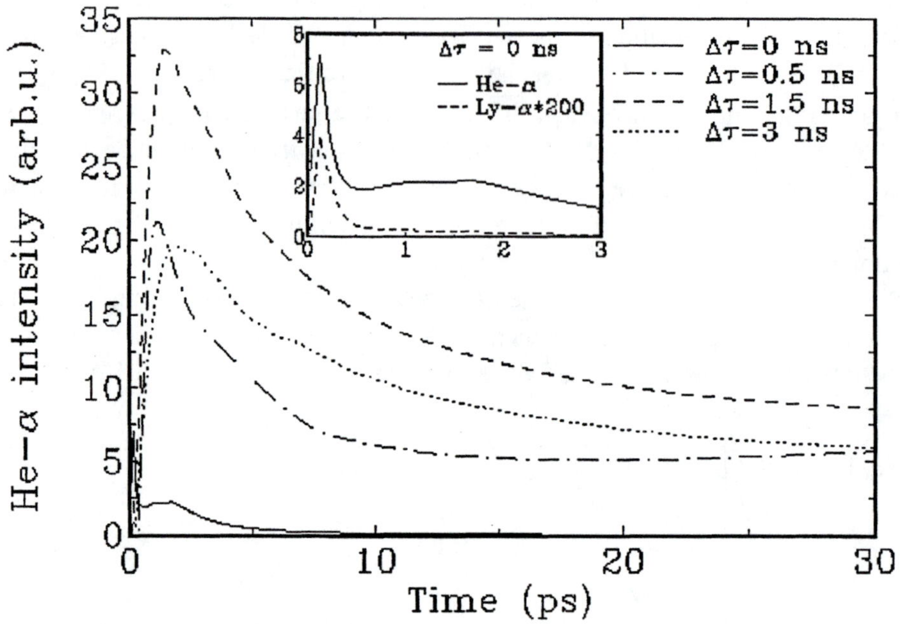

Fig. 3.3.14. Temporal profiles of pulses of He-α, Ly-α emission for various time separations $\Delta\tau$ between pre-pulse and main pulse.

The computed x-ray spectrum near the He-α line emitted in the direction 45° from the target is plotted in the **Fig. 3.3.15** for laser pulse separation $\Delta\tau$=2 ns and compared with the experiment. The ratios of resonance, inter-combination and satellite lines compare favourably. The widths of especially the satellite lines seem to be underestimated to a certain extent in the simulations, which is probably connected to a rather simple model of the line shapes taken used in the post-processor. The figure demonstrates our ability to compute emission X-ray spectra for the assumed experimental conditions with a reasonable accuracy.

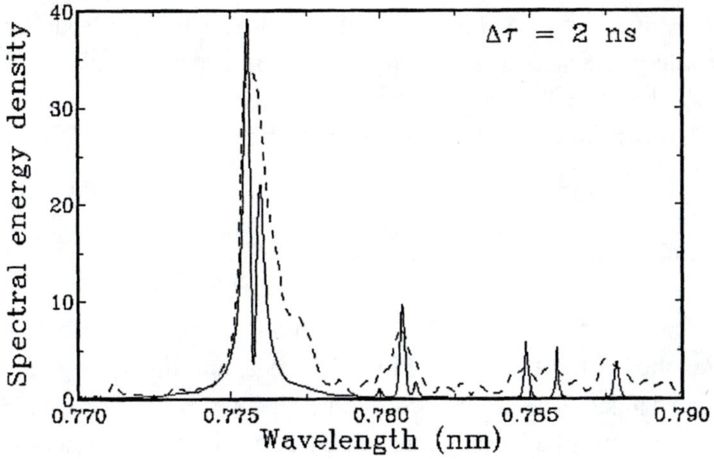

Fig. 3.3.15. The computed x-ray spectrum near the He-α line emitted in the direction 45° from the for laser pulse separation $\Delta\tau$=2 ns and compared with the experiment (dashed line).

Laser pre-pulse is used here in order to improve the laser energy transfer into He-α line emission. The enhancement of the integrated line emission energy by the laser pre-pulse is plotted in **Fig. 3.3.16** versus laser pulse separation $\Delta\tau$. Both experiment and simulations clearly demonstrate maximum He-α line energy at optimum pulse separation $\Delta\tau_m \approx 1.5$-2 ns. A satisfactory agreement between the experimental results and numerical simulations for He-α line emission is also apparent in the shape of the curve.

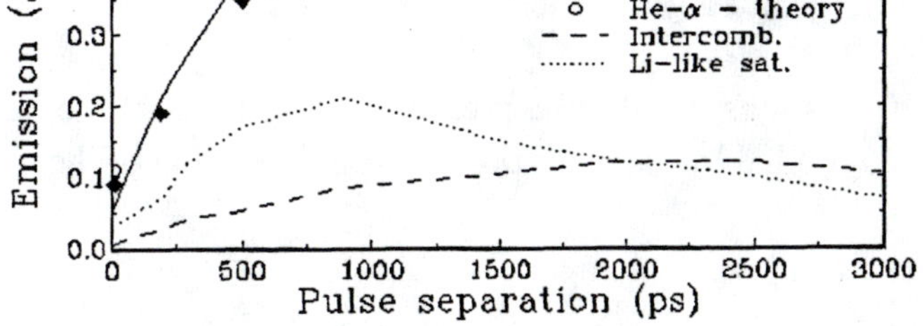

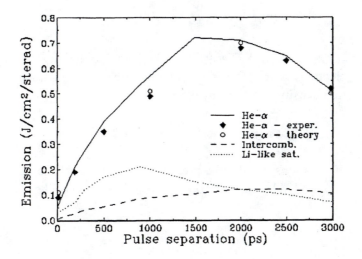

Fig. 3.3.16. The energy emitted by the He-α, intercombination line, Ly-α and the sum of Li-like satellites to the He-α line versus pulse delay for the parameters of the experiment. The laser wavelength λ_0=800 nm, pulse width t_{FWHM}=130 fs, the main pulse intensity $I_0=2\cdot10^{16}$ W/cm^2, and the pre-pulse intensity $I_p=10^{15}$ W/cm^2.

Analytical Model

We suggest an analytical model in order to better interpret our experimental and simulation results. The density scale length L of the expanding plasma is $L/\lambda < 1$ for a laser intensity I_L below 10^{17} W/cm^2 and a short time delay $\Delta\tau$ between the pre-pulse and the main pulse. Therefore, we can use a simple formula for the absorption coefficient in the case of S-polarized light (see sect.3.1):

$$\eta_S \quad 0.17\,(n_e/n_0)^{0.2}\,(L/\lambda_0)^{-0.3}\,(\nu_{ei}/\omega_0)^{0.7}\,(\cos\theta)^{1.3},$$

where $n_0 = 1.6\times10^{23}$ cm^{-3}, ν_{ei} is the electron-ion collision frequency, ω_0 is the laser frequency, θ is the angle of laser radiation incidence, and the average ion charge Z. The formula indicates a decrease in the absorption efficiency η_S with increasing density scale length of expanding plasma. A long pulse delay $\Delta\tau$ leads to a large density scale length L of this plasma. For this case, we have used a more sophisticated formula derived in the work [129,130] for an exponential density profile. The calculations presented in Fig. 3.3.11 are consistent with the simulations and experimental data. There is almost no difference between the absorption efficiencies for S- and P-polarized light. One can conclude that the application of normal laser pulses in this case is preferable because the fast electron generation is minimized and the control of experiments is improved.

In order to describe the plasma emission in lines, the plasma corona produced by the laser pre-pulse is assumed to be uniform during the main pulse interaction. However, we deal with a non-stationary interaction, as the main pulse duration is much shorter than the time for equilibrium to be established. This reasoning has lead us to the following set of equations for the electron temperature, density and populations of respective ion states. The plasma temperature is described by the equation

$$n_e\,\partial T_e/\partial t \quad (k_0\eta_e\,\nu_{ei}/\,\omega_0\sqrt{1-\eta_e})\,I_L + \kappa_T\,(L^{-2}+r_L^{-2})T_e\,n_e, \qquad (3.3.17)$$

where $\eta_e = n_e/n_{cr}$, $k_0 = \omega_0/c$, $\kappa_T = T_e/m_e\nu_{ei}$, r_L is the laser spot radius. We suggest that the total number of electrons is greater than the number of electrons released by Li-like ions. The plasma electron density is defined by the plasma expansion as

$$n_e \quad n_{e0}\,[t_0/(t+\Delta\tau)]^{\chi},\ \chi - \text{a geometrical factor},\ \chi = 1,2,3 \qquad (3.3.18)$$

Only Li-like ions are assumed to be present in the plasma at the initial stage, while the ion density is controlled by the ionization and recombination of the ions:

$$\partial n_i/\partial t = n_e\,n_{i0}\langle\sigma v\rangle_{i0} - \gamma_{0i}, \qquad (3.3.19)$$

Here, $\langle\sigma v\rangle_{i0} \approx \sigma_0 T_e^{1/2} J_{Li}^{-2}\exp(-J_{Li}/T_e)$, J_{Li} is the ionization potential of a Li–like ion and $\gamma_{0i} = 4\ 10^{-27} Z_{nu}^{3} T_e^{-9/2} n_e^{2} n_i$, is given by the rate of electron collision ionization due to the high electron density. For the population N_2 of the He-α upper state, the rate equation is

$\partial N_2/\partial t = -A_{21}N_2 - <\sigma v>_{21} n_e N_2 + <\sigma v>_{12} n_e N_1$ (3.3.20)

The initial populations are found on the assumption that all ions are in the lower state: $N_1 = n_i$. Then the line emission rate for the of 2→1 transition is found as

$\partial W_{21}/\partial t = E_{21} A_{21} N_2$ (3.3.21)

To find the line intensity, one should solve the set of equations (3.3.17-3.3.21). However, at $n_e < n_{cr}$ at $t \geq \Delta\tau$, the heated plasma length is: $x_T \approx L + x_f$, (here x_f is the propagation length of the heat wave), and the following approximate expression is derived from equation (3.3.17):

$(npxf + n_eL)\, T_e \approx \eta s \;\; L$ (3.3.22)

where n_p is solid plasma density. For a long plasma corona with $n_eL > n_p x_f$, one can estimate the density scale length $L \approx c_{s0} \sqrt{t_p \Delta\tau}$ and the electron density $n_e \approx n_{e0}(L_0/L)^{\chi}$. The electron temperature is found from equation (3.3.17) as

$T_e \quad C_T(\Delta t/t_p)^{6\chi/13}(t_L I_L/n_{e0})^{6/13}$,

where t_p is the laser pre-pulse length, C_T - constant. Using formulas (20) and (11), we get the line emission rate $W_{21} \approx E_{21}A_{21}n_i \exp(-E_{21}/T_e)\tau$, where $\tau = t + \Delta\tau$.

An approximate formula for the ion density n_i can be found from equation (3.319) as $n_i \approx n_e/(Z_{nu}-3)\ [1 - \exp(-n_e<\sigma v>_{Li}\ \tau\ ] \approx n_e^2 <\sigma v>_{Li}\ \tau \approx n_e^2\ T_e^{1/2} \exp(-J_{Li}/T_e)\tau$, while the emission rate is $W_{21} \approx W_0 y^{11/6} \exp(-y)$, where $y = E/T_e$ and $E = E_{21} + J_{Li}$. The emitted fluency of the He-α line is then expressed as

$F_x = N_2A_{21}E_{21}Lt_{em} \quad F_m x^{(23\chi-13)/26} \exp[-b(x^{3\chi/13}-1)]$, (3.3.23)

where $b=(23\chi-13)/6$, $x = \Delta\tau_m/\Delta\tau$, $F_m = F_m(\Delta E, I_{pL}, t_p, I_L)$. t_{em} – emission time. It is worth noting that the model described in [128] assumed the plasma corona to have an average density greater than the critical density before the main pulse arrives with a short delay $\Delta\tau$. From the above analytical model, we have derived the following formula for the optimum pulse delay $\Delta\tau_m$ relative to the pre-pulse:

$\Delta\tau_m \quad t_p t_L^{-2/\chi} (n_{e0}/I_L)^{2/\chi}\ [(E_{21}+J_{Li})/C_T b]^{13/3\chi}$ (3.3.24)

It indicates that the line emission increases with laser intensity, as the plasma temperature rises. However, there is a limit to the plasma temperature imposed by the requirement of a sufficiently ionized but not over ionized plasma. Another condition for the optimum x-ray line emission from the target bulk is that the x-ray line absorption length should exceed the hot plasma extension, i.e., the plasma region where the He-like ion density is greatest.

The resulting integrated line emission in the experiment [131] is compared with the simulations and analytical theory in Fig. 3.3.16. The laser is incident normally on a plane bulk Al target. One can see a satisfactory agreement between the experimental data, the analytical curve and the numerical calculations for He-α line emission.

For the relatively low laser intensity in the experiment described in [127] and [126], we have obtained long optimum pulse delays $\Delta\tau_m$ 1 ns, as compared with the results of [128], where this time was found to be approximately 20 ps. One can see from our scaling (3.3.14) that the optimum pulse delay increase with decreasing laser pulse duration and in the pre-pulse intensity. In order to check this dependence, we have calculated the line emission by the plasma under different conditions (see **Fig. 3.3.17**) and obtained an optimum pulse delay comparing well with the analytical scaling formulas.

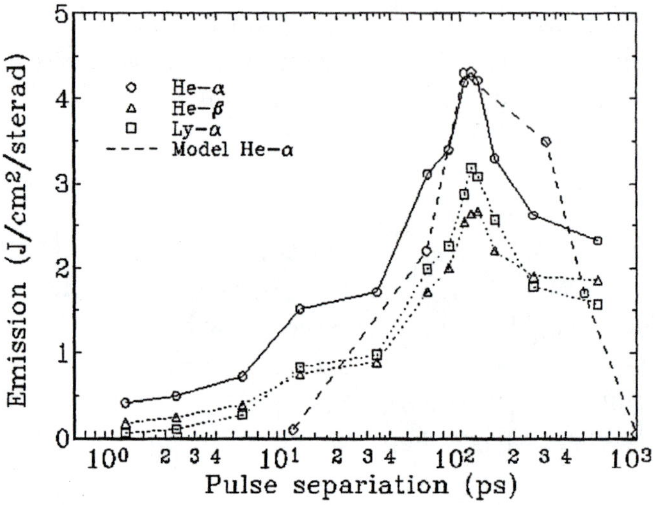

Fig. 3.3.17. The energy emitted in He-α, He-β, Ly-α and sum of Li-like satellites to He-α line versus pulse separation for the parameters of the experiment. The laser wavelength is λ=256 nm, pulse width t_{FWHM}=600 fs, main pulse intensity $I_0=5\times10^{16}$ W/cm^2, pre-pulse intensity $I_p=3\times10^{15}$ W/cm^2.

3.3.5 Hard X-Ray Emission by a Solid Target Nonlinearly Interacting with an Intense Circularly Polarized Laser Pulse

When a solid target nonlinearly interacts with a superhigh power circularly polarized laser pulse, it emits fast electrons and hard bremsstrulung X-rays. The lifetimes of fast electrons and x-ray emission exceed the pulse duration and are found to be several tens of picoseconds at an intensity of 10^{18} W/cm^2. The emitted electrons are sufficiently abundant to be registered experimentally. Generating laser pulses at a certain repeatition rate, one can get a quasi-stationary hard x-ray source suitable for X-ray examination of micro-objects, for excitation of nuclear levels, and for other applications.

The interaction between the laser pulse and the target plasma is described by a set of equations consisiting of the kinetic equation for plasma electrons and the Maxwell equations for electromagnetic fields. This problem is usually solved analytically by expanding into a power series the ratio of the characteristic velocity υ_E of an electron in a laser field to its thermal velocity υ_T (when the electron is in a low field). In the present section, we will consider the inverse limit $\upsilon_T/\upsilon_E<1$ permitting the following assumptions in the initial set of equations.

1. With increasing electron energy, the collision frequency decreases, so the plasma temperature (but not the electron energy) does no longer rise. Therefore, the target plasma can be regarded to be collisionless at very high intensities of the incident laser pulse, exceeding 10^{16} W/cm^2.
2. When the inequality $\upsilon_T/\upsilon_E<1$ holds, which happens at intensities above 10^{17} W/cm^2, an approximate solution to the initial set of equations can be found by solving the hydrodynamic equations. In the zero-th order in υ_T/υ_E, the collisionless kinetic equation transforms to a set of hydrodynamic equations.

A self-consistent solution to the set of hydrodynamic equations and the Maxwell equations can be found analytically (see Sect. 3.1). In this way, we can find the laser field distribution in the plasma, as well as the density and mean velocity of the plasma electrons. This is sufficient for the calculation of quantities varying with the mean electron energy, for example, the bremsstralhung emission intensity. Thus, the approximate self-consistent solution of the set of Maxwellian and collisionless kinetic equations involves a hydrodynamic approximation to self-consistent electromagnetic fields.

We consider here the normal incidence of a circularly polarized electromagnetic wave onto semibound plasma. The plasma electron temperature is T_e and the electron density n_e is assumed to be much higher than their critical concentration n_c. The amplitude E_0 and frequency ω of the electromagnetic wave are taken to be such that the characteristic velocity of an electron, υ_E, exceeds its thermal velocity υ_T . The electron density nonuniformity scale at the plasma boundary is smaller than the skin layer length, since the ion motion at times smaller than 100 fs can be neglected and the plasma boundary remains to be sharp.

It was pointed out above that the plasma electron dynamics is described by a self-consistent set of equations consisting of the collision-less kinetic equation and the Maxwell equations for electromagnetic fields. The solution to these equations will be found using the expansion in the parameter υ_T/υ_E. In the zero-th order, we will have the hydrodynamic equations for the plasma electron motion. In a one-dimensional case, the conservation of the transverse canonical momentum allows one to reduce the self-consistent set of hydrodynamic and Maxwell equations to two non-linear equations in partial derivatives for the vector potential $A(x,t)$ of the electromagnetic wave and the longitudinal electromagnetic field $E(x;t)$ in the plasma (3.1.73). Having found the fields from equation (3.1.73), we can calculate the electron phase path in these fields and the electron distribution function.

One case of a successful analytical solution of the set of equations (3.1.73) is that of the standing circularly polarized wave reflected by the plasma. The steady-state approximation under discussion does not account for the effects associated with the passage of the laser pulse front. These will be analyzed individually.

In the case of linear polarization, the time-oscillating ponderomotive force generates electron fluxes directed inward the target [56]. Note that the effect of the pulse front and tail on the target, (non-stationary effects) will lead to a considerable change in the electron motion, in particular, to the appearance of a longitudinal velocity component. The account of non-stationary effects requires non-stationary solutions of equations (3.1.73), which will be discussed below.

Intensity of Bremsstrahlung Emission of Fast Electrons by the Skin Layer

Hard emission intensity is determined by the characteristic energy of fast electrons. When the condition $\upsilon_T/\upsilon_E \ll 1$ holds, the emission intensity can be calculated in terms of the hydrodynamic approximation alone, in which all electrons have an energy value equal to the mean electron energy $\mathcal{E}_e = \sqrt{1+a^2}$ - 1. The spectral power density of bremsstrahlung emission of a single electron is defined by the well-known Bethe-Heitler formula:

$$\frac{d\mathcal{E}}{dtd\omega_x} = \frac{16}{3}\frac{e^6Z^2n_i}{m^2c^4(1+\frac{\omega_p^2\gamma^2}{\omega_x^2})}\cdot F, \tag{3.3.25}$$

where ω_x. is the frequency of a hard quantum, $F = \ln(\frac{mcd_p}{\hbar}) - \frac{1}{2}$ at $\omega_x < \frac{2c\gamma^2}{d_p}$, $F = \ln(\frac{2\gamma^2 mcd_p}{\hbar\omega_x}) - \frac{1}{2}$ at $\frac{2c\gamma^2}{d_p} < \omega_x < \frac{mc^2\gamma}{\hbar}$, $\gamma = \sqrt{1+a^2(\xi)} \gg 1$ is the Lorenz factor for an electron, and d_p is the Debye length in the plasma. By integrating equation (3.3.25) in the volume occupied by fast electrons and accounting for the electron density dependence on the longitudinal coordinate, we get the emission intensity for the whole skin layer:

$$\frac{d\mathcal{E}^{pl}}{dtd\omega_x} = \frac{16}{3}\frac{\pi D^2}{4}\frac{e^6Z^3n_i^2c}{m^2c^4\omega_{pe}}\cdot\int_0^{\xi*}\frac{(1+\frac{\partial^2}{\partial\xi^2}\sqrt{1+a^2})}{(1+\frac{\omega_p^2}{\omega_x^2}(1+a^2(\xi))(1+\frac{\partial^2}{\partial\xi^2}\sqrt{1+a^2}))}\cdot Fd\xi \tag{3.3.26}$$

Here, $d_p = \dfrac{\upsilon_T}{\omega_p(1+\dfrac{\partial^2}{\partial\xi^2}\sqrt{1+a^2})}$, the upper limit of the integration in the coordinate ξ^*. is defined by the equality of the kinetic energy of an electron and the energy of a hard quantum $\sqrt{1+a^2(\xi^*)}-1=\dfrac{\hbar\omega_x}{mc^2}$, and D is the diameter of the laser spot on the target.

Expression (3.3.26) takes the following effects into account: the electron density variation due to the ponderomotive pressure (the factor $(1+\dfrac{\partial^2}{\partial\xi^2}\sqrt{1+a^2})$, the enlargement of the effective area occupied by fast electrons with increasing field strength (the increase in ξ^* with $a_0 = \max a$), the emission intensity suppression at low frequencies due to the density effect (the factor $\dfrac{1}{1+\dfrac{\omega_p^2\gamma^2}{\omega_x^2}}$), and the ion field screening by plasma electrons (the change in the logarithmic argument at the frequencies $\omega_x > \dfrac{c}{\upsilon_T}\omega_p(1+\dfrac{\partial^2}{\partial\xi^2}\sqrt{1+a^2})(1+a^2(\xi))$).

It is known that a relativistic electron emits along the direction of its motion, so in the situation of interest, quanta will be emitted along the target plane. Let us identify the effects essential for the evaluation of the emission intensity. It was stated above that the maximum value of a_0, at which the reflection mode is steady-state, is $\dfrac{3}{2}\dfrac{\omega_p^2}{\omega_0^2}$. Therefore, the density effect is essential at the frequencies $\omega_x < \omega_p\dfrac{\omega_p^2}{\omega_0^2}$. Under real conditions, the medium becomes optically opaque to quanta of low energy (comparable with the electron temperature) and the quantum spectrum becomes similar to the thermal spectrum, so it can no longer be described by formula (3.3.26). Since the inequality $\hbar\omega_p\dfrac{\omega_p^2}{\omega_0^2} << T_e$ is valid, the range of the density effect lies beyond the applicability of (3.3.26); therefore, this effect can be ignored.

The screening effect is present in the frequency range $\omega_x < \dfrac{c}{\upsilon_T}\omega_p\dfrac{\omega_p^4}{\omega_0^4}$ and is observable against the background of thermal radiation if the inequality $\hbar\omega_p\dfrac{c}{\upsilon_T}\dfrac{\omega_p^4}{\omega_0^4} > T_e$ is valid. In real conditions, this inequality does hold, and the screening effect is essential.

If we impose an additional restriction on the field value $1 << a_0 << \frac{3}{2}\frac{\omega_p^2}{\omega_0^2}$, allowing the neglect of the electron density variation, expression (3.3.26) reduces to

$$\frac{d\mathcal{E}^{pl}}{dtd\omega_x} = \frac{4\pi}{3}\frac{D^2 e^6 Z^3 n_i^2}{m^2 c^3 \omega_p} \cdot \int_0^{\xi^*} F' d\xi \,, \tag{3.3.27}$$

where $F' = \ln(\frac{mc\upsilon_T}{\hbar\omega_p}) - \frac{1}{2}$ at $\omega_x < \frac{2c}{\upsilon_T}\omega_p a^2(\xi)$; $F' = \ln(\frac{2mc^2}{\hbar\omega_x}(1+a^2(\xi)) - \frac{1}{2}$ at $\frac{2c}{\upsilon_T}\omega_p a^2(\xi) < \omega_x$; and $\xi^* = -\xi_0(a_0) + \frac{1}{\sqrt{1-\frac{\omega_0^2}{\omega_p^2}}}\text{arcch}(\sqrt{1-\frac{\omega_0^2}{\omega_p^2}}\frac{1+\frac{\hbar\omega_x}{mc^2}}{\sqrt{(1+\frac{\hbar\omega_x}{mc^2})^2-1}})$.

The order of magnitude of (3.3.27) can be evaluated from the following simple formula:

$$\frac{d\mathcal{E}^{pl}}{dtd\omega_x} \approx \frac{D^2 e^6 Z^3 n_i^2}{m^2 c^3 \omega_p} \ln a_0^2 \frac{mc^2}{\hbar\omega_x} \,. \tag{3.3.28}$$

Let us calculate the frequency-integrated emission intensity. For this, expression (3.3.27) will be integrated over the frequency range from $h\,\omega_x^*$ to $\hbar\omega_x = mc^2\sqrt{1+a_0^2}$. The result has the form

$$\frac{d\mathcal{E}^{pl}}{dt} = \frac{\pi D^2 e^6 Z^3 n_i^2}{mc\hbar\omega_p} \int_0^{\xi_1} \sqrt{1+a^2(\xi)}[\ln 2\sqrt{1+a^2(\xi)} - \frac{1}{3}] d\xi \tag{3.3.29}$$

where $a(x) = \left[\left(1+\hbar\omega_x^*/mc^2\right)^2 - 1\right]^{1/2}$. At $a_0 >> 1$, the intensity (3.3.29) linearly rises with a_0. The coefficient K of the laser radiation conversion to x-ray emission is found by dividing expression (3.3.29) by the incident laser flux:

$$K_x = \frac{64\pi e^8 Z^3 n_i^2}{m^3 c^4 \hbar\omega_p \omega_0^2} \frac{\int_0^{\xi_1} \sqrt{1+a^2(\xi)}[\ln 2\sqrt{1+a^2(\xi)} - \frac{1}{3}] d\xi}{a_0^2 + (1+a_0^2)(2\frac{\omega_p^2}{\omega_0^2}(\sqrt{1+a_0^2} - 1) - a_0^2)} \tag{3.3.30}$$

The function $K_x(a_0)$ for $\omega_p/\omega_0 = 10$, $Z = 13$, and $\hbar\omega_x^* = mc^2$ is shown on **Figure 3.3.18**. The K_x value reaches its maximum of 10^{-5} at $a_0 = a(x_1)$.

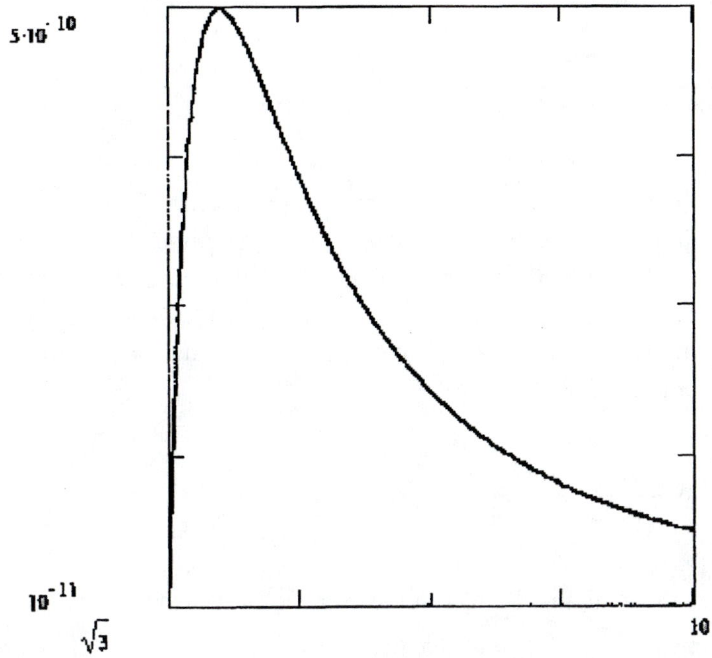

Fig. 3.3.18. The dependence of conversion efficiency $K_x(a_0)$ from laser amplitude a_0 for plasma parameters $\omega_p/\omega = 10$, $Z = 13$.

Contribution of Nonstationary Effects of Fast Electron Generation to X-Ray Emission Intensity

The hard emission intensity has been calculated for the stationary reflection of laser radiation by the plasma when the ponderomotive pressure of the laser light is compensated by Coulomn forces. This reflection mode operates only with a circularly polarized electromagnetic wave, because in that case the ponderomotive pressure does not clearly depend on time (A^2=const). There are no stationary solutions of equations (3.1.73) for a linearly polarized wave. No are there such solutions at $a_0 > \frac{3}{2}\frac{\omega_p^2}{\omega_0^2}$, as shown above. Therefore, the stationary reflection mode permitting a rigorous analytical treatment represents only a particular case.

Let us consider the nonstationary processes and their effect on hard X-ray emission intensity and the conversion coefficient K_x. These processes are due to the passage of the pulse front and tail and to the related acceleration of plasma electrons. When the pulse front enters the plasma, the electrons acquire the longitudinal velocity $v = a_0^2/2(1 + a_0^2/2)$

and transversal velocity $v_\perp = a_0/(1+a_0^2/2)$, moving in a spiral [132]. At $a_0 >> 1$ and $v > v_\perp$, the electrons will be injected into the plasma bulk, until the ponderomotive pressure becomes equalized by the arising longitudinal Coulomn field. The electrons move with the energy $\mathcal{E}_e$ for their free path time:

$$t_e = \frac{3\mathcal{E}_e}{4mc^3 Zn_{nu}\sigma_T \ln\left(\mathcal{E}_e^3 / 2J^2mc^2\right)}, \qquad (3.3.31)$$

where J is the characteristic ionization energy of an atom in the medium and σ_T is the Tompson scattering cross section. The time $t_e = 1$ ps corresponds to the electron energy at $a_0=1$. Therefore, the main contribution to the emission will be made by short laser pulses ($t_L < t_e$) at times exceeding the pulse duration. At $t_L > t_e$, the contribution of nonstationary processes is small. The energy of hard emission in the nonstationary mode will again be defined by equation (3.3.29):

$$\mathcal{E}_x = \frac{32e^6 Z^2 n_i t_{ef} N_{ef}(\mathcal{E}_e)\mathcal{E}_e}{3m^2c^4\hbar}\ln\left(2\mathcal{E}_e / mc^2\right) \qquad (3.3.32)$$

where $N_{ef}(\mathcal{E}_e)$ is the total amount of fast electrons in the nonstationary mode and $t_{ef} = t_L + t_e$. is the lifetime of fast electrons. Expression (3.3.32) describe the stationary mode discussed above if $N_{ef}(\mathcal{E}_e)$ is replaced by the total amount of electrons in the skin layer and t_{ef} is replaced by the laser pulse duration t_L. Then expression (3.3.32) will give the emission intensity. $\mathcal{E}_x / t_L$. coinciding with that of (3.3.29).

Let us find the value of $N_{ef}(\mathcal{E}_e)$ for the nonstationary case. The value $N_{ef}(\mathcal{E}_e)* \mathcal{E}_e$ represents the total energy of fast electrons. The authors of [133] studied the behavior of the conversion corfficient K_e when the total energy $\mathcal{E}_L$ of a short laser pulse is converted to the energy of fast electron, $N_{ef}(\mathcal{E}_e)*\mathcal{E}_e$. At $1 < a_0 < 10$, the value of K_e rises in proportion with a_0^2. The saturation occurs at $a_0 \sim 10$, and we have $N_{ef}(\mathcal{E}_e)* \mathcal{E}_e = \eta\mathcal{E}_L$, where η is the radiation absorption coefficient, i.e., all absorbed energy is converted to the kinetic energy of fast electrons. By dividing $\mathcal{E}_x$ by $\mathcal{E}_L$, we get the coefficient of the laser pulse conversion to hard emission in the nonstationary mode:

$$K_x = \frac{32e^6 Z^2 n_i t_{ef}}{3m^2c^4\hbar}\ln\left(2\mathcal{E}_e / mc^2\right) \qquad (3.3.33)$$

Consider qualitatively the variation of K_γ with laser radiation intensity for short pulses. The value of t_e is proportional to the energy of an electron, $t_e \sim I^{1/2}$. Therefore, at $1 < a_0 < 10$, we have $K_x \sim I^{3/2}$ and at $a_0 > 10$, $K_x \sim \eta\, I^{1/2}$. This dependence is consistent with more rigorous calculations including the distribution function of fast electrons [134]. Let us evaluate K_x at a_0=2 and t_L=0.1 ps. The coefficient K_x is equal to 6 10^{-3} as in [134] and the energy of a fast electron $\mathcal{E}_e$=0.4 MeV. The value of K_x will then be as high as 10^{-2}.

Thus, nonstationary effects increase the amount of fast electrons due to their generation by the laser pulse front. As a result, the hard emission intensity rises considerably. The coefficient of laser radiation conversion to hard emission varies with the radiation intensity in a different way for the stationary and nonstationary mode. The angular distribution of emission in the nonstationary mode has a maximum directed towards the plasma bulk.

3.4 Fast Particle Acceleration in Short Laser Pulse Interaction with Solid Target

As we have seen from the above, an ultra high intensity laser pulse can produce an electric field of amplitude E 0.3 $I_{18}^{1/2}$ [GV/cm] and an electromagnetic pressure of value P 0.3 I_{18} [Gbars]. When such a laser beam is incident upon a solid target, the atoms are strongly ionized, and free electrons gain relativistic momentum: p mc $(I_{18}\lambda^2_{\mu m})^{1/2}$. The production of fast electrons and ions (see for example [135]) is one of the interesting new features of intensive laser-matter interaction. High-energy particles generated in this interaction can be used in many applications from technology to medicine. For example, recently interest has developed in ion acceleration by compact high intensity short pulse lasers with potential applications for the initiation of tabletop nuclear reactions. Critical for ion acceleration is the efficiency of laser-energy conversion into a high-energy electron component, since the latter can produce the strong electrostatic fields needed to accelerate ions. A collimated ion beam can be achieved by focusing an intensive laser beam onto the surface of a solid film. Due to the small diameter of the laser spot and the anisotropy of the hot electron velocity distribution, the ion emittance can be comparable to or even better than that of electrostatic accelerators.

We begin our analysis from the case where fast particles are accelerated into a vacuum by the interaction of a laser pulse with a solid target at oblique incidence and than we consider fast particle acceleration inside over-dense plasmas.

3.4.1 Escape into Vacuum of Fast Electrons Generated by Oblique Incidence of an Ultra Short Super Intense Laser Pulse on a Solid Target

The present section is devoted to a study of the dynamics of electrons in the incidence and reflection regions of an ultra-short laser pulse [149]. Here we assumed that the laser plasma is collisionless and that the motion of electrons is determined by the external electromagnetic field and by the ambipolar potential resulting from charge separation.

Analytical Treatment of the Escape of Electrons from the Target Surface

We consider the oblique incidence (at an angle θ) of a laser pulse of finite width D on a target surface. Here the following obvious distribution of fields develops (**Fig. 3.4.1**): region *i* contains the field of incident wave, *o* that of the reflected wave, *b* a superposition of the incident and reflected waves, and *pl* the field within plasma. We assumed that near the target surface there is a small group of electrons and that those electrons have no effect on the distribution of the fields in the vacuum or in the plasma. This group results from the electron density profile being steeped as a result of pondermotive pressure of the laser pulse and a small number of electrons with a density below critical remaining in the vacuum. This sort of distribution of electron density was obtained via 1D numerical simulations. These electrons will also be examined subsequently. Since the longitudinal and transverse dimensions of the laser pulse are of the order of tens of wavelengths, the laser field consists of an electromagnetic wave with slowly-varying amplitude $E_0(\mathbf{r},t)$. In dimensionless variables: $\tau \Leftrightarrow \omega t$; $y,x \Leftrightarrow (\omega/c)y,x$; $\boldsymbol{E}$, $\boldsymbol{B} \Leftrightarrow e\boldsymbol{E}/m\omega c$, $e\boldsymbol{B}/m\omega c$; $A,\varphi \Leftrightarrow eA/mc^2$; $e\varphi/mc^2$; $\mathbf{v} \Leftrightarrow \mathbf{v}/c$.

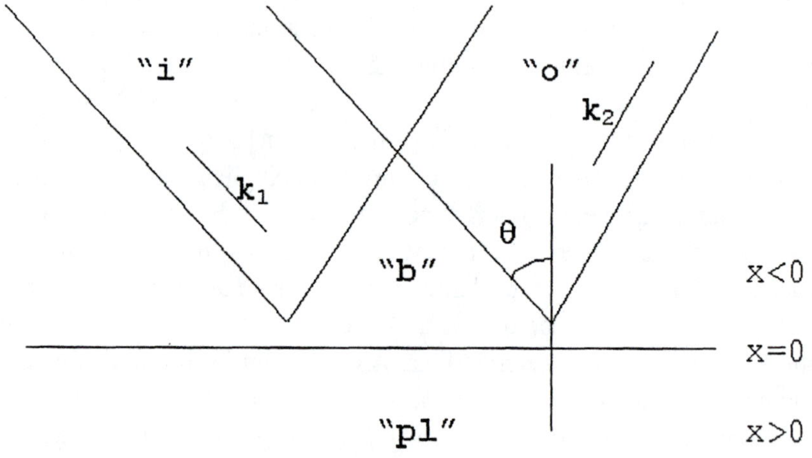

Fig. 3.4.1. Incidence of a laser beam on a plasma surface (x=0): region *i* contains the field of incident wave, *o* that of the reflected wave, *b* a superposition of the incident and reflected waves, and *pl* the fields within plasma.

The laser field corresponding to the various regions of Fig. 3.4.1 are:

for the region *i*

$$B_z^{(i)}=-E_0^{(i)}\cos[x\cos\theta+y\sin\theta-\tau];\ E_x^{(i)}=\sin\theta B_z^{(i)};\ E_y^{(i)}=-\cos\theta B_z^{(i)}, \tag{3.4.1}$$

for region *o*

$$B_z^{(o)}=-RE_0^{(o)}\cos[-x\cos\theta+y\sin\theta-\tau+\psi];\ E_x^{(o)}=\sin\theta B_z^{(o)};\ E_y^{(o)}=-\cos\theta B_z^{(o)} \tag{3.4.2}$$

and for region *b*

$$B_z^{(b)}=B_z^{(i)}+B_z^{(o)}\ E_x^{(b)}=E_x^{(i)}+E_x^{(o)}\ E_y^{(b)}=E_y^{(i)}+E_y^{(0)}. \tag{3.4.3}$$

The reflection coefficient R and phase ψ of the reflected laser wave derive from matching of the electromagnetic fields (3.4.3) with the fields within the plasma in region *pl*. In general, the fields in region *pl* are determined by the self-consistent solution of Maxwell's equations and the kinetic equations for the plasma particles.

Having determined the parameters of the laser fields above the plasma surface, we now proceed to describe electron dynamics. In the fields (3.4.1)-(3.4.3), the trajectory of electrons is two-dimensional, and the equations of motion in region *b* have the form

$$\frac{dp_x}{d\tau}=\left(\sin\theta-\mathrm{v}_y\right)\left(B_z^{(i)}+B_z^{(o)}\right)-\frac{\partial U_{am}}{\partial x} \tag{3.4.4}$$

$$\frac{dp_y}{d\tau}=\mathrm{v}_x\left(B_z^{(i)}+B_z^{(o)}\right)-\cos\theta\left(B_z^{(i)}+B_z^{(o)}\right), \tag{3.4.5}$$

where $B_z^{(i,o)}$ –is the magnetic field in the corresponding region and, $U_{am}(x,t)$ is the ambipolar potential of the plasma ions which we determine below. In the regions *i* and *o* $B_x^{(o)}$ and $B_z^{(i)}$ respectively are absent from Eqs. (3.4.4) and (3.4.5).

We first examine the solution of Eqs.(3.4.4) and (3.4.5) for non-relativistic laser fields, $|B_z^{(i,o)}|< 1$. Then **p=v** in the chosen system of dimensionless variables. We distinguish oscillating and mean component of coordinate and velocity of the electron: $\mathbf{v}=\langle\mathbf{v}\rangle+\delta\mathbf{v}$. Then averaging (3.4.4) and (3.4.5) over the period of oscillations, we obtain the mean components,

$$\frac{d\langle\mathbf{v}\rangle}{d\tau}=\frac{\partial\left(U_{ef}+U_{am}\right)}{\partial\mathbf{r}} \tag{3.4.6}$$

where

$$U_{ef}\left(r,t\right)=\left(E_0^{(i)}\right)^2+R^2\left(E_0^{(o)}\right)^2-R\cos2\theta E_0^{(i)}E_0^{(o)}\left(2x\cos\theta-\psi\right) \tag{3.4.7}$$

is the effective potential, which is the time-averaged square of the laser electric field: $U_{ef}=\langle E^2\rangle/2$. The potential (3.4.7) exhibits three substantially different time scale patterns: the laser wavelength; the transverse extent of the laser beam and the transverse scale of the laser pulse which is the hundreds of wavelengths. The greatest contribution to the force on an electron in (3.4.6) is obtained by differentiating the potential (9) with respect to the very shortest scale length. In this approximation, Eq.(3.4.6) takes the form

$$\frac{d\langle v_x \rangle}{d\tau} = -R\cos 2\theta \cos\theta E_0^{(i)} E_0^{(o)} \sin(2x\cos\theta - \psi) - \frac{\partial U_{am}}{\partial x} \tag{3.4.8}$$

$$\frac{d\langle v_y \rangle}{d\tau} = 0, \quad \langle v_y \rangle = \sin\theta \left(E_0^{(i)}\right)^2 + R^2 \left(E_0^{(o)}\right)^2 - R\cos 2\theta E_0^{(i)} E_0^{(o)} \cos(2x\cos\theta - \psi) \tag{3.4.9}$$

The quantity $<v_y>$ in (3.4.9) follows from the conservation of the y-component of the canonical momentum of the particle, which happens when the dependence of the $E_0^{(i,o)}$ on the coordinate is neglected. The motion of electrons in this approximation can be described qualitatively as follows (see Fig.3.4.2).

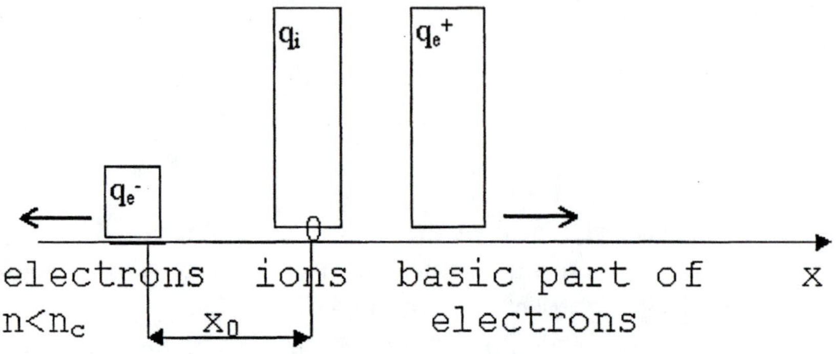

Fig. 3.4.2. Approximated schema of electron charge position.

Initially electrons are near the plasma surface ($x = 0$). They then approach the plane of the nearest minimum in potential (3.4.7), i.e., $2x\cos\theta - \psi = 0$. As they oscillate in the potential well near the plane of minimum, the particles move at velocity $<v_y>$ toward region o, where they are captured by the reflected wave and acquire a mean velocity $R^2(E_0^{(o)})^2/4$ in the direction of the reflected pulse. Taking the transverse structure of the laser pulse into account leads to a modification of the physics of electron motion: large scale potential variation shows up, and the moving electrons tend to occupy positions that minimise the potential energy.

This kind of motion can be approximately represented by a mass point, sliding over a curvilinear surface. Here the kinetic energy of the electron, after it escapes from regions occupied by the field, will equal its initial potential energy,

$$U(x=0) = \left(E_0^{(i)}\right)^2 + R^2 \left(E_0^{(o)}\right)^2 - R\cos 2\theta E_0^{(i)} E_0^{(o)} \cos\psi + U_{am}(x=0) \tag{3.4.10}$$

and the angle of escape will depend on the specific form of the function $U(y,x)$.

We now estimate the angle of escape for a laser beam with a triangular transverse distribution of field amplitudes,

$$E_0^{(i)}(y,x)=E_0\left(1-\frac{|x\sin\theta-y\cos\theta|}{R_L}\right),\quad |x\sin\theta-y\cos\theta|\le R_L \tag{3.4.11}$$

$$E_0^{(o)}(y,x)=E_0\left(1-\frac{|x\sin\theta+y\cos\theta|}{R_L}\right),\quad |x\sin\theta+y\cos\theta|\le R_L \tag{3.4.12}$$

Where R_L is the radius of the laser beam in units of c/ω, i.e. $R_L = D\omega/2c$. Note that estimates of the time the electron moves in region *b*, *o* and *i* for $E_0 > 1$, along with transverse beam size of the order of twenty wavelengths, yield a time much shorter than pulse duration. Thus (3.4.11), (3.4.12) do not contain laser pulse duration. Electrons initially concentrated in the neighbourhood of $x \approx 0$, $y \approx 0$ escape to the vacuum near that point. The equation of motion for electrons escaping to the vacuum in region b has the form

$$\begin{aligned}\frac{d^2x}{d\tau^2}&=2E_0^2\sin\theta\left(1+\left(x\sin\theta-y\cos\theta\right)/R_L\right)/R_L+\\&2E_0^2R^2\sin\theta\left(1+\left(x\sin\theta+y\cos\theta\right)/R_L\right)R_L\end{aligned} \tag{3.4.13}$$

$$\begin{aligned}\frac{d^2y}{d\tau^2}&=-2E_0^2\cos\theta\left(1+\left(x\sin\theta-y\cos\theta\right)/R_L\right)/R_L+\\&2E_0^2R^2\cos\theta\left(1+\left(x\sin\theta+y\cos\theta\right)/R_L\right)R_L\end{aligned} \tag{3.4.14}$$

In regions *i* and *o* we must remove terms respectively containing or not containing *R* in Eqs. (3.4.13) and (3.4.14). In writing (3.4.13), (3.4.14) we have averaged over small scale (of the order of the wavelength). Equations (3.4.13) and (3.4.14) do not contain a force corresponding to the ambipolar field. As will be shown below, this is true for small values of the phase ψ of the reflected wave. Eqs. (3.4.13) and (3.4.14) comprise a set of linear equations with constant coefficients. Their solution in regions *i*, *b* and *o*, and the matching of solutions at the boundary present no difficulties. As a result, for the rightward escape angle $\theta_e^{(+)}$ (between velocity of the electron and *x* axis) of an electron from a laser spot, we obtain

$$\theta_e^{(+)}=arctg\left(tg\theta\frac{\chi(1-R)-\cos 2\theta(1-R)/\left(\mu^2-R^2\right)}{\chi(1+R)+\cos 2\theta(1-R)/\left(\mu^2-R^2\right)}\right), \tag{3.4.15}$$

where

$$\chi = 2^{1/2} E_0 / R_L, \quad \mu^2 = \left(1 + R^2 + \left(\left(1 + R^2\right)^2 - 4R^2 \sin^2 2\theta\right)^{1/2}\right)/2. \tag{3.4.16}$$

Obviously, it is impossible to determine the angle of escape for a laser beam with an arbitrary transverse profile by the method considered above. We now suggest a means of estimating the angle of escape. Since electrons move along the plane of minimum effective potential $U_{ef}(y,x)$, we can assume that they move along the field line of U_{ef} that passes through the neighbourhood of $y \approx 0$, $x \approx 0$ (for a convex transverse laser profile, electrons initially near the maximum of the effective potential will escape to the vacuum). The differential equation for the field line is $dy/dx = (\partial U_{ef}/\partial y)/(\partial U_{ef}/\partial x)$, $y(0)=0$.

Hence, it is easy to find the field line at the boundary of region b, and to estimate the speed, by invoking energy conservation. These quantities will be initial data for the electron's motion in regions i and o. In these regions the field lines are perpendicular to the wave vector of the incident and reflected waves respectively, so the electron acquires an additional component of velocity in these directions, whose magnitude is also determined by energy conservation, i.e. potential U_{eff} at the point of entry into region i or o. Knowing the final values of the velocity components, it is easy to find the angle of escape of an electron for an arbitrary field configuration.

We now proceed to analyse the ambipolar potential $U_{am}(x)$. The ambipolar potential, like U_{eff}, is proportional to E_0^2, since the ambipolar field balances the pondermotive pressure of the laser beam. To find the ambipolar field and the corresponding potential, we assume that the pondermotive force of the radiation acts on the plasma from the vacuum side, pressing the bulk of electrons into the depth of the plasma and not affecting the small part of the plasma lying in the transparent region. Neglecting the effect of this small part on the formation of the ambipolar field, in region pl we have equilibrium between the pondermotive force and the ambipolar field created by the bulk of the electrons:

$$E_{am} = -\left(\sin\theta - v_y\right)\left(B_z^{(i)} + B_z^{(o)}\right) \quad x \succ 0. \tag{3.4.17}$$

Thus, the bulk charge density corresponding to the ambipolar field is

$$\rho = (1/4\pi)\frac{\partial E_{am}}{\partial x} \tag{3.4.18}$$

The charge density is nonzero only within the skin layer, i.e. in the very narrow region near the surface. For example, under actual conditions the scale length of the skin layer is of the order of one tenth of the wavelength of the incident radiation. Thus, at distances from the plasma boundary comparable to the wavelength, but less than the laser beam diameter, the charge configuration (3.4.18) is equivalent to a plane with surface charge density σ given by

$$\sigma = \int_0^{\infty} \rho dx = (1/4\pi) E_{am}\big|_{x=0} = -(1/4\pi)(\sin\theta - \mathrm{v}_y) B_z\big|_{x=0} \tag{3.4.19}$$

The ambipolar field in vacuum, which affects the escape of particles, can be assumed to be the field due to the surface charge σ, i.e. $E_{am}=4\pi\sigma$. As a result, the ambipolar potential takes the form

$$U_{am}(y,x) \approx \frac{\partial E_{ef}}{\partial x}\Big|_{x=0} \cdot |x| = -R\cos 2\theta \cos\theta E_0^{(i)} E_0^{(o)}\big|_{x=0} \cdot |x|. \tag{3.4.20}$$

At distances comparable to the physical dimensions of region b, the ambipolar field falls off in accordance with Coulomb's law. Thus, Eq.(3.4.20) holds for small x, and is actually the Taylor series expansion of the ambipolar potential. At small x the ambipolar field balances the force due to pondermotive pressure up to the linear term of the Taylor series. This leads to a reduction in the dependence of the energy of an escaping electron on the laser intensity for $E_0 < 1$.

To conclude this section we consider relativistic laser intensities, for which $E_0 > 1$. The averaged equation of motion then takes the form

$$\frac{d\mathbf{p}}{d\tau} = -\frac{\nabla\langle \mathbf{E}^2\rangle}{2\left(1+\mathbf{p}^2+\langle \mathbf{E}^2\rangle\right)}, \tag{3.4.21}$$

where **p** is the electron momentum. This equation can be solved analytically in regions i and o. Solving Eq.(3.4.21) in region o yield the angle of escape of an electron from the region occupied by the field,

$$\theta_e^{(+)} = \theta + arctg\frac{\left(2(\gamma/\gamma_b - 1)(1+\mathrm{v}_b)\right)^{/2}}{\gamma - \gamma_b(1-\mathrm{v}_b)}, \tag{3.4.22}$$

where $\gamma_b, \gamma, \mathrm{v}_b$ and v are the Lorentz factor and velocity upon entering and leaving region b. One important consequence of Eq.(3.4.22) is that acceleration of an electron by the reflected laser pulse ($\gamma_b < \gamma$) causes it to escape in the specular direction, i.e. $\theta_e^{(+)} \to \theta$ at $\gamma \to \infty$.

Summarizing the analytical treatment of electron motion in incident and reflected laser fields, we find that for non-relativistic fields the parameters of the trajectory are determined by the form of the effective potential (i.e., the spatial distribution of the laser fields). At relativistic intensities, capture of the electron by reflected beam, induces it to move in the direction of the latter.

Numerical Simulation of Electron Motion

The analytical model for the motion of the electron above the plasma constructed in the preceding section contains many approximations, but it provides a qualitative description of the escape of electrons from the surface. A more rigorous examination of this problem requires a comprehensive solution of the electron equation of motion and Maxwell's equations in 2D geometry. This complicated computation problem will be solved by the method described below. To simplify the problem we assume that the transverse field above the plasma surface ($x<0$) is given, and we calculate the ambipolar field and electron trajectories self-consistently.

The superposition principle is used to calculate E and B in the region where the incident and reflected laser radiation intersect. For $x > 0$ and outside the beam, all components of the electric and magnetic fields are assumed equal to zero. The ambipolar field is calculated at each time using the Poisson equation and is added to the fields from the laser pulse.

We take the parameters of the laser pulse and target to be close to the possible experimental values. Consider a plane laser beam of width D=10 μm incident on a target at an angle $\theta = 30^\circ$ (Fig. 3.4.1). We take the reflection coefficient to be 0.8 and the phase to be θ=0. This is consistent with our data for Nd laser pulse of duration t_L =100 fs and intensity $I = 10^{18}$ W/cm^2.

The target plasma with which the laser radiation interacts has initial electron density: $n_e(x) = \beta n_c \exp(-x/x_c)$ at $x < 0$ and $n_e(x) = \beta n_c$ at $x > 0$, where β=3 and x_c = 0.3 μm is the scale of inhomogenity at the critical point. The mean electron temperature is T_e=1 KeV. The density and temperature dependences are taken from estimates of the expansion dynamics of plasma exposed to a picosecond preheating laser pulse.

To describe the motion of plasma electrons we have used the particle in cell method. Three dimensional particle motion is described as relativistic one, including Lorentz force. The difference scheme employed here is a second order approximation in space and time. To improve the accuracy of the fields at the particles, they were calculated separately for each particle at each time step. All particles were injected into the laser spot region at the initial time in accordance with the density profile and initial energy of the target plasma, and they continue to move until the pulse terminated. The distribution of electrons $n_e(\alpha)$ with respect to the angle of emission α is determined by summing the accumulated information over all particles, and the mean electron energy is calculated. In this model the ions are also assumed to move. The ion and electron density profile initially coincide, and the ions subsequently move under the influence of the same fields as the electrons.

We begin our examinations of the results by analysing a triangle pulse with intensity 10^{18} W/cm^2, **Fig. 3.4.3, 3.4.4** shows the spatial distribution of electron density at two times, t=40 and 308.9 fs.

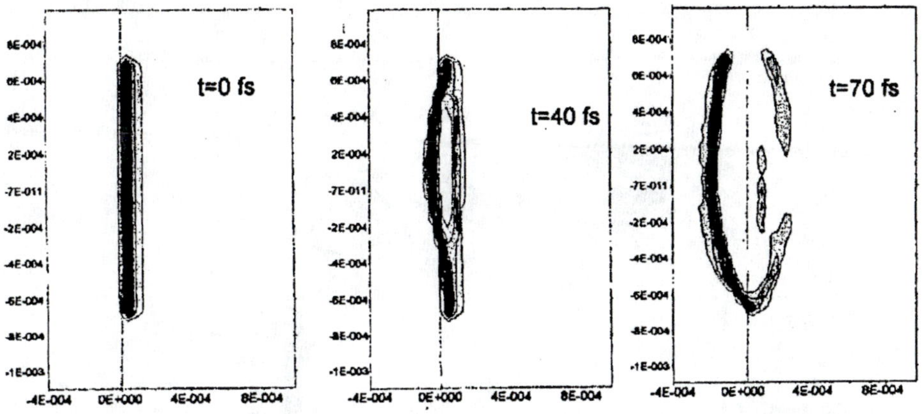

Fig. 3.4.3. Spatial distribution of electron density for a laser pulse with a triangular transverse profile at various times. The laser intensity is 10^{18} W/cm^2, electron density is given in units 10^{19} cm^{-3}. The abscissa is the position normal to the plasma surface and the ordinate is the position along the surface. The origin of the abscissa corresponds to the position of the target surface at the initial time, and that of the ordinate to the centre of the laser spot.

The abscissa in both plots of Fig. 3.4.3 is the linear position x of an electron above the plasma surface, and ranges from –4 μm to 10 mm. The density scales are in unit electrons/cm^3. The calculations show that the laser radiation is already "pressuring" electrons into the plasma by 40 fs. The resulting pondermotive potential repels some electrons from the point $x=x_c$. Those that are "turned back" by the laser light and ambipolar potential subsequently form two broad beams and emerge to the right.

Comparison of the distributions at time t= 40 fs and 308.9 fs indicates that a beam of electrons that does not move along the surface normal is actually formed from the sub-critical plasma. This is illustrate most clearly by Fig.3.4.5 which shows the angular distribution $n_e(\alpha)$ of electrons at x >0 at time t=308.9 fs. Analysis of **Fig. 3.4.5** indicates the presence of a beam of electrons emitted at an angle of –58° to the normal. The results of this calculation for intensity 10^{17} W/cm^2 are shown **in Fig. 3.4.6.** At lower intensities the angular distribution of the emitted electrons becomes narrower (they emerge close to normal), and the angular asymmetry of the distribution is reduced. Note that the energy of the emerging electron increases slowly as the laser intensity rises. This effect is related to compensation of the accelerating force by the ambipolar field, as noted before.

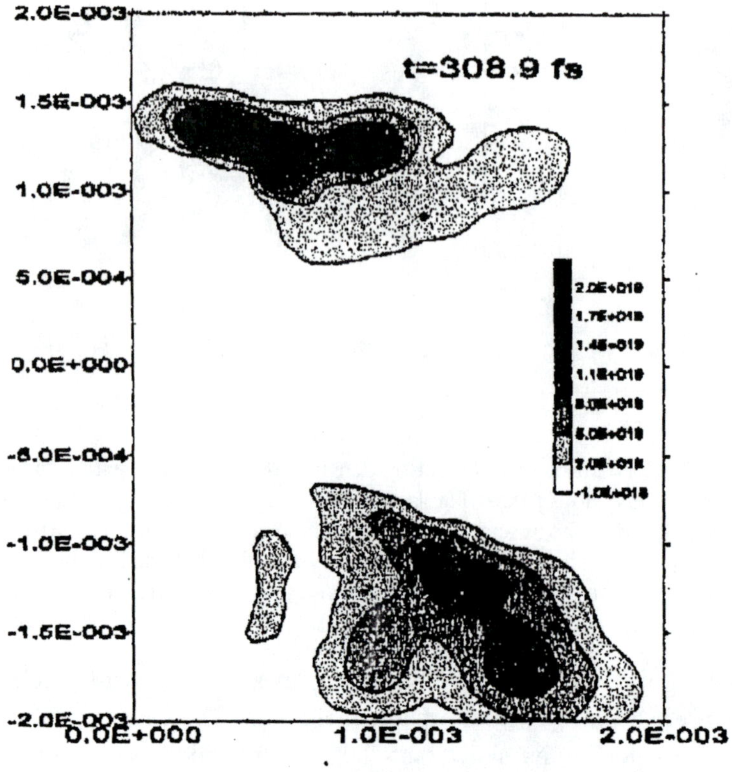

Fig. 3.4.4. Spatial distribution of electron density for a laser pulse with a triangular transverse profile at t= 308.9 fs. The laser intensity is 10^{18} W/cm^2, electron density is given in units 10^{19} cm^{-3}. The abscissa is the position normal to the plasma surface and the ordinate is the position along the surface. The origin of the abscissa corresponds to the position of the target surface at the initial time, and that of the ordinate to the centre of the laser spot.

Ions also leave the target, but unlike electrons, they leave normal to the surface. The ions move because of the ambipolar potential. The mean ion energy is the same order of magnitude as the electron energy for Z=1. We also analyse the calculations for rectangular spatial and temporal distributions of the laser intensity at 10^{18} W/cm^2. **Figure 3.4.7** shows two calculated two dimensional electron density distributions at time 50 fs and 240 fs, as done previously for triangle distribution. The abscissa in both plots of Fig. 3.4.7 is the linear position x, which ranges between the same limits as in Fig. 3.4.3. The calculations show that the laser radiation also "pressures" electrons into the plasma at 40-50 fs. The resulting pondermotive potential repels some electrons from the point $x \approx x_{cr}$. In contrast to the case of triangle pulse shape, the part that is "turned back" falls into a minimum of the pondermotive potential (3.4.7) and retained there for the duration of the laser pulse, undergoing oscillations at the minimum of the potential $U_{eff} \propto \cos(2x\cos\theta -$

ψ). Fig. 3.4.7 shows that the electron burst has a transverse size of 0.8 – 1 μm at 50 fs and is stretched out along the entire of laser spot.

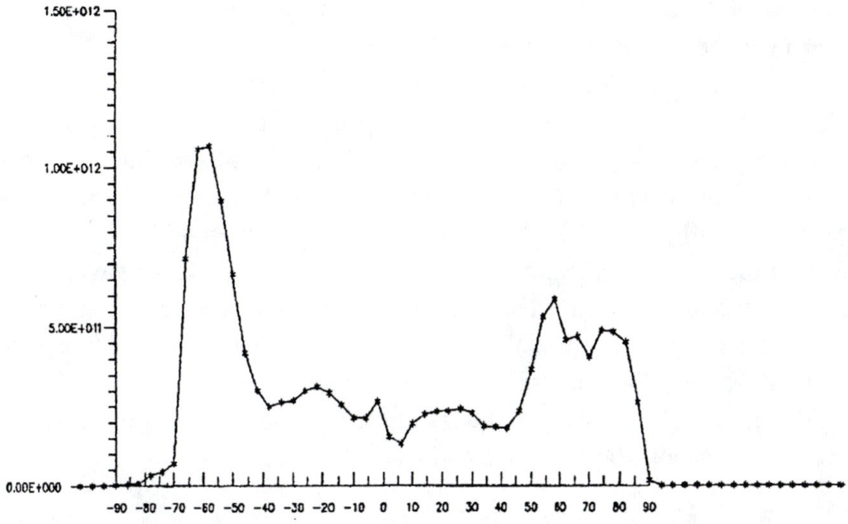

Fig. 3.4.5. Angular distribution of escaping electrons for a laser pulse with a triangular transverse profile and intensity 10^{18} W/cm^2. The abscissa is the emission angle relative to the surface normal and the ordinate is the number of electrons.

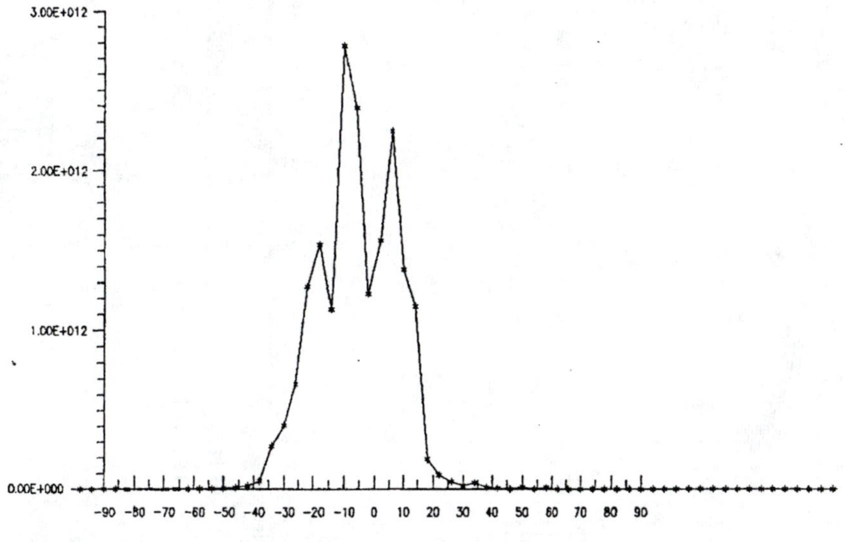

Fig. 3.4.6. Angular distribution of escaping electrons for a laser pulse with a triangular transverse profile and intensity 10^{17} W/cm^2. The abscissa is the emission angle relative to the surface normal and the ordinate is the number of electrons.

After laser pulse end (>100fs), the resulting electron bunch escapes the target at characteristic angle of 34°. Comparison of the results at equal intensities but different transverse profile shows that the mean electron energy is insensitive to pulse shape. The spatial structure of electron density and the angle at which electrons are ejected into vacuum, however, differ substantially for rectangular and triangular transverse laser profiles. Assuming that the contribution of the ambipolar field decreases as the laser intensity increases, we have done some additional calculations of the range of emission angles for triangle distributions at intensities 10^{19} W/cm^2 and 10^{20} W/cm^2, neglecting the ambipolar field. The corresponding emission angles are 48 – 57° and 42 – 51°, while the energies are 450-570 KeV and 4.6 – 5.2 MeV. These data imply that electrons energy at angles between the specular direction and the direction of the wave electric field, while the electron energy increases in proportion to the intensity of laser radiation in according with the relativistic dynamics of an electron in the field of monochromatic wave [132]. To evaluate the influence of the initial conditions on the angle at which electrons are ejected from the laser field region, we have done some calculations with special initial conditions. It was assumed that electrons leave the plasma (x=0) either at a fixed angle to the normal or isotropic. The initial energy was 2 KeV and the laser intensity was 10^{18} W/cm^2. In either case, the angular distribution and energy of emitted electrons were the same as in Fig. 3.4.5. Thus, the parameters of the emitted electrons are determined by the laser field configuration above plasma surface.

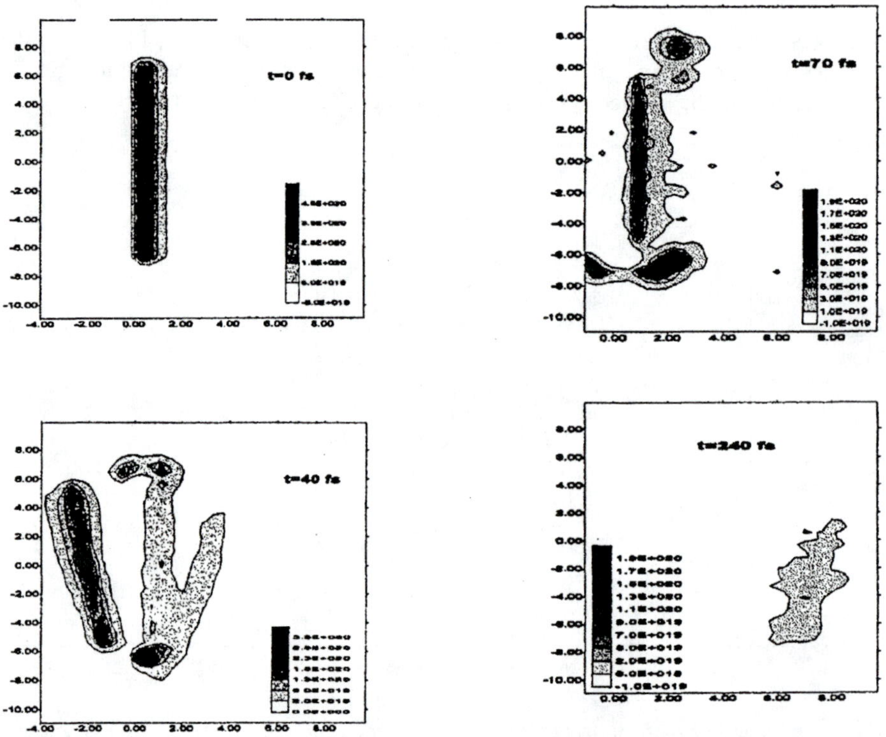

Fig. 3.4.7. Spatial distribution of electron density for a laser pulse with a rectangular transverse profile at various times. The laser intensity is 10^{18} W/cm^2, electron density is given in units 10^{19} cm^{-3}. The abscissa is the position normal to the plasma surface and the ordinate is the position along the surface. The origin of the abscissa corresponds to the position of the target surface at the initial time, and that of the ordinate to the centre of the laser spot.

Comparison of Numerical and Analytical Results

We now compare the results of the numerical and analytical models. Above all, we note the qualitative agreement of the calculations with the theory of electron motion in the effective potential. Comparing Figs. 3.4.3 and 3.4.7, where the intensities and energies of the laser pulses are the same, while the transverse profiles are different, we see that in Fig.3.4.3 for a triangle pulse shape, electrons are pushed out of the high field region to form two lateral bunches. In Fig. 3.4.7, where the pulse profile is rectangular, there is a single electron bunch that is uniform along the surface of the plasma at times 0-100 fs. Figures 3.4.3 and 3.4.7 imply that during initial interaction of the laser pulse with electrons (i.e. at times of 0 – 40 fs), the particles are divided into two groups, moving respectively into vacuum and into plasma. This is also explained by the structure of the effective potential (3.4.7) at $2x\cos\theta = -\pi$. At $\theta = 30^\circ$. And wavelength 1.06 μm, x is 0.3 μm. The initial concentration of electrons falls off exponentially with a characteristic scale length $x_c = 0.3$ μm. Electrons are repelled from the region $x \approx x_c$ as they approach the minimum of effective potential. The subsequent minima of U_{eff} along the x-axis are weekly populated by electrons, owing to the exponential drop in initial electron density and (cf. Figs. 3.4.3 and 3.4.7) are absent up to 10^{18} cm^{-3} on the chosen density scale. If the density resolution is increased, it is possible to observed a "striped" structure in electron density along the x-axis. After formation of the initial electron bunches over times of 0 – 70 fs under the influence of the strong gradients of U_{eff} over scale lengths of the order of the wavelength, these bunches begin to move under influence of the weaker gradient of U_{eff} on scale lengths of the order of the transvers size of the laser beam. The time over which this motion takes place is of order $D/(U_{eff}/m)^{1/2}$, which for the numerical parameters chosen here is less than the pulse duration. An electron burst can therefore cover the entire region occupied by the field. For non-relativistic intensities, the function $U_{eff}(y, x)$ ends up symmetric with respect to the normal to an accuracy of the order of the difference between the reflection coefficient and unity. Thus, the angular distribution of the emitted electrons in Fig. 3.4.6 is almost symmetric.

When the ambipolar field is taken into account at intensities 10^{17} W/cm^2 - 10^{18} W/cm^2, the mean energy of the emitted electrons is not proportional to U_{eff} (i.e to the intensity of the radiation), but has a weaker dependence on U_{eff}. Finally, when the intensity rises to relativistic levels, the number of electrons emerging in the direction of the reflected pulse increases (cf. Fig. 3.4.6, and Fig. 3.4.5). For ultra-relativistic intensities, the emission angle approached the specular value.

We now compare the emission angle and electron energy for a triangle laser beam with intensity 10^{17} W/cm^2 [Eqs.(3.4.15) and (3.4.16)] with the numerical results of Fig. 3.4.6. A calculation using Eqs.(3.4.15) and (3.4.16) yields $\theta^{(i)} \approx 17^\circ$ and $\theta^{(o)} \approx 28^\circ$; the numerical simulations yields $\theta^{(i)} \approx 10^\circ$ and $\theta^{(o)} \approx 20^\circ$. Finally, estimations of emission angle from the slope of the field lines the effective potential yield $\theta^{(i)} \approx 6^\circ$ and $\theta^{(o)} \approx 16^\circ$. The electron energy, according to estimates of U_{eff} at $x \approx 0, y \approx 0$, is 12 keV and the numerical calculation yields 7.9 keV. Given the approximations entailed in the analytic estimates (the lack of an ambipolar field) we can say that the numerical and analytical results are in agreement. This analysis implies that by varying the angle of incidence, the intensity, and the transverse profile of a laser beam, it is possible to control the motion of the electron bunch produced by the interaction of a laser pulse with a solid target.

3.4.2 Production of Fast Electrons by High-Power Laser Pulse in Dense Plasmas

Let the plane linearly polarized electromagnetic wave comes along the axis x normal to the semi limited plasma. Plasma temperature is T_e, the density grows linearly from zero to n_e on the distance L. The wave with the parameters E_0,ω is chosen in such a way that $n_e \gg n_c = m\omega^2/4\pi e^2$; $(T/m)^{1/2} \ll eE_0/m\omega$, and the length of the transient region L is much less than the skin layer l_s. The reason is that during the laser pulse the movement of ions is negligible and the plasma edge preserves its sharpness. Movement of the plasma electron component is described by the self sustained set of equations, consisting of the collision-less kinetic Boltzman equation ($0<x<l_s$) and Maxwell equations for electromagnetic fields and of the kinetic Fokker-Plank equation for $x>l_s$ (see Sect.3.1). Ambipolar field E_a is determined from the zero current along the axis X condition. In the **Fig.3.4.8a** are shown the levels of the distribution function f calculated for various moments of time. The calculation was carried for the irradiating laser radiation intensity of 10^{18} W/cm^2 and plasma with the starting temperature 10 keV and concentration $n_e = 25n_c$ (by KINET code). One can see that f is significantly distorted comparison with the starting Maxwell distribution.

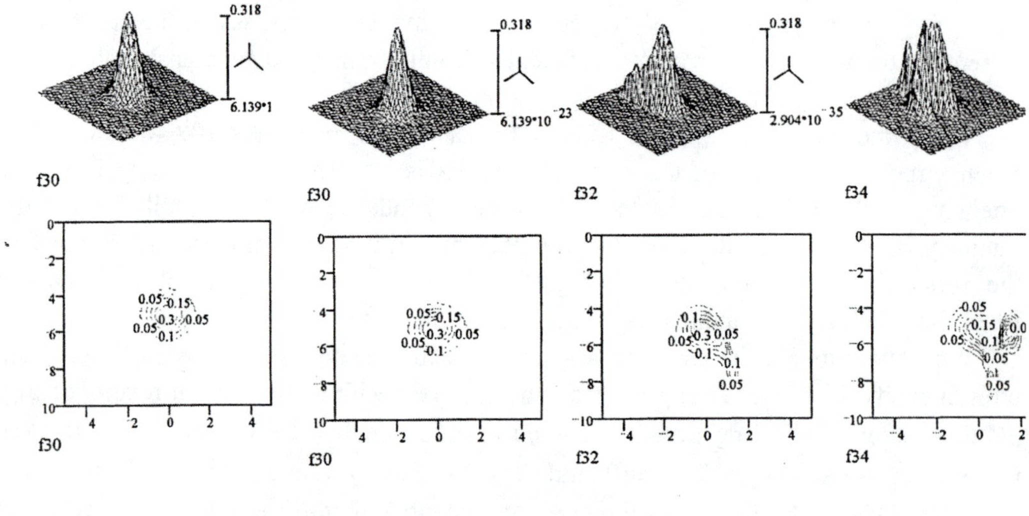

Fig. 3.4.8a. The levels of the distribution function f calculated for various moments of time. The calculation was carried for the laser intensity of 10^{18} W/cm^2 at normal incidence on over-dense plasma with initial temperature 10 keV and concentration $n_e = 25n_c$.

Distribution function can be analytical modeled by evaluation of the electron movement in the fields and taking Maxwell distribution as the starting one:

$$f(x;p_x;p_y;t) = \frac{n_e(x)}{2\pi mT}\exp\left[\frac{mc^2 - (m^2c^4 + p_x^2(0)c^2 + p_y^2(0)c^2)^{1/2}}{T}\right] \tag{3.4.23}$$

Solution of the relativistic movement equations looks like:

$$P_y(0) = p_y + e/c\left[A_y(x - V_x t;0) - A_y(x,t)\right]$$

$$P_x(0) = p_x + e\int_0^t E_x(x(\tau);\tau)d\tau + \frac{e(p_y - e/cA_y(x;t)}{mc\gamma}\int_0^t \frac{\partial A_y\left(x(\tau_0);\tau_0\right)}{\partial x}\partial\tau + \tag{3.4.24}$$

$$+\frac{e^2}{2m\gamma c^2}\int_0^t \frac{\partial A_y^2(x(\tau);\tau)}{\partial x}d\tau,\ A(x;t) = \frac{cE_0\cos(\omega t)}{\omega(1 + x^2/l_s^2)},\ \gamma = \left(1 + \frac{p_x^2 + p_y^2}{m^2c^2}\right)^{1/2}.$$

According to the numerical calculation, the electromagnetic field in plasma can be approximated by the following equations [136]:

$$E_y(x;t) \cong \frac{E_0\sin(\omega t)}{1 + x^2/l_s^2} \qquad B_z(x;t) \cong -\frac{2cE_0 x\cos(\omega t)}{\omega l_s^2(1 + x^2/l_s^2)^2} \tag{3.4.25}$$

$$E_x(x) \cong E_0\left[\Theta(x)\exp(-x/l) - \Theta(-x)\exp(x/l)\right]$$

Here $\Theta(x)$ is the Heavyside step-like function, $l_s = (c^2V_T/\omega\cdot\omega_p^2)^{1/3}$, $l \approx L$, $E_{x0} \cong E_0\left(\frac{cL}{\omega l^2}\right)\left(\frac{\omega^2}{\omega_p^2}\right)\left(\frac{eE_0}{m\omega c}\right)(1+R)$. The lengths l_s and l are much shorter than the length of free flight of electron in plasma; hence the trajectory x(τ) can be approximated by the straight line : $x(\tau) = x + V_x\tau = x - V_x(t-\tau)$. Hence, the equations (3.4.23), (3.4.24) and (3.4.25) make it possible to determine the analytical expression for the distribution function. In the **Fig. 3.4.8** are shown these functions for L=0.02λ, $l = \upsilon_T/\omega$, $eE_0/m\omega c = 1$ and for various t. Qualitative explanation of its shape is as follows. For the electrons, moving under the action of the transverse field E_y parallel to the plasma edge the ponderomotive force deepens them into the plasma, excluding out of the interaction region. So, twice during the period of laser wave oscillation is produced the flux of fast electrons, moving inside the plasma. In the Fig. 3.4.8 this flux reveals itself as the "tail" of the distribution function. The average energy of hot electron is ≈ 0.5 MeV.

As was shown in Sec 3.1 in the case of oblique incidence it is convenient to calculate the intensity of X-ray emission in the coordinate set, moving with the speed V=$c\sin(\theta)$ (θ -incidence angle) along the plasma edge. In this system the incidence is normal.

Longitudinal field E_x recalculation to this set of coordinates results in the additional constant magnetic field and, hence, in the following field configuration (3.7.2).

Solution of such set can be drawn out of the solution of (3.4.24) by replacement of γ in (3.4.25) by $\gamma = \left[1 + p_x^2 / m^2c^2 + (p_y / mc + \sin(\theta))^2\right]^{1/2}$. Analysis of the obtained distribution function shows that the constant magnetic field in the set of normal incidence results in acceleration of the fast electrons flux in longitudinal (X) direction during one half of laser oscillation period and in its slowing during another half of period (see **Fig. 3.4.8b**).

Let's now estimate the parameters of high-speed electrons produced by a laser pulse. When the radiation intensity amounts to 10^{19} W/cm^2, the radiation absorption involves nonlinear plasma effects. Numerical simulations have shown that, in the 10^{19} – 10^{21} W/cm^2 intensity range, the absorption coefficient becomes independent of the angle of incidence, and the absorption is about 10% without pre-pulse. Laser pre-pulse increase scale of plasma inhomogenity L and absorption coefficient η. It was numerically shown [134] that in the range 10^{18}-10^{20} W/cm^2 there is $\eta \propto L$.

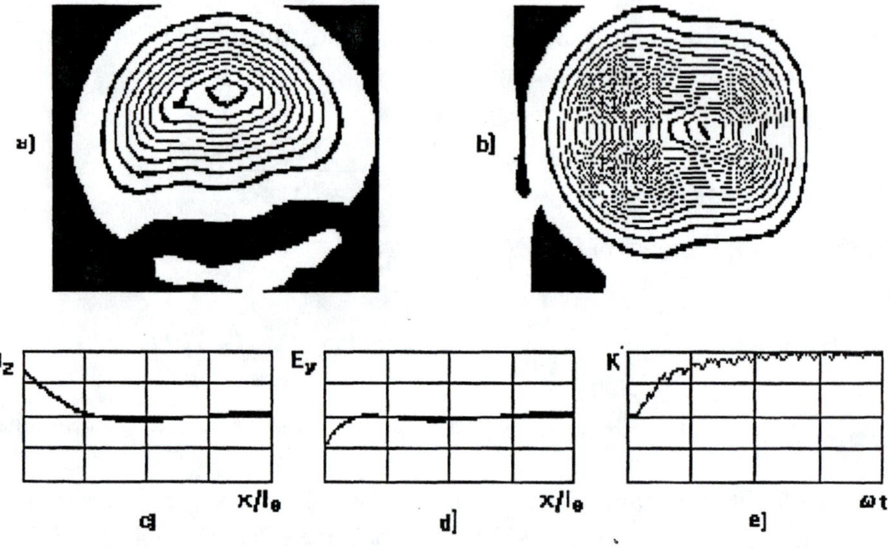

Fig.3.4.8.b. The levels of electron distribution function (a) at t = 30 fs, averaged in time (b). The calculation was carried for the laser intensity of 10^{18} W/cm^2 at oblique incidence on over-dense plasma.

The same dependence was obtained in the part 3.1.5, where analytical model of pondermotive absorption was considered. Follow (3.1.89) we will use the scaling of $\eta(L,I)$ as:

$$\eta(L,I_{18})=(0.1+0.01\mathrm{L})I_{18}/(30+I_{18})^{0.7}, \qquad (3.4.26)$$

where I_{18} - radiation intensity in units 10^{18} W/cm^2, L = $L\omega/c$, $L \approx c_s\, t_{pl}$. Here t_{pl}.- pre-pulse duration. At pre-pulse intensity 10^{12} W/cm^2 we have L=10 at t_{pl}=10 ps and we will use L = 0, 10 in our father numerical simulations.

At the beginning of this section, we considered the physical mechanism of formation of fast electron jets due to the ponderomotive light pressure, whereby an electron oscillating in an electric field is driven into plasma by the ponderomotive pressure force. When a laser pulse is normally incident, the electrons are injected twice per period. Altogether, there are $n_c l_s S$ electrons in in a skin layer, where n_c is the critical density, l_S is the typical skin-layer size, and S is the area of a laser spot. During laser pulse length t_L, $\omega t_L/\pi$ half-cycles of oscillations occur. Therefore, N_{ef} = $n_c l_s S\omega t_L/\pi$ electrons will be accelerated during the laser pulse. It should be emphasized that the pulse length has to be short enough to allow for ignoring the motion of ions. It has been proved in Ref. [42] that this condition is satisfied for t_L < 100 fs. Furthermore, as electrons move inward a plasma, the charge separation will occur, which will give rise to the electron current from the surrounding plasma into the skin-layer region and will retard the output of the high-speed electrons from the skin-layer. Therefore, in the strict sense, the N_{ef} estimate is the upper bound of the number of high-speed electrons. To verify this statement, we make an alternative estimate as N_{1ef}=$K_e(I)\mathcal{E}_L/\mathcal{E}_e$, where $\mathcal{E}_L$ is the laser pulse energy, $K_e(I)$ is the transformation coefficient of laser energy into fast electron energy, and $\mathcal{E}_e$ is the energy of an electron. It was shown [133] that in the range 10^{18}-10^{20} W/cm^2 $K_e(I)$ has linear rise from 0.03 up to the η for the subpicosecond laser pulses. At laser intensity more then 10^{20} W/cm^2 $K_e(I) \approx \eta$. This means that all the absorbed energy being assumed to be transferred to the motion of high-speed electrons. The energy of an individual electron is obviously specified by the wave field strength inside the skin-layer [137,138]

$$\mathcal{E}_e/mc^2 \approx (1+\eta I_{18})^{1/2}-1. \qquad (3.4.27)$$

Now we numerically estimate N_{ef} and N_{1ef} for the following plasma and laser pulse parameters: I_{18}=10^2, λ=10^{-4}cm, t_L = 100 fs, n_c = 10^{21}cm^{-3} and S = 10^{-6}cm^2. We take length l_S of the skin layer for the anomalous skin effect, l_s=$c/\omega_p(\upsilon_T\omega_p/c\omega)^{1/3}$, which takes place because the background plasma temperature is ~ 10 keV and $\upsilon_T/c > \omega/\omega_p$, the intensity being I_{18}=100. For the chosen parameters N_{ef}=9.6 10^{10} and N_{1ef}=1.3 10^{10}, which are fairly consistent allowing for the fact that the estimates are rough and N_{ef} is the upper bound of the effective number of high-speed electrons.

It is well known [132] that the electron kinetic energy within the first part of a laser pulse at the vicinity of its maximum can be as high as $\mathcal{E}_e/mc^2 = (\upsilon_E/c)^2/2$ but the second part of this pulse will damp electron because it can not obtain any energy in vacuum from a plane wave if it was at rest initially. Anyway this electron can get energy from electromagnetic wave if it inertial penetrates a target and the laser pulse reflects from this target. Pre-plasma with a length approximately equal to the laser pulse length can significantly increase electron energy. If we have initial electron momentum in pre-plasma P_{e0} one can easily estimate its maximum energy

$$\mathcal{E}_e/mc^2 \approx (P_{e0}/mc)\,(\upsilon_E/c)^2 \approx (P_{e0}/mc)I_{18}$$

From this formula we see that electron energy can be tens of MeV level for relativistic electrons generated in pre-plasma by the Brunel effect for example [40].

Radiation Losses of Fast Electrons and Limit of Laser Intensity for Acceleration

It is well known that electron interacting with an electromagnetic wave cannot get an infinite energy from this wave because of radiation losses [132]. Fast electron with energy $\mathcal{E}_e$ (in the units mc^2) emits radiation intensity:

$$I_{rad} = (4e^2\omega^2/3c)\ \mathcal{E}_e^2 I_{18}$$

Here the intensity of electro magnetic wave is I_{18}. The condition for small radiation losses of fast electron is found as

$$(4e^2\omega^2/3c)\ \mathcal{E}_e^2 I_{18} < mc^2\,(\omega/2\pi)\ \mathcal{E}_e$$

Take into account the estimation of electron energy in the field of a high intensity wave in vacuum: $\mathcal{E}_e \approx 0.25\, I_{18}$, we can rewrite this condition as

$$I_{18} < (3\lambda/4\pi^3 r_0)^{1/2}$$

Here r_0 – is electron classical radius. For laser wavelength $\lambda = 10^{-4}$ cm we obtain that the maxim value of laser intensity should be $I < 3\cdot10^{21}$ W/cm^2.

For the interaction of laser wave with fast electron in plasma skin layer we have $\mathcal{E}_e \approx \sqrt{I_{18}}$ and then max $I_{18} < (3\lambda/4\pi^3 r_0)^{2/3} = 1.5\cdot10^4$. It means that there is no sense in increasing laser intensity more than the limit $I_{lim} = 10^{23}$ W/cm^2 for acceleration of particles.

3.4.3 Ion Acceleration at Moderate Laser Intensity

Acceleration Model

In this section we have assumed an electrostatic acceleration of ions by the ambipolar field in the expanding plasma at the plasma-vacuum boundary. The spectrum of fast ions is found from the electron spectrum in terms of the model [139]. This model assumes quasi-neutrality, and the time of fast electron roundtrip in the corona is taken to be shorter than the laser pulse duration. Then the electron distribution is symmetric in the longitudinal velocity υ, and the electron concentration is governed by the electrostatic potential

$$Zn_i = n_e(\varphi) = 2\int_0^{\infty} f_A\left(\sqrt{\upsilon^2 + 2e\varphi / m_e}\right) d\upsilon \tag{3.4.28}$$

where f_A is the above distribution function of electrons accelerated for example by a P-polarized laser wave. The electrostatic potential is set to $\varphi = 0$ at the critical surface and at the vacuum side $\varphi \to \infty$ for $x \to -\infty$. The evolution of the ion density and velocity is then described by collision-less hydrodynamics. A self-similar solution is applied to calculate the time-integrated energy distribution of ions. The electron distribution f_A is averaged over the laser pulse duration to obtain a constant electron distribution for the model of ion acceleration. Then the ion density n_i, the velocity u and sound velocity c_s are functions of the self-similar variable $\upsilon = x/t$. The ion sound velocity c_s as a function of the electrostatic potential φ is expressed through the ion concentration $n_i(\varphi)$, and the electrostatic potential φ is found implicitly as

$$u = \upsilon + c_s(\varphi), \quad c_s^2 = \frac{eZn_e(\varphi)}{M_i}\left(\frac{dn_e}{d\varphi}\right)^{-1},$$

$$-\upsilon = c_s(\varphi) + \frac{eZ}{M_i}\int_0^{\varphi} \frac{d\varphi'}{c_s(\varphi')}.$$

The ion velocity distribution is expressed via the following transformation of the variables

$$\partial N/\partial u = n_i\,|\partial x/\partial u| \approx n_i\, t_L |\partial \upsilon/\partial u|$$

where symbol N represents the number of ions emitted per unit surface of the target.

In situations with a large ratio of the fast-to-thermal electron temperatures (T_h/T_c >10), the function $\upsilon(\varphi)$ is not necessarily monotonous; consequently, two values of φ as a function of υ can be found [140]. Physically, this is associated with the formation of a narrow layer inside the expanding plasma, where the assumption of quasineutrality does not hold any more. A strong electric field in this layer leads to a nearly stepwise acceleration of ions. Therefore, there is a minimum in the energy spectrum of ions, corresponding to the velocities within the region of fast acceleration. In order to meet all the conditions of the present model, we introduce a step in the $N(\upsilon)$ function, which provides the solution $f_i(u) = 0$ in a certain range of ion velocities. Notify that to overcome this problem the use of a truncated bi-Maxwell distribution with the limited maximal energy has been suggested in [141].

Results of Simulations and Discussion

We performed simulations for the experimental conditions described in the literature [16,35,48,142] and compared the resulting spectra of fast ions with the measurements. To facilitate the comparison with our simulations, a standard transformation of the ion blow-off current traces presented in **Figure 3.4.9** of the paper cited above [48] to the ion spectra was made. The experimental ion spectrum and the simulation result are presented in Figure 3.4.9. The position of the fast ion maximum and the slope of the tail in the ion spectrum compare well with the experimental data. Fast ions are confined within an angle of about 5° from the target normal in the experiment of [48]. Therefore, the discrepancy at low velocities for a detector positioned normally to the target can be interpreted as being due to the geometrical factor, since the angular distribution of thermal ions is much broader. The total energy of fast ions measured in this experiment is about 20 % of the incident laser energy, while the computed value is 9.6 %. The computed laser absorption efficiency $\eta = 39\%$ is lower than the experimental efficiency $\eta \cong 63 \pm 10$ %, which can partly account for the difference in the energy transfer to fast ions.

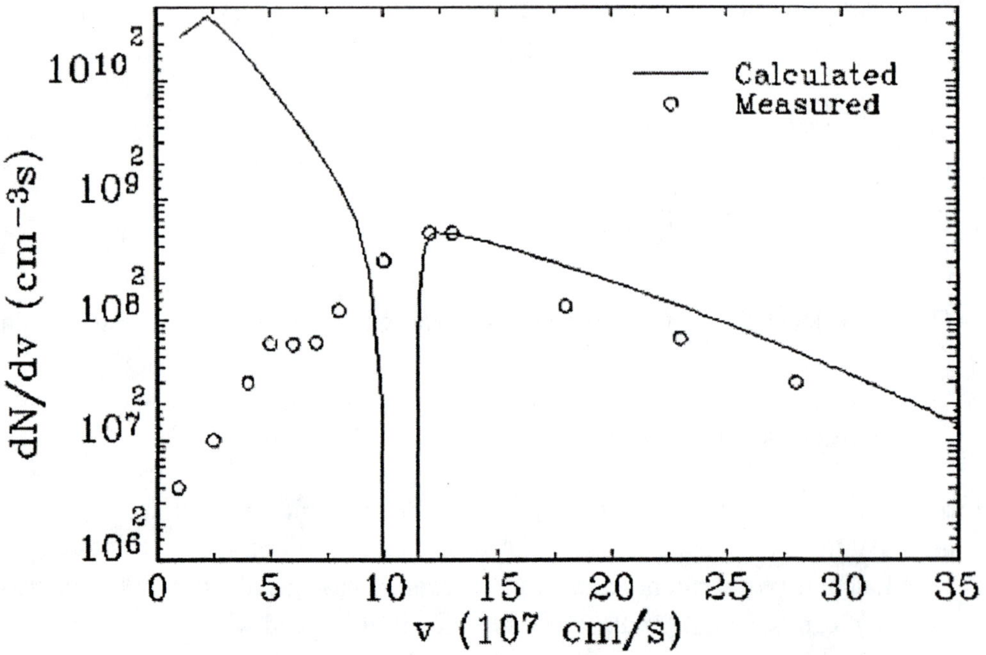

Fig. 3.4.9. The spectrum of ions emitted normally from a solid Al target by P-polarized 1 ps FWHM Gaussian Nd - laser pulse with a peak intensity of 8 10^{15} W cm^{-2} incident at 57° angle. The calculated data (curves) compared with the experimental spectrum (circles).

It should be noted that the entirely depleted gap in the computed ion velocity spectrum is associated with the employed model of ion acceleration. It is present when the ratio of the fast electron temperature T_h to the thermal electron temperature T_c exceeds a certain limit, which is $T_h/T_c > 9.6$ for a bi-Maxwell distribution of electrons [140]. This effect is not observed in the simulations of experiments of [48] with maximum laser intensities below 5.10^{15} W/cm^2.

We have also carried out simulations for the experimental conditions described by [35]. A fast ion detector was placed at 45° to the target normal in this experiment. Our one-dimensional model cannot describe this measurement correctly. However, the basic parameters of the computed fast ion spectrum, shown in **Fig. 3.4.10**, are similar to the experimental parameters. Since the experimental spectrum is not completely calibrated, i.e. the total energy of fast ions is unknown, we have normalized it, so that its maximum is equal to the maximum of the computed fast ion distribution. The computed spectra are qualitatively quite similar to the measured spectra. The number of ions at high energies in the experiment is reduced because the observation angle is 45° to the target normal.

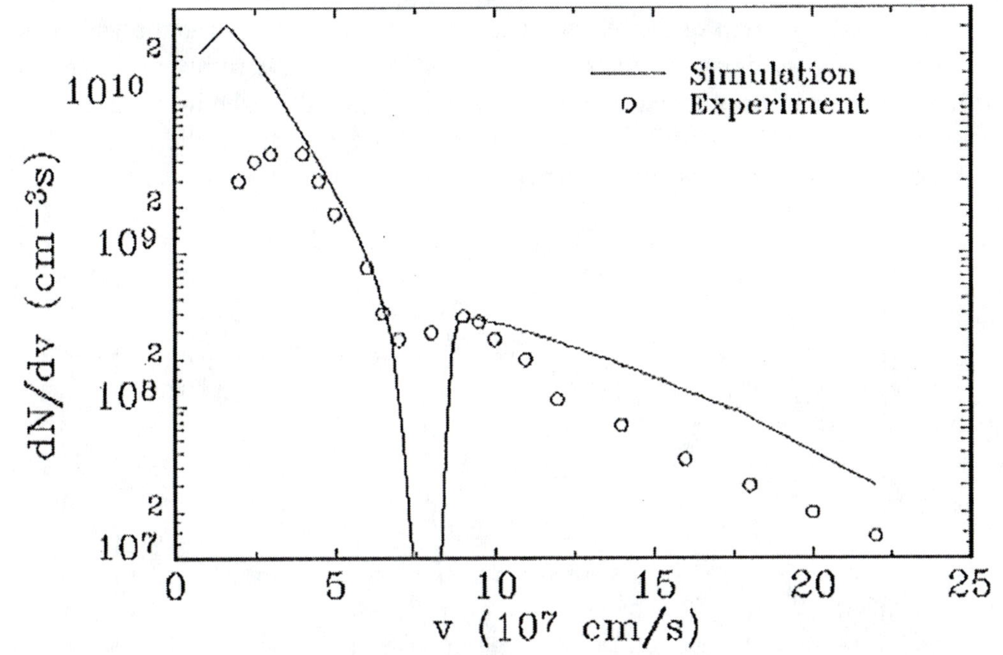

Fig 3.4.10. The spectrum of ions emitted from a solid Al target, irradiated by a 1.5 ps FWHM by a P-polarized Gaussian Nd - laser pulse with a peak intensity 10^{16} W cm^{-2} incident at 57° angle. The calculated data (curves) are compared with the experimental spectrum, recorded at 45° angle from the target normal (circles).

The spectrum of ions emitted in the direction normal to the target was measured in recent experiments by [142]. In these experiments, a 1.5 ps FWHM Nd-laser pulse was incident at 45° onto a plane Al target. The laser light intensity was varied between 10^{16} and 10^{17}W/cm^2 and the intensity contrast was ~ 10^6 - 10^7. The details of the simulation of absorption coefficient in these experiments are presented before and below.

The computed profiles of the plasma density and electric fields at the laser pulse maximum are plotted in **Figures 3.4.11** and **3.4.12** for the laser intensities 10^{16} and 10^{17} W/cm^2. Note that the small scale oscillations in the density profile in Figure 3.4.11 can be considered as a numerical artefact, since they disappear when an artificial viscosity is introduced. On the other hand, the density profile in Figure 3.4.12 shows an inward shock wave initiated by the ponderomotive force of the laser radiation. The density maximum of about 20 n_c in Figure 3.4.12 is only ≅ 0.01 λ wide. Similar shock waves were observed earlier in hydrodynamic simulations of the interactions of intense picosecond pulses with a solid target [141]. A comparison of Figures 3.4.11 and 3.4.12 shows that the increase in the laser intensity leads to a considerable shortening of the electron density scale length at the critical surface. While the density profile is near to the optimum for resonance absorption at 45° incidence for the lower intensity, the resonance absorption is less efficient for $I = 10^{17}$ W/cm^2. The structure of the electric fields at this intensity should lead to electron acceleration normal to the target surface. A much broader angular distribution of fast electrons should evolve for the higher intensity, as the longitudinal field is decreased by wave-breaking and the density profile modification (Figure 3.4.12). A broad angular spectrum of fast electrons leads to fast ion emission into a wide cone. The computed ion spectrum is presented in **Figure 3.4.13** for the respective intensities. The normalization of the experimental ion spectrum measured for the higher intensity is the same as in Figure 3.4.10. Thus, there is an excellent agreement of the computed and experimental fast ion energies and spectra.

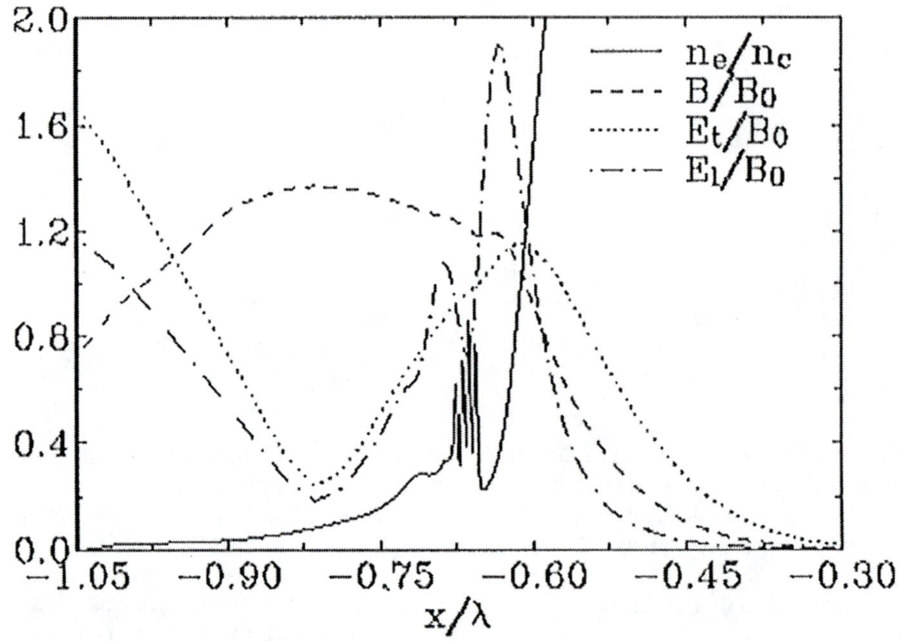

Fig. 3.4.11. Profiles of the plasma density and of the laser fields at the peak $I = 10^{16}$ W cm^{-2} of a 1.5 ps Gaussian pulse of P-polarized Nd – laser radiation, incident at 45° angle on a solid Al target.

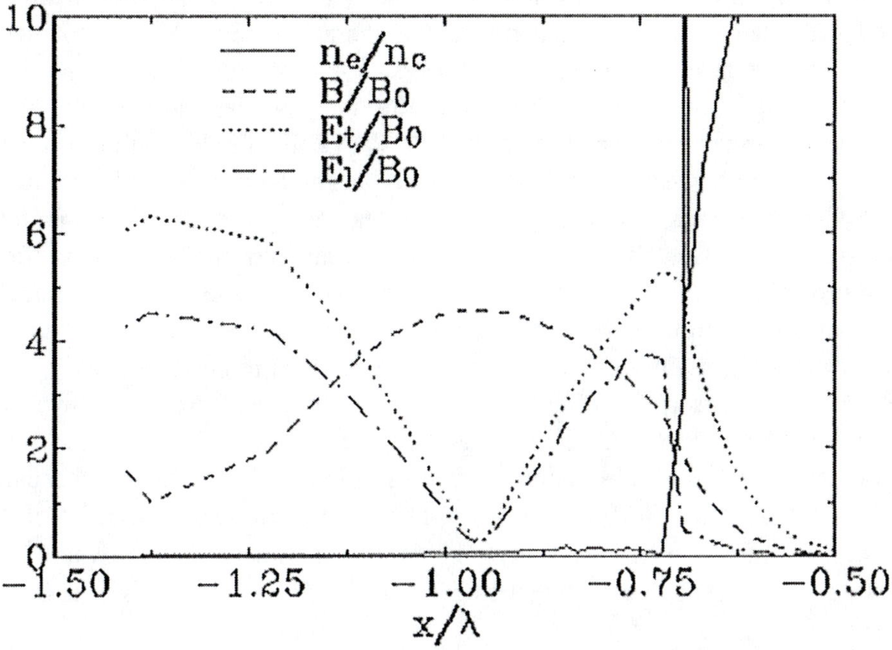

Fig. 3.4.12. Profiles of the plasma density and of the laser fields at the peak $I = 10^{17}$ W cm^{-2}.

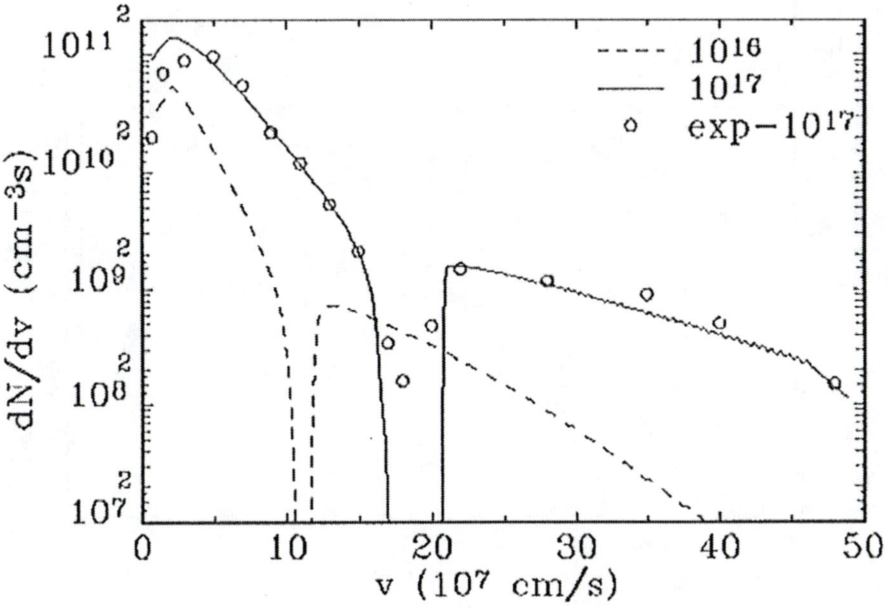

Fig. 3.4.13. The spectrum of ions emitted from a solid Al target, irradiated by a 1.5 ps FWHM P-polarized Gaussian Nd - laser pulse incident at a 45° angle. The peak laser intensity is 10^{16} W cm^{-2} (dashed curves) and 10^{17} W cm^{-2} (solid curves). The data are compared with the experiment (circles).

Since the applicability of our model has been demonstrated, we investigate the general dependence of basic characteristics of the interaction on the laser parameters. The laser absorption efficiency plotted in **Figure 3.4.14** decreases slowly with the laser intensity. The enhanced density profile due to the rising ponderomotive force leads not only to a decline in the collision absorption but also to a slight reduction in the resonance absorption. For the highest intensity studied, the plasma corona dynamics is, however, very complicated; there are periods when the plasma density profile changes rapidly and cavitons are formed. This also leads to a fast time variation in the absorption efficiency, so that the overall absorption efficiency is affected. In this case, there is a remarkable variation of the overall absorption efficiency with the details of the simulation parameters. This uncertainty in the absorption efficiency is represented by error bars in Figure 3.4.14. The absorption efficiencies calculated here are somewhat lower than in experiments and in some PIC simulations. This implies that it is rather difficult to account precisely for all the absorption mechanisms of ultra-short laser pulses incident on a solid target if one uses the 1D hydro-code. The same figure depicts the rise of the hot electron temperature with laser intensity. The deduced scaling is $T_h \sim I^{\alpha}$, where $\alpha \cong 0.6$, in good agreement with the scaling observed in the PIC simulations and the above experiment [48].

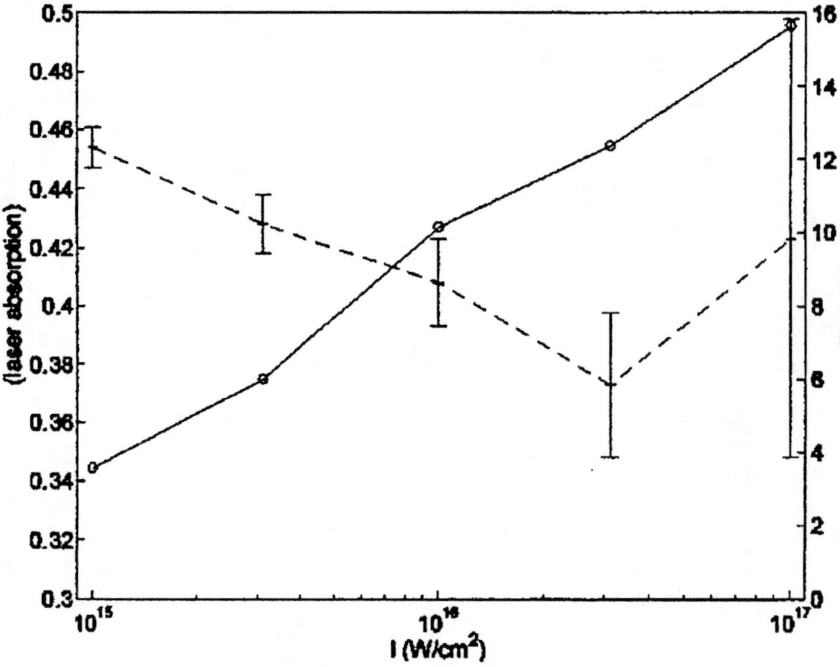

Fig. 3.4.14. Laser absorption efficiency (dashed line) and hot-electron temperature (solid line) versus peak laser intensity. P-polarized 1.5 ps FWHM Gaussian Nd – laser pulse is incident at 45° angle onto a solid Al target.

The energy of the fast ion emission grows with the laser intensity, but the efficiency of laser energy conversion to fast ions displayed in **Figure 3.4.15** decreases with light intensity. This behaviour is due to the fact that the energy of fast ions is proportional to the hot electron temperature, $T_h \sim I^{\alpha}$, with $\alpha < 1$. However, if the energy conversion to a group of very fast and rather mono-energetic ions is preferable, the energy conversion increases very fast with laser intensity. This is demonstrated in Figure 3.4.15 for Al ions in the energy range 800 - 1000 keV. Therefore, high light intensities are preferable for the generation of intense beams of energetic ions in applications. However, the validity of our model is limited to laser intensities $I\lambda^2 < 10^{18}$ W cm^{-2} μm^2. For greater intensities, the ponderomotive pressure is expected to induce an inward motion of the critical surface and of the corona plasmas during most of the pulse time. It should significantly reduce the ion acceleration by the ambipolar field to vacuum. Thus, one can conclude that the optimum intensity for the ambipolar acceleration of high- energy ions in the bulk target is around $I \approx 10^{17}$ W/cm^2 and that the efficiency of laser energy conversion to the emission of energetic ions may reach ~1%.

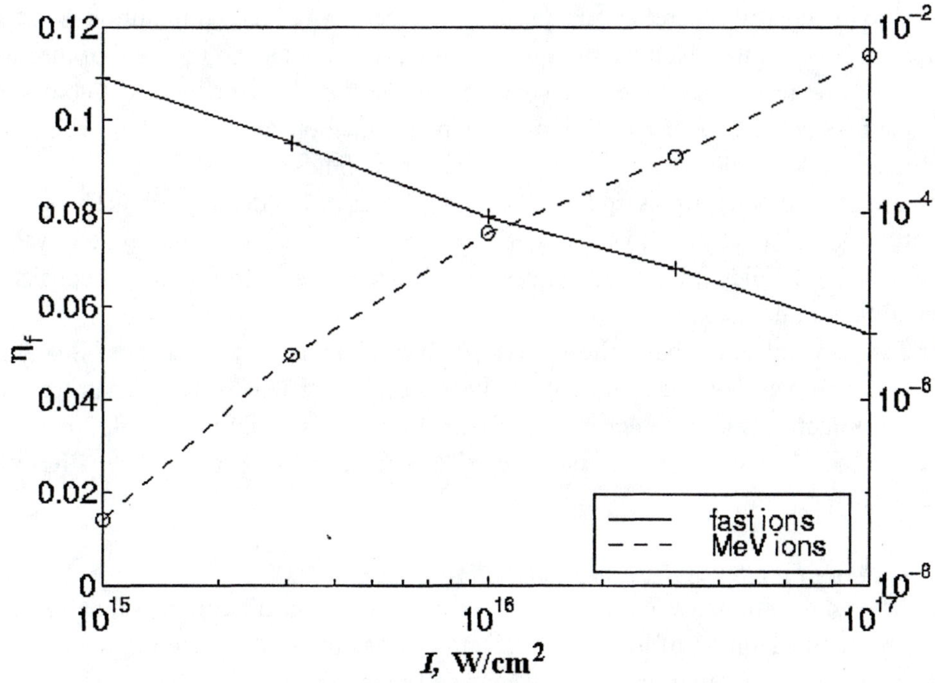

Fig. 3.4.15. The efficiency η_f laser energy transformation to the energy of fast ions (solid line) and the efficiency η_M of energy transformation to ions in the energy range 800 – 1000 keV (dashed line) versus peak laser intensity. A P-polarized 1.5.ps FWHM Gaussian Nd – laser pulse is incident at a 45° angle onto a solid Al target.

3.4.4 Ion Acceleration when an Ultra Intense Laser Pulse Interacts with a Foil Target

In this section we consider ultra high intensity laser pulse acceleration of fast ions in dense plasmas. The most important point is the efficiency of this fast ion production. We analyze - with help of an analytical model and PIC code simulations - the different acceleration mechanisms and compare the efficiency of electrostatic ion acceleration, at the front and rear of a foil target, the ponderomotive mechanism and acceleration by the shock wave in detail.

Basic Equations

We consider acceleration of ions from a plasma layer with a thickness of several laser wavelengths. Such plasma can be formed by laser illumination of a thin foil. To describe the acceleration of ions in laser plasma we formulated a set of equations. First we consider the given ion distribution and movement of an electron subsystem. The electron motion can be described with the help of kinetic or hydrodynamic equations. The electrons play the role of intermediary through which the energy of the laser pulse is converted to ion energy. Therefore, the purpose of describing of electron motion is to find a connection between the electrostatic field operating on the ions and the field of the laser pulse. In the kinetic approximation, plasma electron motion describe by a self-consistent set of equations consisting of a collision-less kinetic equation for a distribution function of electrons, and Maxwell equations for electromagnetic fields (see Sec.3.1).

In these equations we are using the analytical profiles of ion density $n_i(x)$. The system of equations theoretically allows us to express electrostatic potential by the vector potential of a laser field and the density profile of ions, and thus to find a self-consistent force operating on the ions.

For ion movement we have the system of hydrodynamic equations (see Sect.3.2). However, it is impossible to solve analytically this self-consistent system. Therefore, we simplify the situation and consider two approximate methods of finding a solution of the system. To check the used methods we also simulate the problem by PIC-code calculations.

Approximation of a Given Electron Distribution Function

As the first approximate method we set a given electron distribution function and investigate the acceleration of ions in this electron distribution. We describe the electron distribution function as a two-temperature Maxwell function:

$$f_{e0} = f_c \exp(-u^2 / \mathrm{v}_c^{\ 2}) + f_h \exp(-u^2 / \mathrm{v}_h^{\ 2}), \tag{3.4.29}$$

where

$$u^2 = \mathrm{v}^2 - \mathrm{v}_0^{\ 2} - 2e\varphi / m; \quad \mathrm{v}_{c,h} = (T_{c,h} / m)^{1/2}. \tag{3.4.30}$$

Temperatures of hot and cold electrons is considered to be known. Selection of a such distribution function is natural and takes into account the basic groups of electrons, both accelerated in a skin layer, and cold electrons of background plasma.

In **Fig. 3.4.16** show, calculated with the help of PIC - code simulations [143], electron distribution functions for the interaction of a laser pulse of duration 1 ps and intensity 510^{18} W/cm^2 with Al foil target. From this figure we see two reference slope angles of the distribution function, corresponding to the two indicated temperatures. Energy of hot electrons can be derived from laser intensity I_L as $T_h \approx mc^2(\sqrt{1+\eta I_{18}}-1)$ (see 3.4.27). The electron density is expressed by a distribution function:

$$n_e = 2\int_0^{v_{\max}} f_{e0}(\sqrt{v^2 - v_0^2 - 2e\varphi / m_e})dv\,. \tag{3.4.31}$$

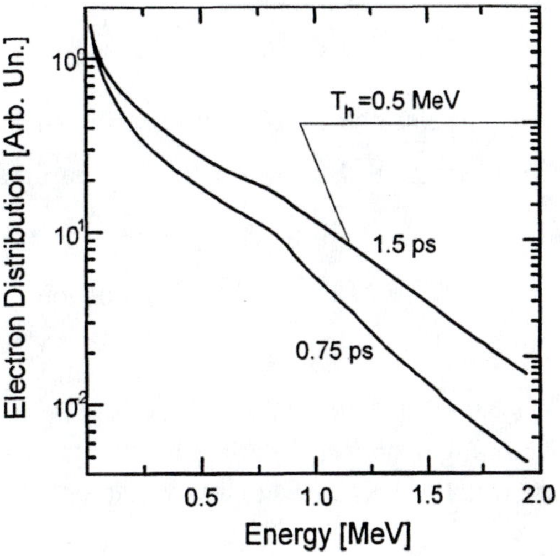

Fig. 3.4.16. Calculated electron distribution functions for the interaction of a laser pulse of duration 1 ps and intensity 510^{18} W/cm^2 with Al foil target at different moments of time.

System hydrodynamic equations of plasma motion transforms to:

$$\frac{\partial n_i}{\partial t} + \frac{\partial}{\partial x}(n_i v_i) = 0$$

$$\frac{\partial v_i}{\partial t} + v_i \frac{\partial}{\partial x} v_i + \frac{1}{M n_i} \frac{\partial \sum_{j=c,h} T_{ej} n_{ej}}{\partial x} = -\frac{Ze}{M}\frac{\partial \varphi}{\partial x}, \tag{3.4.32}$$

$$\frac{\partial^2 \varphi}{\partial x^2} = -4\pi e(Z n_i - \sum_{j=c,h} n_{ej})$$

where

$$n_{ej} = n_{0ej} \exp(-2e\varphi / m_e \mathrm{v}_j^{\,2}) \tag{3.4.33}$$

The system (3.4.32) is complete and allows to find $n_i, \mathrm{v}_i, \varphi$. For analysis of the system we rewrite this in dimensionless variables:

$$\begin{aligned}
&\frac{\partial \eta_i}{\partial \tau} + \frac{\partial}{\partial \xi}(\eta_i \mu) = 0 \\
&\frac{\partial \mu}{\partial \tau} + \mu \frac{\partial}{\partial \xi}\mu + \frac{1}{\eta_i}\frac{\partial}{\partial \xi}(\exp(-Z\psi)) + p\frac{1}{\eta_i}\frac{\partial}{\partial \xi}(\exp(-Z\Lambda\psi)) = -\frac{\partial \psi}{\partial \xi} \\
&\delta \frac{\partial^2 \psi}{\partial \xi^2} = \eta_i - Z\beta \exp(-Z\Lambda\psi) - Z\exp(-Z\psi)
\end{aligned} \tag{3.4.34}$$

The dimensionless variables are selected as follows: $\eta = n_i / n_0$ - profile of ion density, normalized on density in over dense plasma, $\xi = x / L$ - coordinate, the scale of plasma inhomogenity $L = c_s t_L$, $\tau = t / t_L$ - time, normalized on duration of laser pulse, $\mu = \mathrm{v}_i / c_s$ (where $c_s = \sqrt{ZT_{eh} / M}$) - Mach number of ions in relation to velocity of an ion sound,
$\psi = Ze\varphi / Mc_s^{\,2}$; $\Lambda = T_h / T_c$; $\beta = n_{eh} / n_{ec}$; $p = n_{ec}T_c / n_{eh}T_h$; $\delta = r_D^{\,2} / L^2$ - dimensionless Debye screening distance, Z - effective charge of the ions. Obviously, the following inequalities are true for the dimensionless variables:

$$\Lambda \approx 1, \beta \quad 1, \delta \quad 1, p \quad 1 \tag{3.4.35}$$

Quasi-neutrality is assumeed for the solution of the Poisson equation. The quasi-neutral status is valid if the process time is great than the time of fast electron oscillation about the ion:

$$t_L[\mathrm{ps}] \geq \left(\frac{m_e}{4\pi n_h e^2} \right)^{1/2} \sim (10^{-3} n_{cr} / n_h)^{1/2} \tag{3.4.36}$$

In a quasi-neutral approximation $\delta = 0$. The maximum velocity of ions we present as [141]:

$$\mu_{\max} = 1 + \varepsilon_{\max} = 1 + \frac{m_e \mathrm{v}_{e,\max}^2}{2T_h}. \tag{3.4.37}$$

By expansion of the dimensionless function Ψ into a series, the coordinates of the asymptotic values of these functions are found to be as follows:

$$\Psi \quad 0.5(\xi-\xi_t)^2;\ \mu \quad 0.5(-\xi+\xi_t),\ \eta \quad (1/\ \ 2\pi)(-\xi+\xi_t)\,, \tag{3.4.38}$$

here coordinate ξ_t is defined from the requirement $\psi(\xi_t)=0$.

There is the next physical picture: fast electrons carry away ions behind themselves; the average electron moves together with ions, however precise electron motion represents an oscillation about ions in an electric field during a time like (3.4.36). In the one-dimensional case the electrons can not leave the ions, since the potential increases ad infinitum. We define $\mathcal{E}_{\max}$ from the requirement that the time of an electron returning from vacuum is about the laser pulse duration:

$$t \approx 2\mathrm{v}_{e,\max}(Lm/e\varphi_{\max}) \geq t_L \tag{3.4.39}$$

From here

$$\mathcal{E}_{\max} \approx \mathrm{v}_h t_L/2\sqrt{2} \approx (AmT_h/Zm_pT_c)(t_L/t_p)^2 \approx 10, \tag{3.4.40}$$

where t_p - duration of laser pre-pulse, producing an initial scale of plasma inhomogeneity. We can estimate in this case the ratio of ion velocity to speed of light as: $\mathrm{v}_i/c \approx 6\cdot 10^{-2}$. In a real situation the potential is restricted. We take for the estimation the potential of a charged disk of diameter $2r$ (about the size of a laser spot) on the surface of the plasma, then

$$\varphi = \frac{2q}{r^2}(\sqrt{r^2+z^2}-z); \quad q = e\pi r^2 l\cdot \mathrm{n}_{h,\max} \tag{3.4.41}$$

where

$$n_{h,\max} = \int_{\mathrm{v}_{\max}}^{\infty} \frac{dn_e}{d\mathrm{v}} d\mathrm{v} \tag{3.4.42}$$

and l is the distance which fast electrons move out from the target during the laser pulse. The magnitude of this charge is estimated, thus, from the assumption of quasi-neutrality. Then $\mathcal{E}_{\max} \approx \ln\psi \approx \ln(\omega_{p,h}{}^2 lr/\mathrm{v}_T{}^2)$. The calculation for $T_h \approx$ 100 keV, $n_{e,h} \approx 10^{-3} n_{cr} \approx 10^{-18}$ cm^{-3}, $r \approx 3\cdot 10^{-3}$ cm gives $\mathcal{E}_{\max} \approx 6$ or $\mathrm{v}_i/c \approx 3\cdot 10^{-2}$. Thus the one-dimensional model gives the same order of magnitude of ion velocity.

We consider now the estimations for very short pulses, when there is no time for quasi-neutrality to be reached: $t_L[\mathrm{ps}] \leq (10^{-3} n_{cr}/n_h)^{1/2}$. In this case [144]

$\mu_{\max} \approx 2(1+\ln(Z\omega_{pi}t_L))$

$$\eta_i = \begin{cases} 4\varepsilon_i^2 + (1/\varepsilon_i)\exp(-\varepsilon_i^2/2), & \xi < \xi_s \\ 0, & \xi > \xi_s \end{cases}. \quad (3.4.43)$$

$$\varepsilon_i = \exp(\xi - \xi_s); \quad \mu = 1 + \varepsilon_i + \xi$$

At distance, where the scale of plasma inhomogeneity is about the Debye screening distance, we assume that ion concentration is equal to zero. Then

$$r_D = c_s t_L;\ n_e = n_{0e}\exp(-\xi), \text{ from this } \xi_s = 2\ln(Z\omega_{pi}t_L) \quad (3.4.44)$$

$$\mu_{\max} \approx 2(1+\ln(Z\omega_{pi}t_L)) \approx 10$$

for ion velocity $v_i/c \approx 6\cdot10^{-2}$ again is obtained.

For plasma with a sharp boundary the value of velocity can be determined from the value of the ambipolar field

$$E_a \approx T_h/r_D. \quad (3.4.45)$$

Velocity of an ion then $v_i = \dfrac{Z}{M}E_a t_L$, and their energy

$$\varepsilon_{\max} \approx (mn_hT_h/2n_cm_i)(\omega_0 t_L)^2 \approx 10T_h, \quad (3.4.46)$$

at $n_h = 10^{-2}n_c$ and t_L = 1.5 ps. Thus, within an order of magnitude, the same ion velocities are derived from the different models. The connection between ion velocity, duration of laser pulse and pre-pulse is determined by the following: for short pre-pulse (the initial scale of inhomogeneity is less than the Debye screening distance of fast electrons) ion energy does not depend on pre-pulse duration, but for longer pre-pulse the ion energies less than that given in (3.4.40). For short $t_L[\text{ps}] \le (10^{-3}n_{cr}/n_h)^{1/2}$ basic pulses the energy of an ion is weakly (see (3.4.43)) dependant on pulse duration. For longer pulses the energy of an ion starts to increase with t_L^2. This increase over the above is restricted by the fact, that the energy and number of fast electrons are limited by the finite duration of the laser pulse. The pulse duration, for which there is saturation, is much more than 1 ps, and pulses of such duration are not considered.

We compare these scalings with the results of our numerical simulations **Fig. 3.4.17.** Figures 3.4.17, where plasma foil densities profiles are shown, illustrate that an intense laser pulse generates a shock wave with considerable of density 3-4 times that of foil density.

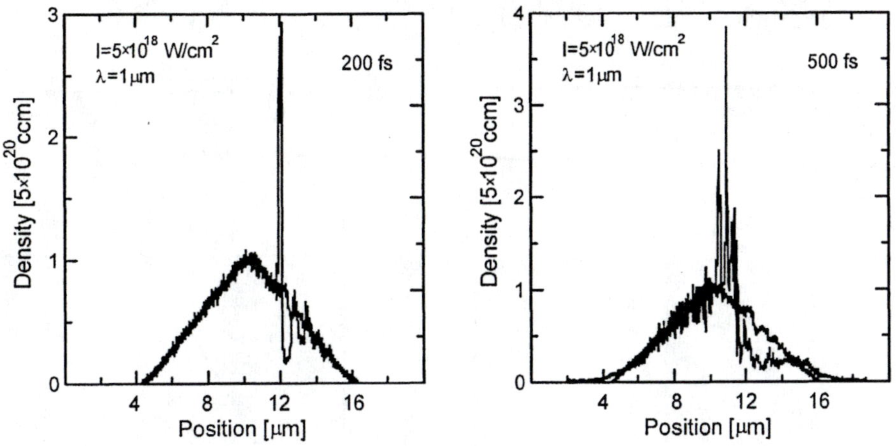

Fig.3.4.17. Calculated plasma foil density profiles at different moments of time (here initial density profile is black solid line).

From **Figure 3.4.18**, where are the ion velocity distributions, we see that the ions are accelerated by fast electrons on two surfaces of the thin foil, where the electrons leave this target and carry behind themselves ions. The acceleration happens also at a critical density $n = n_c$, where ponderomotive pressure forms the shock wave. Part of the ions take off into a vacuum in the opposite direction. The scaling from the given section are valid only for ions on the left-hand boundary, the opposite boundary we do not consider. From figures 3.4.17, 3.4.18 we see that the values of ion velocity are in the interval 0.02-0.04 c, that well coincides with our scaling.

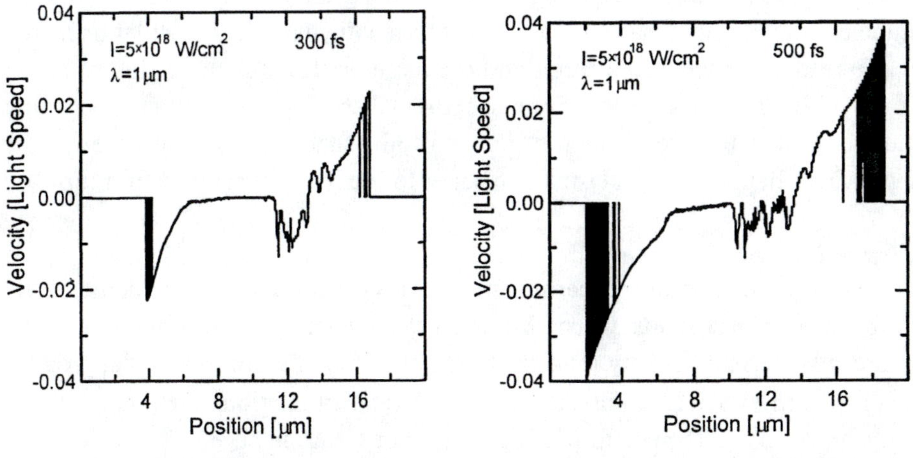

Fig. 3.4.18. The ion velocity distributions at different moments of time (here laser pulse of duration 1 ps and intensity $5 10^{18}$ W/cm^2 interacts with Al foil target).

It also follows from the graphs, that the maximum velocity of ions increases with time. So at 300 fs and 500 fs the velocity increases from 0.02 till 0.04, that approximately corresponds to quadratic dependence (3.4.34, 3.4.40) of ion energies from duration of laser pulse. We analyze also the dependence of ion energy from a scale of plasma inhomogeneity and now consider the results of simulations for plasma with a sharp density profile (such case is implemented without prepulse): **Fig. 3.4.19.**

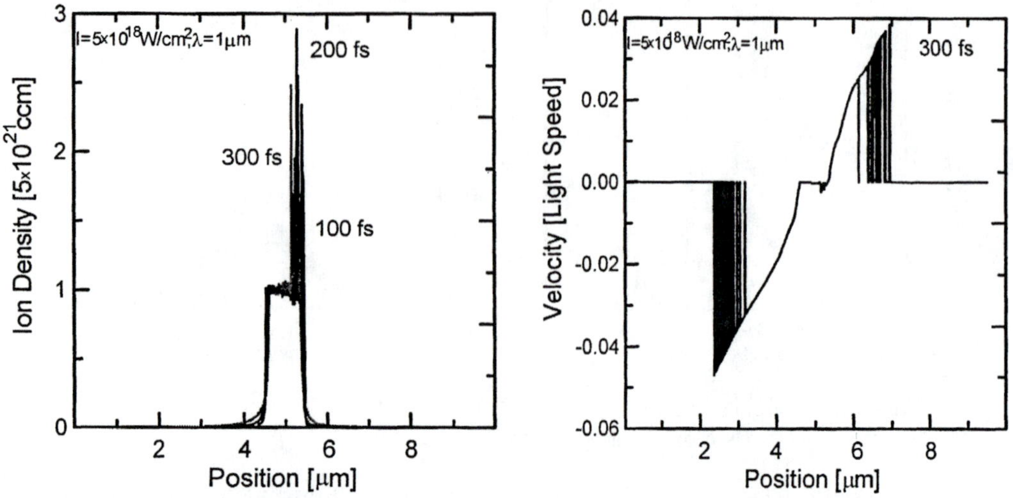

Fig. 3.4.19. Density and velocity distribution at the different time for the same laser pulse parameters and Al foil target with sharp boundary.

The graphs from Fig. 3.4.19 of ion velocity for time 300 fs show: that on the back side of the foil the sharp plasma profile provides approximately twice as high ion speed compared with the smooth profile case. The explanation is as follows: on a sharp boundary, fast electrons travel a smaller distance from ions, as an ambipolar field is stronger. During the laser pulse the electrons make more cycles of oscillation about ions, and the rate of acceleration thus is higher. The scaling under the formula (3.4.46) gives magnitude of velocity of ions $v_i \sim 0.05c$, this agrees with numerical calculations.

So, the reason for ion acceleration on both target surfaces is the action of ambipolar field from fast electrons. The given electron distribution function approximation accurately describes the acceleration of ions on foil boundaries, but for an exposition of ion shock wave shaping inside plasma it is better to use hydrodynamic approximation.

Hydrodynamic Approximation

The hydrodynamic approach permits us to analyze processes inside dense plasma. It is valid for intense laser fields, when plasma electron velocity is increased by a field, and greatly exceeds thermal velocity. The electron distribution function in this approximation can be considered as a δ-function. As in the numerical simulations there is the following inequality between the temperature of cold and hot electrons $T_c << T_h$, we can use the hydrodynamic approximation, in which $T_c = 0$. We show, that in hydrodynamics there is an additional mechanism of acceleration of ions when propagating a shock wave in a nonuniform medium.

We find the solution of the self-consistent set of equations by expansion on parameter v_T/v_E. In the zero order of approximation, we have instead of collision-less kinetics equation - hydrodynamic equations of motion of plasma electron component. The conservation law of transverse canonical momentum of electrons allows to reduce a system of hydrodynamic equations of electrons motion and Maxwells equations to two nonlinear partial equations for vector potential $\overline{A}(x;t)$ of electromagnetic wave and longitudinal electric field $E(x;t)=-\partial\varphi(x,t)/\partial x$ in plasma (see Sect. 3.1). Using the quasi-neutrality approximation, we have one equation for vector potential amplitude ($\vec{A}(x,t)=\vec{A}(x)e^{i\omega t}$):

$$\left(\frac{\partial^2}{\partial\xi^2}+\frac{\omega^2}{\omega_p{}^2}\right)a(\xi)=\left(\eta(\xi)+\frac{\partial^2}{\partial\xi^2}\sqrt{1+a^2}\right)\frac{a}{\sqrt{1+a^2}} \tag{3.4.47}$$

and one equation, connecting longitudinal electric field with vector potential:

$$E=\frac{\partial}{\partial\xi}\sqrt{1+a^2}+\delta_T{}^2\frac{\partial}{\partial\xi}\ln\eta(\xi). \tag{3.4.48}$$

In a system (3.4.47-48) the following dimensionless variables are used:

$\xi=\frac{\omega_p}{c}x$; $\vec{a}=\frac{e\vec{A}}{mc^2}$; $E=\frac{eE_x}{mc\omega_p}$; $\delta_T=(T_e/mc^2)^{1/2}$; $\eta_i(\xi)=\frac{n_i(\xi)}{n_i(\xi=\infty)}$. Now we return to system (3.4.42), by adding in the equations viscosity and by calculating an field E from (3.4.48). For the analysis of a strong shock wave parameters in plasma it is convenient to consider equations in the following dimensionless variables: instead of velocity v_i we again enter a Mach number $\mu=v_i/\sqrt{ZT_{ec}/M}$, and ion time τ_i we measure in terms of $\frac{c}{\omega_p}\sqrt{\frac{M}{ZT_{ec}}}$, i.e. time an ion sound takes to transit a skin layer. Then the dimensionless hydrodynamical equations for ions have the form:

$$\begin{gathered}\frac{\partial\eta_i}{\partial\tau_i}+\frac{\partial}{\partial\xi}(\eta_i\mu)=0\\ \frac{\partial\mu}{\partial\tau_i}+\mu\frac{\partial}{\partial\xi}\mu+\frac{\partial}{\partial\xi}\ln\eta_i(\xi)-\frac{1}{\eta_i(\xi)}\frac{\partial}{\partial\xi}\tilde{\tau}_i\frac{\partial}{\partial\xi}\mu=\frac{1}{\delta_T^2}\frac{\partial\sqrt{1+a^2}}{\partial\xi}\end{gathered}, \tag{3.4.49}$$

where the dimensionless viscosity $\tilde{\tau}_i$ depends only on temperature of ions:

$$\tilde{\tau}_i=\frac{64}{\Lambda_c}\frac{Z^2T_e{}^2\omega_p}{e^4n_{0i}c}\left(\frac{T_i}{ZT_e}\right)^{5/2}, \tag{3.4.50}$$

and Λ_c – Qulomb logarithm.

The system (3.4.49) is non-complete, and it is necessary to add an equation for temperature. We consider the process of shock wave propagation as adiabatic:

$$\frac{\partial}{\partial \tau_i}\left(\frac{T_i}{ZT_e}\right)\eta_i^{1-\gamma} + \mu\frac{\partial}{\partial \xi}\left(\frac{T_i}{ZT_e}\right)\eta_i^{1-\gamma} = 0. \tag{3.4.51}$$

The adiabatic exponent γ we take as equal 5/3, since the equation of plasma state differs a little from a ideal gas state.

If a laser pulse is shorter than $l_s/(ZT_e/M)^{1/2}$, then laser field pressure is equivalent to instantaneous shock, and the right side of the second equation of (3.4.49) is equivalent to the boundary conditions of the homogeneous system (3.4.49). The self-similar solution of the problem of strong shock is known [100]. From the second equation (3.4.49) the velocity of a fast ion is

$$\frac{\mathrm{v}_i}{c} = \frac{u_{\mathrm{w}}}{c} = \sqrt{\frac{Zmn_{cr}(\sqrt{1+a^2}-1)}{Mn_e}}, \tag{3.4.52}$$

that coexists with [101]. The obtained formula gives the velocity of fast ions in the neighbourhood of a skin layer, i.e. behind the shock wave front. It is known that for a strong shock wave, the velocity of front and velocity of particles behind front connected by the factor 4/3, if an adiabatic exponent γ = 5/3. Then a velocity of front we estimate as 4/3 u_{w}, $\mu = \frac{4}{3}\mathrm{v}_i\sqrt{\frac{M}{ZT_e}}$.

Thus, instead of a solution (3.4.49, 3.4.51) in a medium with a given density profile, we divided the problem into two parts: first we found the parameters of a shock wave in the region of a skin layer, where it is shaped, and then we considered the exit of given shock wave in the region of low density (rear side of foil). In this region in (3.4.49) there is not a laser pulse force and it is possible to neglect viscosity.

Let's select the next law of plasma density dependence: $n_i(x) = bx^{\delta} = n_i(x_1)\left(\frac{x}{x_1}\right)^{\delta}$.

Here x_1 is the distance at which the ion density starts to decrease. A density in this point is $n_i(x_1)$. At x=0 the density equals zero. Thus the shock wave goes from the position x_1 and at the position x=0 goes out the rear side of the foil target. To get the solution of a homogeneous system (3.4.49) in this region we transfer to the self-similar variable $\chi = x/C(-t)^{\alpha} = x/X(t)$. Details of solution of equations (3.4.49) are described in [100]. For our purposes, we must know from [100] only one parameter α= 0.6 and after that, we get the velocity of a shock wave: $\dot{x} \sim |t|^{-(1-\alpha)} \sim X^{-(1-\alpha)/\alpha}$. On the surface at x = 0, velocity is going to infinity. Energy (per unit mass) $\varepsilon \sim \dot{x}^2 \sim X^{-2(1-\alpha)/\alpha}$, pressure

$\wp \sim Mn_i\dot{x}^2 \sim X^{\delta-2(1-\alpha)/\alpha}$. Thus the law of energy gain when a shock wave transits from a point x_1 to a point x is as follows:

$$\mathcal{E}(x) = \mathcal{E}(x_1)\left(\frac{x}{x_1}\right)^{-2(1-\alpha)/\alpha}. \tag{3.4.53}$$

Further it is necessary to find a limit point x, where the ion (thermal ion before front) free path exceeds the width of plasma stratum, and plasma density becomes zero. After this point the hydrodynamic approximation is incorrect. The ion free path depends on its energy and density of plasma:

$$\Lambda_T(z) = \frac{12\pi^{3/2}T_i^2}{Z^4 e^4 n_i(z)\Lambda_c}. \tag{3.4.54}$$

We find this point x from the equation $x = \Lambda_T(x)$:

$$\frac{12\pi^{3/2}T_i^2}{Z^2 e^4 n_i(x_1)\Lambda_c x_1}\left(\frac{x}{x_1}\right)^{-\delta} = \left(\frac{x}{x_1}\right) \tag{3.4.55}$$

hence

$$\left(\frac{x}{x_1}\right) = \left(\frac{12\pi^{3/2}T_i^2}{Z^4 e^4 n_i(x_1)\Lambda_c x_1}\right)^{\frac{1}{1+\delta}} \tag{3.4.56}$$

Then ion's energy behind the front is:

$$\mathcal{E}(x) = \mathcal{E}(x_1)\left(\frac{12\pi^{3/2}T_i^2}{Z^4 e^4 n_i(x_1)\Lambda_c x_1}\right)^{\frac{-2(1-\alpha)}{\alpha(1+\delta)}} \tag{3.4.57}$$

In this formula $\mathcal{E}(x_1)$ - energy of an ion behind the front, which corresponds to velocity (3.4.52), $n_i(x_1)$ - ion concentration inside plasma.

Coordinate x_1 is necessary to find the law of density dependence

$$bx_1^{\ \delta} = n(x_1) \tag{3.4.58}$$

Now the formula of energy gain can be presented as

$$\mathcal{E}(x)=\mathcal{E}(x_1)\left(\frac{x_1}{\Lambda_T(x_1)}\right)^{\frac{2(1-\alpha)}{\alpha(1+\delta)}} \tag{3.4.59}$$

In depth of plasma $\Lambda_T < x_1$, $\alpha < 1$ and the energy really grows.

From this formula we see that the energy gain is the ratio of the hydrodynamic scale to free path and varies as an exponent whose power is determined by the self-similar index α. The strong density dependence on coordinates reduces the rate of energy gain, for $\delta >> 1$, but at $\delta < 1$, i.e. for smoothly varying medium, the energy gain is much better, though it is necessary to understand that these two parameters not absolutely independent and $\alpha = \alpha(\delta)$. In the given model, the free path of a thermal ion in plasma is assumed as small in comparison with any hydrodynamic lengths.

Now we fulfil the numerical estimations. First we estimate free path of thermal ion, by expressing it in terms of laser wavelength:

$$\frac{\Lambda\omega}{c}\sim\frac{\omega}{Z^{*4}e^4 n_i L_c c}(T_i)^2=\left(\frac{1}{Z^{*3}\Lambda_c}\right)\left(\frac{\omega}{\omega_{pe}}\right)^2\left(\frac{c}{\omega r_0}\right)\left(\frac{T_i}{mc^2}\right)^2, \tag{3.4.60}$$

where r_0 - classical radius of an electron. We choose $Z^* = 10$, $n_e = 10n_{cr}$, then the free path is less than a wavelength at $T_i < 1.2$ keV.

The width of shock wave front is close to ion free path length because the viscosity of ions: $\tilde{\tau}_i = n_i T_i \tau_i$ and typical scale of viscosity $l_\eta \propto \tau_i T_i / m_i \upsilon_{Ti} \approx \Lambda_T$, from it width of front and ion free path, thus, are within one order of magnitude. At $\alpha = 0.6$, $\delta = 1$:

$$\mathcal{E}(x)=\mathcal{E}(x_1)\left(\frac{x_1}{\Lambda(x_1)}\right)^{\frac{2(1-\alpha)}{\alpha(1+\delta)}}=\mathcal{E}(x_1)\left(\frac{x_1}{\Lambda(x_1)}\right)^{2/3} \tag{3.4.61}$$

From this formula we see that a free path ~ 0.1λ is necessary for raising ion energy for plasma width 3λ ($x_1 = 3$) and temperature of ions for this purpose should be less than 450 eV.

So parameters of plasma for ion acceleration can be as follows: $n_e = 10n_{cr}$, scale of plasma density $x_1 = 3\lambda$, $T_i = 450$ eV. If we want to increase temperature k times more, scale should raise k^2 times. After laser pre-pulse plasma temperature decreases and scale of inhomogeneity grows - so this case is realizable.

Let's compare these results with PIC simulations. In the previous figures of numerical simulations, the shock wave propagates in the region of increasing density. Now we depict a shock wave in the plasma region with decreasing density.

From **Figure 3.4.20**, we can conclude, that the shock front in plasma with decreasing density breaks up on some fronts.

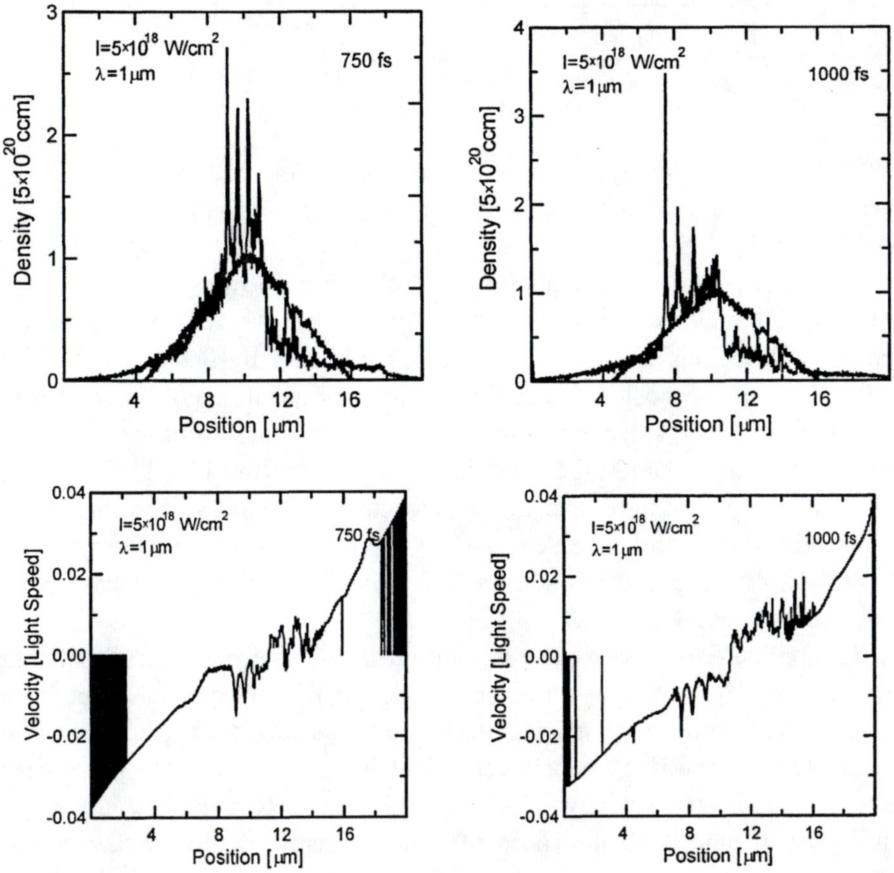

Fig. 3.4.20. Decay of the shock front in plasma with decreasing density and a formation of solitons.

For normalized laser intensity $a^2 < \delta_T^2$ and pulse duration $t_L < c/c_s\omega_p$ we can transform the system (3.4.49) to KdV equation for density perturbation $\delta\eta$ at initial condition (see Sect. 3.2):

$$\delta\eta\Big|_{\tilde{\tau}=0} = \frac{c^2}{4v_{Te}^2}\frac{\partial a^2}{\partial\xi}\tau_L = f(\xi)$$

According to [84] the number of solitons determined by the following formula:

$$N = \frac{1}{2\pi}\int_0^\infty \sqrt{\frac{c^2}{4v_{Te}^2}\frac{\partial a^2}{\partial\xi}\tau_L}\,d\xi \quad N \approx \frac{1}{2\pi}\sqrt{\frac{c^2}{2v_{Te}^2}a^2\tau_L}$$

At $a^2 >> \delta_T^2$:

$$N \approx \frac{1}{2\pi}\sqrt{\frac{\varepsilon_L}{\varepsilon_s}\frac{u_w n_e}{c n_{cr}}\frac{2(\sqrt{1+a^2}-1)}{a^2}}, \qquad \varepsilon_s = \frac{n_e m_i u_w^{\ 2} l_s}{Z},$$

where ε_L – laser pulse energy, ε_s – energy of one soliton with dimension near skin depth(per unit of area). Put in this formula the data of Fig.3.4.20 we obtain $N = 4$, and it coexists with our simulation result.

We compare the data of acceleration of a shock wave following from Figures 3.4.20 to the result of (3.4.61). According to this law the shock wave velocity, on propagation from the point 9 microns to the point 8 microns increases by $[(9\text{-}4)/(8\text{-}4)]^{(1-\alpha)/\alpha} = 1.2$ times. (4 microns is the position, where the density is equals zero). Refering on Fig. 3.4.20, from the graph of velocities in the vicinity of the shock front, we see the visual correspondence to this value. In the numerical calculations from Fig. 3.4.20 we see that the ion free path is ~ 1 micron and the scalelength is x_1 ~ 6мкм therefore, analytical coefficient of the ion velocity increase is ~ $6^{0.6}$ ~ 3 times as in the simulations.

The process of ion acceleration from foil target by a laser pulse has shown the following futures.

The light pressure of a laser pulse is transmited to ions by means of electrons and ambipolar fields. A shock wave is formed in the vicinity of a critical density point (where the pressure is maximum) and propagates deep into the plasma. The hydrodynamic model gives adequate (confirmed by numerical calculations) exposure of this process. A characteristic shock wave velocity in Al target for intensity 5 10^{18} W/cm^2 and pulse duration 1.5 ps is about 0.02 the speed of light. Intensity of the shockwave and ion acceleration have a maximum in plasma with a scale of density gradient about a wavelength, as there are the maximum of absorption, amount of fast electrons and their energy in this case.

A strong shock front disintegrates on several fronts, causing an order of decrease of their intensity when driving in inhomogeneous plasma from one surface of the foil to other.

Ion velocity increases as plasma density decreases. It produces acceleration of ions, and magnification of their energy proportionaly to the ratio of plasma density gradient to the length of ion free path. Gradient of rear side plasma density of the target should be smooth to realize this effect.

Culomb mechanism of electron-ion acceleration from two surface, in contrary to that from a shockwave, is most effective in plasma with a sharp boundary and exceeds the acceleration of ions by a shockwave. At short initial scale of plasma inhomogeneity ion energy does not depend on this scale. For longer scale, ion energy decrease is inversely proportionally to square of plasma gradient. For short basic pulses the energy of an ion is weak (logarithmic) depending on pulse duration. For longer pulse, the energy of an ion begins to increase proportionally with the square of pulse duration down to a condition of saturation, at which energy flows of fast electrons and accelerated ions are comparable.

From the point of view of maximum ion acceleration, the optimal target should have a few micron thickness with smooth density gradient (scale of density inhomogenity about a wavelength) at rear and sharp density gradient front.

3.4.5 Analysis of the Experimental Data

The plasma in the experiment [145] had sharp gradient, $L < \lambda_0$, because of the small laser pre-pulse. Laser field amplitude decreased by ω_p / ω_0 times inside the plasma with a sharp gradient. For this reason, for absorption coefficient $\eta = \int_0^{\tau_i} jE\, dV\, dt \,/\, \mathcal{E}_L$ we can use the formula for collision-less absorption in SIB regime (see Sect. 3.1) to estimate η:

$$\eta \quad \eta_{SIB} = \frac{4}{\sqrt{\pi}} \left(\frac{\upsilon_T}{c} \right)^3 \frac{\omega_{pe}^2}{\omega_0^2}. \tag{3.4.62}$$

This formula is accurate for $\upsilon_T \omega_p / c\omega_0 \leq 1$ (here ~ 1), and shows that $\eta \sim Z$.

At very high laser intensity $\eta \sim$ (0.3 - 0.4) [133,146] and week depends on θ and I. At I 10^{18} W/cm^2, electron velocity in the skin layer is close to the speed of light, so electron current density $\vec{j} = en_e \vec{\upsilon}_e$, in η, is higher for materials where electron density is a maximum of one. **Fig. 3.4.21** shows that absorption for Sn is approximately 1.6 times more than for Al for laser intensity $1 \cdot 10^{18}$-$5 \cdot 10^{18}$W/cm^2.

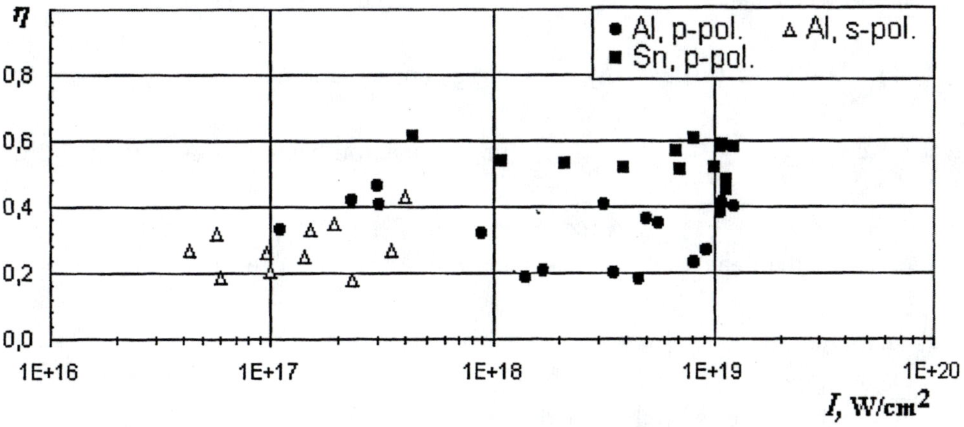

Fig. 3.4.21. Experimental data of absorption coefficient dependence from laser intensity for different laser radiation polarization and target materials.

We estimate effective ion charge in these materials taking into account that plasma temperature is some keV under these conditions [147]. In this case, for Al we have full ionisation and for Sn $Z \approx 25$, as $Z \approx 2/3(AT_e)^{1/3}$ (see. Sect.3.1). So, for $I \leq 5 \cdot 10^{18}$ W/cm^2 Sn gives ~2 times more absorption coefficient for these conditions, compared with Al.

Let's analyse the electron distribution function from the X-ray spectrum [145]. The bremsstrahlung intensity is determined by the following formula:

$$\frac{d\mathcal{E}_X}{d\mathcal{E}_\gamma} = \frac{8\sqrt{2}e^6 Z^2 n_i n_{eh}}{3\pi\sqrt{\pi m_e T_e} m_e c^3 \hbar} \exp\left(\mathcal{E}_\gamma / T_e\right) \tag{3.4.63}$$

From this formula and experimental data we estimate density of fast electrons as n_{eh}~10^{18} см$^{-3}$, and total number of electrons during laser pulse - 10^{11}. Fast electron temperature is $T_h \approx m_e c^2 \left(\sqrt{1+(2-\eta)I_{18}} - 1\right)$. There is another group of electrons with much lower temperature [145], and the explanation of this group is connected with return plasma current, to compensating for the current of the fast electrons [148]. Without this compensation, fast electron current exceeds the limit of Alfven current.

In our experiment most of the fast electrons are accelerated by pondermotive force inside the plasma, but the direction of the fast electrons, escaping from the boundary of the over-dense plasma, corresponds to the specular reflected light direction, as we discussed in Sect.3.4.1.

Photons give a momentum $n_p \hbar k_0$ and energy $n_p \hbar \omega_0$ to electrons escaping into vacuum and ions moving inside the target in the direction normal to a plasma surface. From the laws of conservation we obtain the following equations:

$X\gamma\Lambda\sin\theta_e \approx (1\text{-}R)\sin\theta$ - transverse momentum conservation;

$X(\gamma\text{-}1) \approx (1\text{-}R)\text{-}\beta^2 Z\gamma(m/M)/(1\text{-}R)$ - energy conservation;

$R + X\gamma\Lambda\cos\theta_e + \cos\theta \approx \beta$ - longitudinal momentum conservation,

where $X = n_e/n_p$, $R = 1\text{-}\eta$, γ - Lorentz factor, $\beta = (X/Z)(P_i/\hbar k_0)$, $\Lambda = m_e c^2/\hbar\omega_0 >> 1$.

Than the angle of electron escape to vacuum is close to the specular one:

$\theta_e \approx \arcsin[(1 - 1/\gamma)\sin\theta] \approx \theta$,

at $\gamma >> 1$, and this agrees with the experimental data.

Fast electron energy in laser field:

$$\mathcal{E}_e = m_e c^2(\gamma - 1) \approx (p_0/mc)(\upsilon_E/c)^2, \tag{3.4.64}$$

where $\upsilon_E = eE/m\omega$, and initial momentum along laser beam $p_0 \approx m\upsilon_E$, because it determined by acceleration in laser field over a distance $\lambda=2\pi c/\omega$. From here for $I \approx 10^{19}$ W/cm^2 we obtain $\mathcal{E}_e \approx 10$ MeV, and this agrees with experimental data.

During the laser pulse duration the relativistic electrons (at laser intensity $I \approx 10^{19}$W/cm^2) fly out from a target to distance greater than laser spot size on the target. As a result, the boundary area of the target obtains a positive charge and an ions are accelerated into the target and to the vacuum by the electrostatic field (see **Fig. 3.4.22**).

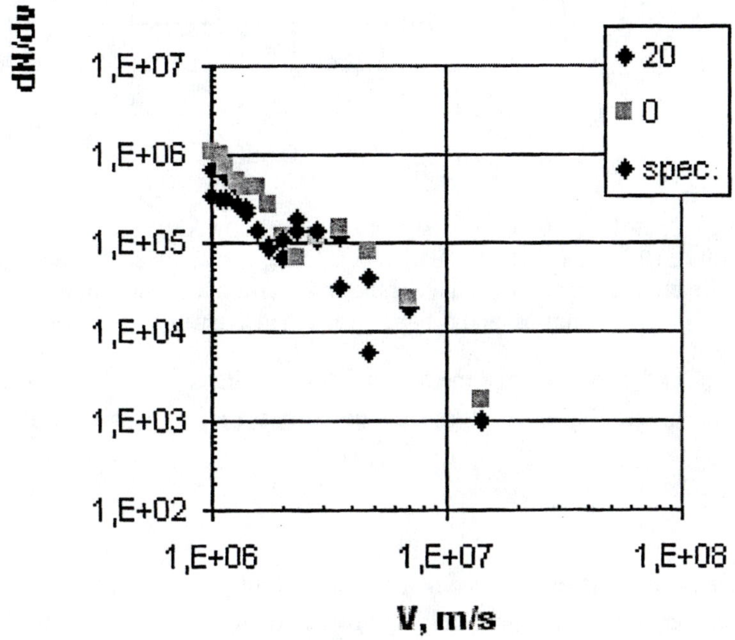

Fig. 3.4.22. Experimental data of the dependence of ion distribution function from ion velocity for the different angles of observation.

We analyse this process by simulations with help of kinetic relativistic code PM2D [149]. To check the simulation results for ion acceleration into target we also calculated the parameters for ions flying out into vacuum and compared these results with experimental results because in our experiments only this part of ions has been recorded. **Fig. 3.4.23** shows spatial distribution of energy flux density of ions flying out into vacuum for laser intensity 10^{18} W/cm^2. In simulations as in experiments, ions fly out at an angle diagram ~$15^{\circ}$to normal with target surface. The difference between theoretical and experimental results is connected with collision-less simulation model.

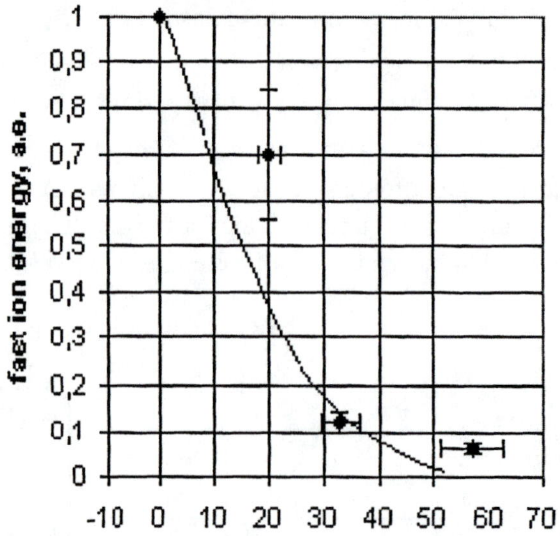

Fig. 3.4.23. Spatial distribution of fast ions escaping from Al target to vacuum at laser intensity $5 10^{18}$ W/cm^2. The simulation data (solid curve) are compared with the experiment (circles). The ordinate is the emission angle relative to the surface normal and the abscissa is the ion energy.

Let's compare the ion average energy and the number of fast ions. From momentum conservation we obtain the momentum flow equality for ions and electrons:

$$n_i m_i \upsilon_i^2 = \cos^2\theta_e n_e \mathcal{E}_e \upsilon_e^2 / c^2,$$

here $n_{e,i}$ – density of ions and electrons (in transparent plasma with scale L, $n_e \approx n_{cr}/Z$). Fast electron energy in the fields of incident and reflected laser pulses: $\mathcal{E}_e = m_e c^2 \sqrt{1+(2-\eta)I_{18}}$, then:

$$\mathcal{E}_i = \cos^2\theta_e m_e c^2 Z \frac{n_e}{n_{cr}} \frac{(2-\eta)I_{18}}{\sqrt{1+(2-\eta)I_{18}}}. \tag{3.4.65}$$

At $I_{18}>1$ from (3.4.65) we get $\mathcal{E}_i \sim I_{18}^{0.5}$, and this correlate with experimental dependence (see Fig.3.4.23) $\mathcal{E}_i \sim I_{18}^{0.46}$. To take Zn_e/n_{cr} in under-dense plasma layer we used X-ray data, and we let $n_e = n_{eh}$, then $Zn_e/n_{cr} \sim 0.1$. For η=0.4, in **Fig. 3.4.24** it is clear that theoretical graph is close to experimental one.

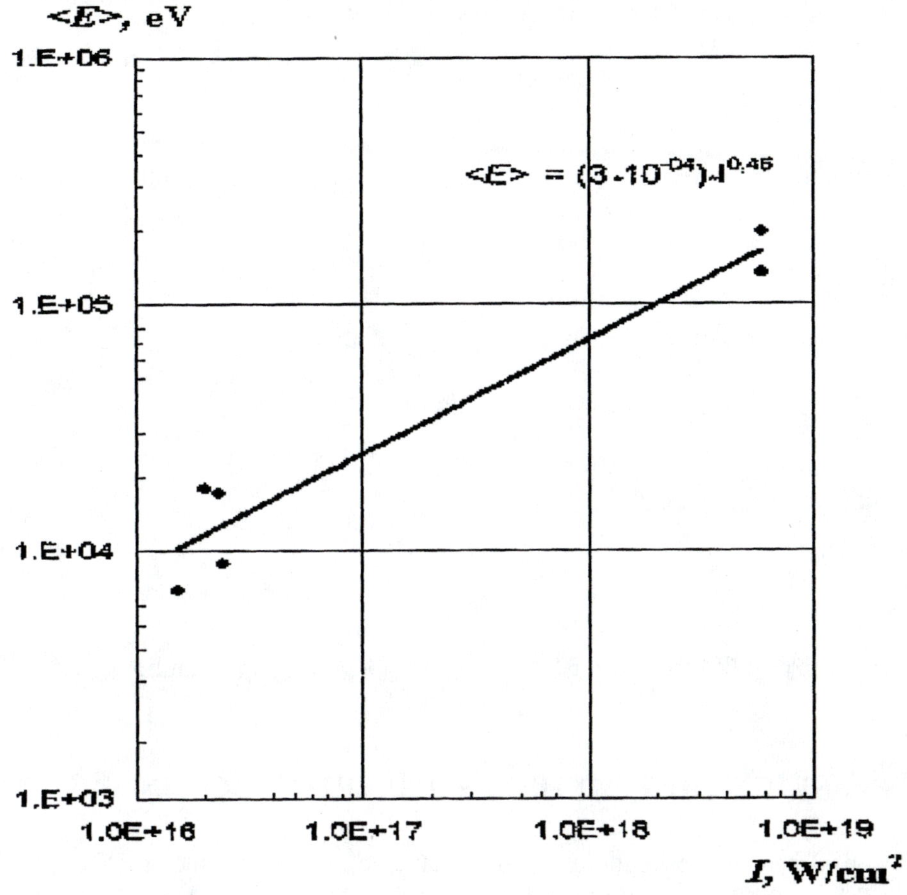

Fig. 3.4.24. Proton energy dependence from laser intensity. The simulation data (solid curve) are compared with the experiment (circles).

From simulation data in Fig. 3.4.24 we can conclude that fast ion energy equals $1.2 \cdot 10^{16}$ eV at laser intensity 10^{18} W/cm^2. In experiment from Figs. 3.4.24 and 3.4.23 we obtain the experimental result $7 \cdot 10^{15}$ eV and this is close to our estimations.

The main part of fast ions there were protons and carbon ions. The reason for this is water vapour on the target surface and polishing procedure, so if we exclude H and C on the surface target we would get Al ions with energy $\mathcal{E}_i \geq 1$ MeV at $Z^* \geq 6$. The number of such ions of intensity $5 10^{18}$ W/cm^2 is of the order of $\sim 10^7$ for Al target, and conversion coefficient into fast ion energy is near ~1%.

According to our simulations, approximately the same (in angle distribution) ion beam propagates inside dense plasma from target boundary area with volume $V = \pi(d/2)^2 L$. This movement is connected with electrostatic potential from charge separation

$$e\varphi = \mathcal{E}_{eh}$$

and this separation is connected with electrons of energy

$$\mathcal{E}_{eh} \approx \eta I / c n_{eh}$$

leaving from this target area. Suppose that fast ion takes energy equal to this potential

$$Ze\varphi \approx \mathcal{E}_{eh}$$

and that fast electron energy exceeds electrostatic potential

$$\mathcal{E}_{eh} \geq e^2 n_{eh}\, \pi L d^2/(4L^2+d^2)^{1/2}$$

we obtain the following formula for ion energy:

$$\mathcal{E}_{ih} \approx 3Z\, mc^2(\eta I_{18}\, dL/\lambda_0^{\,2})^{1/2}$$

This agrees with (3.4.65) and at $(d/\lambda_0)^{1/2} \approx n_e/n_{cr}$ we obtain the same result.

3.5 "Fast Ignition" Method for Inertial Confine Fusion

We consider now the conditions which have to be met in the method of "Fast ignition" (FI) of the DT fuel [26] for Inertial Confine Fusion problem. Thermonuclear fuel is prepared in a shell DT target, i.e. conditions close to but insufficient for fusion reaction are established then a high-energy nanosecond laser pulse is applied [150]. Then a high-power laser pulse of very high intensity is applied to the target so as to ignite DT in a small part of the target. The energy density in the plasma formed in the part of the DT fuel ignited (i.e. the conditions needed for the thermonuclear reaction to proceed) governs the energy which the laser ignition pulse must supply. The expansion time of this region determines the laser pulse duration, since it should be sufficiently long to transfer energy to the plasma. If successful, the fast ignition scheme drastically reduces the difficulty of compression and can lead to high gain with relatively small energy of implosion driver.

3.5.1 Fast Ignition of ICF Target by Hot Electrons

The energy is transferred from the laser radiation to supra-thermal electrons, then to thermal electrons, and finally it is converted into the kinetic energy of the expanding fuel. Each of these stages has its own characteristic time: $t_{\gamma e}$ – is the time for the transfer of the laser energy to electrons, which is the same order of magnitude as one laser radiation period; t_{ee}, - is the time for transfer of energy from the fast to the thermal electrons; t_{ei} – is the electron – ion relaxation time.

The duration of the laser ignition pulse t_i should exceed t_{ei}, but should be less than hydrodynamic expansion time of the hot region $t_s=R_s/C_s$ [R_S is the size of the hot region and $C_s=(T_f/m_i)^{1/2}$ – is the velocity of sound] in order to ensure that the target does not disperse before ions are heated to the fusion temperature T_f.

The task is to establish, at the moment of collapse of the shell target, such conditions in the compressed DT fuel that α-particles maintain the reaction. This can be ensured if $\rho_k R_k \geq 0.3$ g/cm^2 (ρ_k is the density of the DT target at the moment of collapse and R_k is the radius of the target at this moment) and if the temperature at which the collapse takes place is $T_k \geq T_f = 10$ keV [151,152]. We shall assume that the first condition is satisfied throughout the region occupied by the fuel $V_k = 4\pi R_k^3/3$ and the second is obeyed in a small region of size R_s. Then, when the rest of the fuel enters the reaction, the energy gain is by a factor $(R_k/R_s)^3$ (such a large gain applies only in relation to the energy of the ignition pulse, but since it is less than the total energy of the laser pulse used to compress, the net gain is naturally less). It is assumed that at the moment of collapse t_k the shell is practically burnt through. The condition for this is $\Delta R=C_T t_L$ (where ΔR is the thickness of the shell whose density is ρ_p , C_T– is the rate at which the shell is burnt, and t_L – is the duration of the compression pulse). This condition is quite difficult to satisfy for a simple gas-filled target, since we then have $\rho_k \cdot R_k=3\rho_p\Delta R$. Let us therefore consider a "two-stage" target and a profiled pulse, which makes it possible to reduce the shell thickness by the factor P_m/P_0 (R_0/R_1) [152]. Here P_m/P_0–is the ratio of the maximum laser power for what is known as a Teller pulse [152], R_1 is the initial radius of the DT fuel and R_0 – is the shell radius. If we select $P_m/P_0=10^2$ and $R_0/R_1=3$, which are realistic values, we find that $\Delta R=1$ μm for the shell burn-through condition. Let us assume that the degree of compression does not exceed 10^3, so that the hydrodynamic instabilities can be ignored. Then, if the initial density is $\rho_0=0.3$ g/cm^3 (frozen DT), we find that $\rho_k=300$ g/cm^3 [26]. The absorbed power, needed to heat the fuel in such a two-stage target to a given temperature, is $P_n \sim T_k^{3/2}$ [152]. Therefore, if we select $T_k=3$ keV, then the laser energy needed to compress the target can be reduced by a factor of 5 compare with the usual ignition case ($T_k=T_f=10$ keV) when the required laser energy is $E_L \approx 1$ MJ.

It follows from the condition $\rho_k R_k \geq 0.3$ g/cm^2 with $\rho_\kappa= 300$ g/cm^2 that $R_k=10$ μm. Consequently, the initial radius is $R_1=100$ μm and R_0 may be equal to 300 μm. The total mass of DT is $M_f=10^{-6}$ g and the total energy released by "combustion' of this mass in the thermonuclear gain exceeds unity if fuel ignition is performed by a short laser pulse (solid curve in **Fig. 3.5.1** [26]).

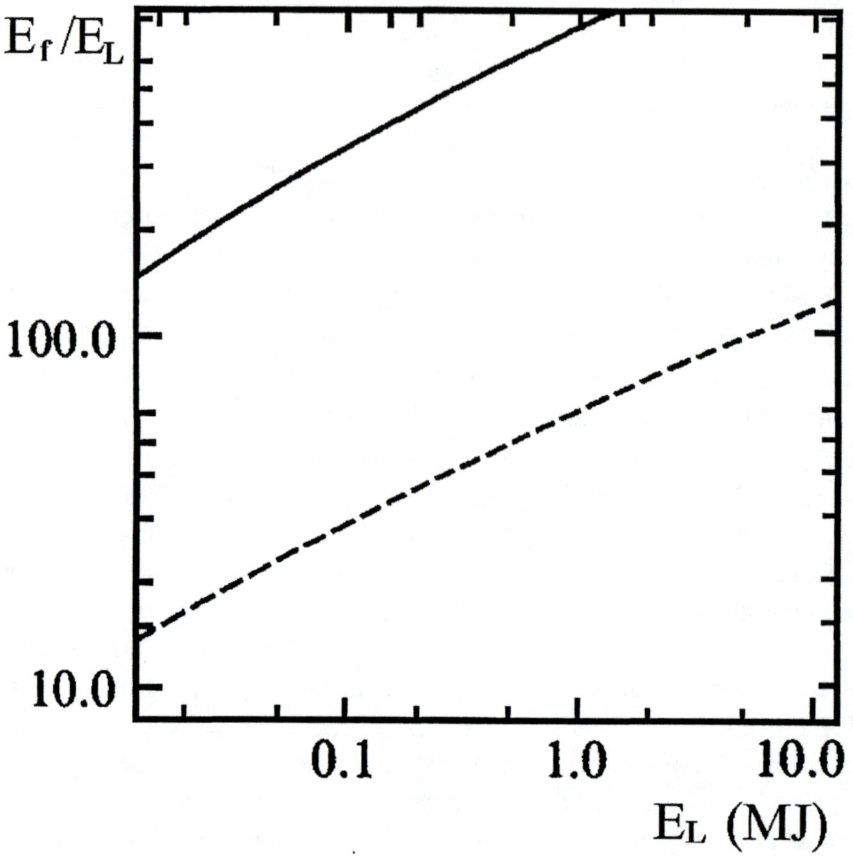

Fig. 3.5.1. Dependence of the thermonuclear energy gain G on the laser energy E_L plotted for the long (dashed curve) and short (solid curve) laser pulse.

Since the velocity of sound corresponding to 3 keV is $C_s \sim 10^8$ cm/s, the expansion time of the compressed fuel is $t_s \approx 10$ ps. Consequently, the duration of the additional pulse t_s which need to be applied to the target should be less. For this reason, a short laser pulse of duration $t_i < t_s$ of energy E_s and intensity $I_s = E_s/t_s \pi R_s^2$ is applied at the moment of collapse in the "fast ignition" method. This short pulse reaches the compressed fuel if the optical pressure of the pulse is equal to the pressure of the DT fuel at the moment of collapse, and the pressure in the plasma corona into which the shell is converted is somewhat less, which is true if $I_p/c = r_k T_k$. We can also apply a different laser pulse which satisfies these conditions for $E_p \approx 10\, E_s$, $t_p \approx 100\, t_s$, $R_p = (E_p/I_p t_p \pi) < R_s$.

Since the laser pulse pushes a plasma along the path formed by a pulse, there may be a considerable degree of conversion to plasma waves and then to hot electrons (see previonse section) [153]. The temperature of the hot electrons is $T_h \sim 1$ MeV and, for the laser intensity under consideration, these electrons should release their energy in a small (compare with R_k) region of radius R_s: $R_s \sim l_h \sim T_h^2/\rho_k$ (under our conditions, $l_h \cong 2.5$ μm), so that the temperature $T_s \sim E_{hl}/M_s$ in this region containing a mass $M_s = 4/3 \cdot \pi R_s^3 \rho_k$

becomes equal to the fusion temperature T_f=10 keV. Hence, for R_p= 2 μm, we find that $R_p=R_s$ and, consequently, $M_s \approx 0{,}03 \cdot M_f$. Since in this region the fuel ions are heated by the laser pulse to 10 keV, the deposited energy is $E_s \approx E_{hl} \approx 10^8 \rho^{-2} \approx 4$ kJ and the intensity of the short laser pulse should be $I_s \approx 10^{21}$ W/cm^2. Ignition in this small region is followed by a spread of the reaction to the whole fuel in the pre-compressed target. We can therefore expect a considerable energy gain (by a factor at least 5), compared with the usual compression, because the energy of ultra-short laser pulse is low compared with the energy needed to condition the target.

It follows from the above discussion that the laser power needed for "fast ignater" is governing by the efficiency of energy transfer and by require ignition energy. Therefore, one of the key features of the method is the need for a strong absorption of the laser radiation and conversion of the energy of the short laser pulse into the energy of fast particles. We analysed this question in Sec.3.1.

Fig. 3.5.2 gives the dependences of the fraction of the absorbed laser radiation on laser intensity obtained in the cited experimental and theoretical investigations. For a given value of intensity, the absorption varies in a fairly wide range and it depends on the maximum plasma density, the characteristic length of the plasma, and the polarisation and angle of incidence of the laser pulse on the target. An analysis of the experimental and calculated results shows that a considerable fraction of the laser energy absorbed by resonant and non-linear mechanisms is acquired by electrons and is then used to accelerate ions.

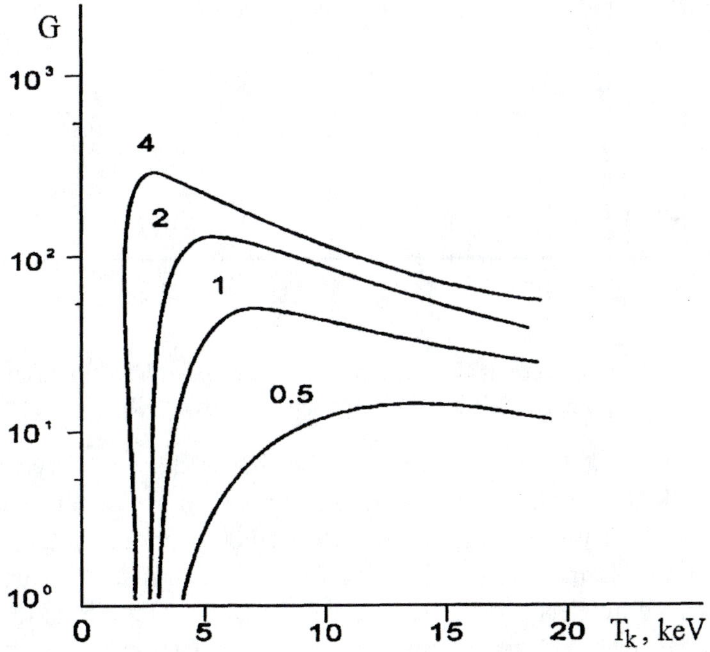

Fig. 3.5.2. The energy gain of homogeneous target as a function of temperature T_k for different ρR_k. The curves are marked with corresponding values of ρR_k in g/cm^2.

Fig. 3.5.3 gives the dependence of the temperature of such fast electrons on intensity. The results were obtained by numerical simulations and in experiments (see for example [42]). The results show that the laser radiation of intensity in excess 10^{19} W/cm^2 can generate effective electrons of energy reaching 1 MeV for which the mean free path is comparable with the mean free path of the α- particles. Multiple scattering and generation of electric and magnetic fields can reduce the size of the region heated by these particles. One can thus heat effectively a small region (3 μm) of the compressed fuel (its core) by the fast electrons generated by a laser pulse.

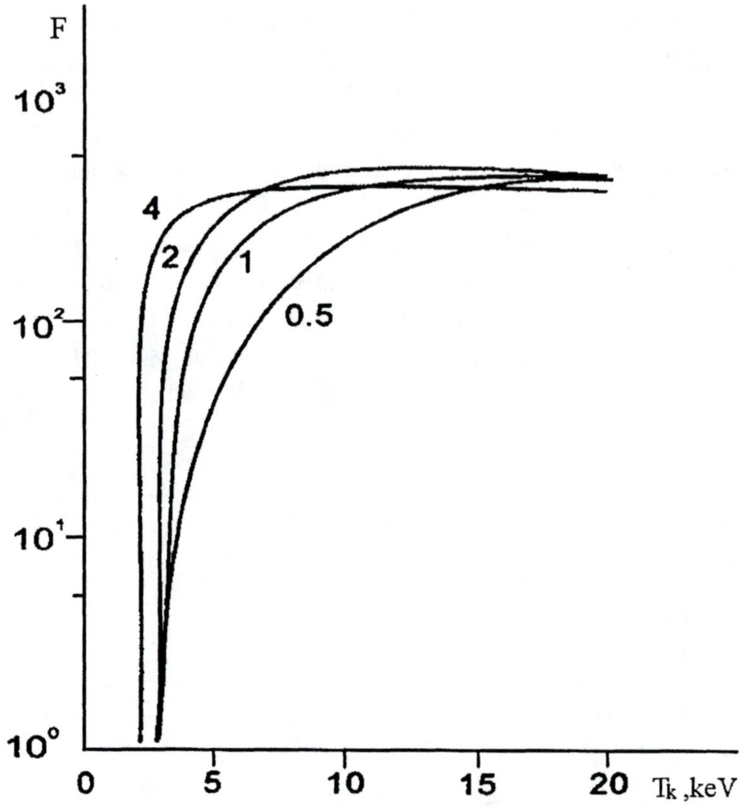

Fig. 3.5.3. Normalized energy gain F as a function of temperature T_k at different ρR_k. The curves are marked with corresponding values of ρR_k in g/cm^2.

Efficient transport of the hot electrons to the fuel is the next key problem in FI method, because the fuel is first compressed to a high density by a longer laser pulse. If the energy of the short pulse is converted into the hot electrons in the region of the critical density of the plasma corona, which is located quite far from the compressed core, the energy deposited by electrons in the core will be small. The aim is to deliver high-intensity radiation much closer to the compressed nucleus, where its energy is transformed into supra-thermal electrons with a short mean free path.

This can be done because of the following two circumstances. First, the relativistic increase in the electron mass means that the critical density [$n_c = n_{c0} / \gamma$, (γ- relativistic factor] increases g-fold, where $g \approx (1 + I_{18}\lambda^2)^{1/2}$, so that radiation can penetrate deeper into the target. Second - this over-dense plasma can be displaced by the pondermotive pressure. If an intense beam reaches a plasma of under critical density such an optical "piston" drives the plasma into the target and the optical pressure on the target surface is then $P=(1+R)\cdot 3.3\cdot 10^{-16} I$ [Мбар], where R- reflection coefficient. For I=3·10^{20} W/cm^2 we have P= 100 Gbar, which is sufficient to create a crater in plasma near critical density. For this reason it is recommended that the over-dense plasma in the corona should be depressed by a longer pulse before a picosecond laser pulse which generate fast electrons. An increase in the laser radiation intensity from 10^{17} W/cm^2, to 10^{19} W/cm^2 can depress the corona to a depth of hundreds of microns, even when the plasma density rises to about 100 n_c. In recent experiments [154] involving measurements of the Doppler shift of the radiation reflected by the critical surface it was found that the pondermotive pressure shifted this critical density surface into the target.

Filamentation and defocusing of the incident beam should be considered particularly in the method under consideration because they are factors that limit the capabilities of the method. Moreover, stimulated scattering mechanisms, leading to the loss of energy by the main pulse, may develop [155]. An analytic model [156,157] demonstrate an instability, similar to the Rayleigh-Taylor instability, may occur when light interacts with the plasma on the critical surface whose reciprocal increment amounts to tens of femtoseconds. According to [156,157], the Taylor formula can be used to estimate the acceleration of this surface by the optical pressure acting across the skin layer thickness, yielding the hydro instability increment $\gamma_g \sim I^{1/2} \lambda_p^{-1/2} / n_i^{1/4} \approx I^{1/2} \cdot 10^5$ [s^{-1}], where $\lambda_p \approx l_s$ (l_s- skin layer) is the perturbation wavelength. It is thus evident that a strong turbulence may develop over a period of several tens of femtoseconds when an ultra-relativistic laser pulse interacts with plasma. The length representing turbulent mixing of light with plasma can be estimated from [147]

$$L_t = 0.2\left(\int_0^t dt\gamma_g\right)^2,$$

which for our parameters gives L_t=50 μm. Naturally, this can complicate implementation of this method under discussion. This effect is predicted also by numerical simulation [26] (as formation and propagation of bubbles) over a relatively short distance, but calculation [25] show that beam filamentation exists for a certain time and then the beam propagates without significant refraction.

Simulation over a greater distance has demonstrated saturation of the hydrodynamic instability, so that one could expect a positive effect of the delivery of a laser pulse to the fuel. This is related to the generation of magnetic fields, which can give rise to stable configurations in the target that propagate into the dense plasma and generate fast electrons. The transport of fast electrons formed outside the critical density surface is modified significantly by self-consistent macroscopic electric and magnetic fields, which are generated by the electron flux. The plasma quasi-neutrality also gives rise to electric potential comparable with the temperature of the fast electrons. Two-dimensional effect generates a magnetic field that can facilitate the propagation of an electron flux [25]. On the over hand, the magnetic fields of the electron-generating filaments attract electrons and form a single channel. Simulation [101] predict magnetic fields order of 100 MG in a plasma with a density equal to several critical densities when $I\lambda^2$=10^{19} W/cm μm^2. This simulation show that, if the laser intensity is 10^{21} W/cm^2, a hole may be bored in high density plasma and a laser pulse may be delivered to a distance of 50 μm from the target core. Calculations of the transport of fast electrons, carried out in the 1D approximation taking account of self-consistent electric fields, showed that energy deposition occurs in the high-density plasma of core. The additional heating increases the temperature of the fuel ions to 10 keV, so that a significant thermonuclear yield should be possible for reasonable laser energy.

3.5.2 Efficiency of the Thermonuclear Burning in Laser Targets with Fast Ignition

In this section we consider some aspects of the gain model, which describes the fast ignition approach to inertial fusion. The study of thermonuclear (TN) burning of inhomogeneous heated targets allows us to calculate the igniter parameters, which provide high TN gain, and as a result to evaluate the required laser pulse parameters both of main driver and of ignition laser.

Physical-Mathematical Model

The ignition energy is delivered rapidly in the fast ignition scheme and the gain is calculated from the isochoric model where the hot spot and main fuel are out of pressure equilibrium [158]. In the original proposal [24] the other parameters of igniter and main fuel were chosen using isobaric model. We feel that parameters of both the igniter and of the main fuel should be reviewed based on the gain model. First, conditions of plasma self-heating are not the same for isolate plasma and for the hot spot in cold plasma because the igniter is cooled by contact with the surrounding plasma due to electron heat conductivity. On the other hand the initial increase in igniter temperature T_s is not a necessary condition for TN flash. Numerical calculations [159] show that under certain initial conditions a burn wave goes through two stages: the initial "subsonic" stage is characterized by a dropping or slightly rising temperature behind the wave front. The "smoldering" igniter expands preparing the conditions for the second "supersonic" stage of intense burning which gives rise to the TN flash. It is sufficient for spark ignition that the duration of "subsonic" stage is far less than the target hydrodynamic lifetime. The critical minimum igniter parameters depend on the parameters of the main fuel. One of the main purposes of this paper is to obtain these dependencies.

Second, on the assumption that the energy of the main fuel is dominated by electron Fermi-energy, the density dependence of the total internal energy in the compressed target takes the form: $E_0 = A\rho^{-2} + B\rho^{2/3}$. It easy to see that if the gain is optimized then igniter energy ($E_s = A\rho^{-2}$) and energy of the main fuel are comparable i.e. the extra laser pulse energy is of the same order as the energy of the main driver in contradiction with the fast ignition concept. However, we doubt that it is possible to realize target implosion with degenerated final stage, because this supposes low temperature. Here the main fuel is supposed to be in the non-degenerate state whose temperature T_0 is one of the free parameters of the problem. The gain is reduced in comparison with the optimal value but the igniter energy is limited to a few percent of total energy. We consider these conditions as more realistic.

The theoretical analysis of TN burning in non-degenerate fuel is drastically different from that of original scheme [24]. In spite of the apparent complexity of having an additional parameter (main fuel temperature) the situation is in some ways simple. It is obvious that in this case, fixing area densities and temperatures of igniter and fuel, the target density ρ plays a major role in determining the order of magnitude of other physical quantities. For example the mass is proportional to $(\rho r)^3/\rho^2$; the target lifetime $\Delta t \sim r/c_s \sim (\rho r)/\rho T^{1/2}$ where c_s is the mean speed of sound; $E \sim MT \sim \rho^{-2}$ and so on. The process of TN burning is approximately scale invariant in terms of variables: $r^* = \rho r$; $t^* = \rho t$; $E^* = \rho^2 E$; $M^* = \rho^2 M$. In particular, given real densities and temperatures, the target gain is independent of ρ and hence of initial target energy E in contrast to the original scheme [24]. The exact scale invariance is violated only by density dependencies of Coulomb logarithms. Our numerical calculations are in good agreement with this conclusion.

The calculations of energy gain are carried out by means of TN burning simulation from the moment of isochoric hot spot generation up to the target destruction. The mathematical model of the TN burning of an inhomogeneous spherical target is described by the equations system of motion, energy balance, continuity, state and kinetic equations for fast TN particles. A single-fluid, two-temperature plasma approximation with allowance made for ion-electron energy exchange and for electron and ion heat conductivity is used in the solution of the hydrodynamic equations. The kinetic processes of fast TN particles and thermal X-rays are characterized by sharp gradients of density and temperature with scales of the order of particle range, substantially by anisotropy of particle angle distribution, by complicated energy dependence of charged particles energy loss, and by presence of a set of correlated channels of primary and secondary TN reactions. Under these conditions a Monte-Carlo method for kinetic processes simulation seems to be more convenient than other standard computational schemes. As the characteristic time of target hydrodynamic processes is much greater than the TN particle transit time, the plasma parameters may be considered as a constant under kinetic process simulation on each time step of a well-known difference non-stationary scheme for differential equations. These approximations are realized in TERA code used here for self-consistent solution of the kinetic and hydrodynamic equation system [159]:

$$\frac{\partial u}{\partial t} = -r^2 \frac{\partial P}{\partial m} + V\mathrm{F}$$

$$\frac{\partial V}{\partial t}=\frac{\partial}{\partial m}\left(r^2U\right)+V^2S$$

$$S=\sum_{i,k}n_in_k\langle\sigma_{ik}\mathrm{v}\rangle\left(m_i+m_k\right)+\sum_i N_im_i$$

$$c_e\frac{\partial T_e}{\partial t}+P_e\left(\frac{\partial V}{\partial t}-V^2\mathrm{S}\right)+\frac{\partial q_e}{\partial m}+\frac{T_e-T_i}{\rho\tau}=Q_e+Q_f$$

$$c_i\frac{\partial T_i}{\partial t}+P_i\left(\frac{\partial V}{\partial t}-V^2S\right)+\frac{\partial q_i}{\partial m}-\frac{T_e-T_i}{\rho\tau}=Q_i$$

$$Q_{e,i}=2\pi m\int\left(\frac{\partial \mathrm{v}}{\partial t}\right)_{e,i}f\mathrm{v}^3d\mu d\mathrm{v},\quad Q_f=2\pi c\int k_\mathrm{v}I_\mathrm{v}d\mathrm{v}d\mu$$

$$F=2\pi m\int af\mathrm{v}^2\mu d\mathrm{v}d\mu\,,\ N_j=2\pi\lim_{\mathrm{v}\to 0}\left(a_e+a_i\right)\mathrm{v}^2\int f_jd\mu$$

$$\mathrm{v}\left(\mu\frac{\partial}{\partial r}+\frac{1-\mu^2}{r}\frac{\partial}{\partial\mu}\right)f+\frac{\partial}{\mathrm{v}^2\partial \mathrm{v}}\left(\mathrm{v}^2af\right)=W_0+W_s-\mathrm{v}f\sum_k n_k\sigma_k$$

$$\mathrm{v}\left(\mu\frac{\partial}{\partial r}+\frac{1-\mu^2}{r}\frac{\partial}{\partial\mu}\right)f_n=W_0+W_{ns}-\mathrm{v}f_n\sum_k n_k\sigma_{nk}$$

Here $V(r,t) = 1/\rho(r,t)$, $P(r,t) = P_i + P_e$, $W_0(r,T,v)$; $W_s,W_{ns}(r,v,m,(f_n,f))$ – sources of primary TN particles and recoil nuclear from elastic scattering of TN neutron, σ_{ik} – cross-sections of reactions between plasma nuclei of sort *i* and *k*, N_i – number of fast particles of sort *i*, n_i – nuclear concentration of sort *I*, a_i, a_e – Coulomb coefficient of de-acceleration of fast particles by electrons and ions, $\sigma_k(r,v)$, $\sigma_{nk}(r,v)$ – cross-sections of secondary TN reactions between fast particle and plasma nuclear with elastic scattering of neutron from plasma nuclei, κ_v, $\kappa_{v\sigma}$ - spectral coefficient of Bremsstrahlung absorption and coefficient of scattering, I_{vp} - equilibrium radiation intensity.

The target at the end of implosion is supposed to be homogeneous with the exception of a relatively small central hot spark (igniter). The initial conditions of isochoric igniter are described by two parameters: characteristic dimension (area density) ρR_k and temperature T_f. (Let's consider product ρR as separate physical value marked by two letters, then $\rho_k R_k=(\rho R)_k = \rho R_k$ and so on). It turns out in agreement with [159] that the target gain is practically independent on ignition origin if a TN flash takes place. As a consequence, the analysis of burn efficiency is reduced to determination of the critical (minimum) values of igniter parameters. As for initial parameters of the main fuel ρR_k, T_k, we suppose that there is no TN flash in a homogeneous target. In the opposite case the concept of fast ignition has no physical meaning. Preliminary calculations carried out for different homogeneous targets make it possible to establish the upper bound for target temperature and dimensions.

Thermonuclear Burning of Homogeneous Target

The study of homogeneous target TN burning is starting point to obtain the limits of initial target parameters in the fast ignition problem. It may be supposed that in the case of effective TN flash the process of burning is self-regulating and the target gain is practically independent on the ignition origin. The results of calculations set forth below confirm this prediction. That's why TN burning of homogeneous target may be used as a point of reference in analytical treatment of burn efficiency of targets with arbitrary temperature and density distributions. The general measure of target TN burning efficiency is the TN gain $G = E_{tn}/E_0$, where E_{tn} is the released TN energy, and E_0 is the initial internal energy coupled to the target. The results of TERA calculations of the gain of homogeneous targets with different ρR_k, T_k are presented in **Figure 3.5.2.**

The sharp increase in the gain within narrow interval of target temperature T_k corresponds to generation of self-sustaining TN burning. If a TN flash takes place the released energy depends weakly on the initial temperature T_k, as a result of further temperature increase the gain varies in inverse proportion to temperature: $G \sim 1/T_k$.

The most interesting result consists of the fact that the TN flash of thick targets (ρR_k>1g/cm^2) may occur at initial temperatures T_k ~ 3-5 keV far lower than the temperature of maximum TN reaction rate ($T \sim$ 15-20 keV). As a result, the energy gain ranges into the hundreds. It is due to target heating by α particle in the initial stage of burning. (a cold plasma with ρR_k>1g/cm^2 is impenetrable to fast TN particles). The transparency of thin targets with ρR_k<1g/cm^2 leads to high values of critical temperature $T_k \sim$ 15-20 keV and as a consequence to low values of gain ($G \leq 1$).

Let us consider in detail the energy gain of volume ignited targets as a function of ρR_k, T_k. By definition, the value of G may be presented in the form:

$$G = E_{tn} / E_0 \sim \langle \sigma v \rangle \rho \Delta t / T_k, \qquad (3.5.1)$$

where $\langle \sigma v \rangle$ is a time-averaged TN fusion rate, Δt is a characteristic target hydrodynamic lifetime. The TN fusion rate as a function of temperature shows a pronounced maximum at T ~ 20 keV and changes relatively little with further temperature increase up to T ~1000 keV [160]. As a result, the value of $\langle \sigma v \rangle$ in the range of active TN burning is virtually independent of the target initial parameters. The target lifetime may be presented as the ratio of target radius and mean speed of sound $v \sim \langle T \rangle^{1/2}$: $\Delta t \sim R_0/\langle T \rangle^{1/2}$. Our calculations show that in the process of active TN burning of a thick target with ρR_k > 1 g/cm^2 temperature runs into hundreds of keV. The corresponding lifetime reduction due to target self-heating decreases the index of a power in ρR_k dependence of the gain.

In fact, during the thick target TN flash the fraction of released TN energy causing plasma self-heating is essentially greater than initial energy E_0, so the TN flash mean temperature $<T>$ and target lifetime Δt are determined by target gain: $<T> \sim E_{tn} \sim GT_k$, $\Delta t \sim G^{-0.5}$. As a result the energy gain may be presented in the form:

$$G = F(\rho R_k, T_k)\ \rho R_k^{2/3} / T_k \qquad (3.5.2)$$

where the coefficient F weakly depends on ρR_k, T_k.

The function $F(\rho R_k, T_k)$ may be obtained from the data presented in **Figure 3.5.3**. It is seen that the value of coefficient F is approximately constant with accuracy of the order of 10% provided that the conditions of volume ignition are satisfied ($G>>1$). Under these conditions the TN energy gain may be evaluated as:

$$G = 370 \rho R_k^{2/3} / T_k \qquad (3.5.3)$$

In eq. (3.5.3) T_k and ρR_k are measured in keV and g/cm^2, respectively.

As may be seen from their derivation, equations (3.5.2), (3.5.3) are applicable only if $G >> 1$. There is another factor limiting the application scope of these equations. During TN burning, the density ρ changes not only due to hydrodynamic processes but also partially due to fuel depletion. If a significant fraction of the fuel is burned the effective density in eq. (3.5.1) as well as the mean temperature $<T>$ depend on the target gain. Let us define burn efficiency g as the fraction of fuel depleted: $g = \Delta M/M$. The released TN energy can be expressed in terms of a number of deuterium and tritium nuclei undergoing the DT fusion reactions ΔN: $E_{tn} = \Delta N<E>/2$, where $<E> \cong 17.6$ MeV. The initial energy E_0 is connected with the total number of D- and T-nuclei N: $E_0 = 3NT_k$. Taking into account that $g = \Delta M/M = \Delta N/N$, Eq. (3.5.3) may be rewritten in terms of burn efficiency:

$$g = 6GT_k / <E> = \rho R_k^{2/3} / 7.9 \qquad (3.5.3a)$$

The real value of fuel depletion is approximately ten percent greater than that given by Eq.(3.5.3a) due to the contribution of DD fusion reactions and so on. In spite of this it is convenient to use Eq. (3.5.3a) as a specific measure of efficiency of DT burning. In accordance with the above mentioned reasons Eq.(3.5.3) is valid only at relatively small values of the burn efficiency g. The results of TERA numerical calculations of the gain of homogeneous targets with different ρR_k, expressed in terms of DT burn efficiency g, are presented in **Figure 3.5.4**.

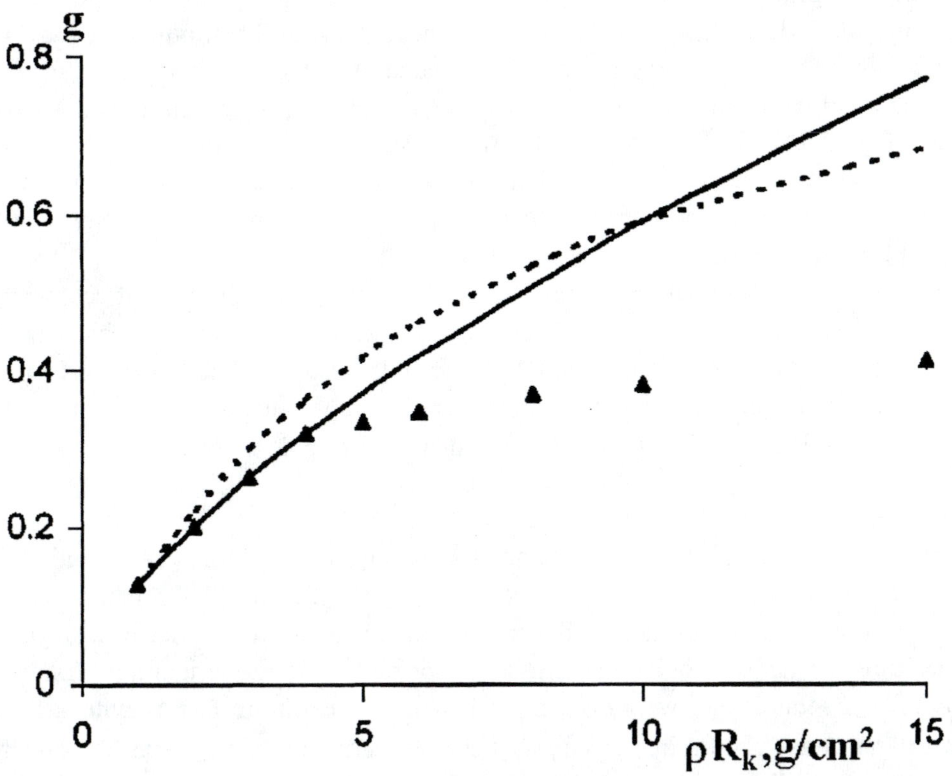

Fig. 3.5.4. The DT burn efficiency g of a homogeneous target as a function of ρR_k. Solid and dashed lines correspond to $g = \rho R_k^{2/3}/7.9$ and $g = \rho R_k/(\rho R_k+7)$ respectively. Markers show the results of TERA numerical calculations.

Comparing these results with the values given by Eq.(3.5.3a) we obtain the range of applicability of eq.(3.5.a): $g \le 0.4$. According to foregoing arguments the Eqs. (3.5.2), (3.5.3) may be used under the conditions: $G >> 1$, $g \le 0.4$. The corresponding range of area density and temperature ($1\ g/cm^2 \le \rho R_k \le 5\ g/cm^2$; $T_k < 10$ keV) is rather appropriate to the fast ignition targets.

If the effect of lifetime reduction due to target self-heating is neglected (it is correct for thin targets with $\rho R_k < 1\ g/cm^2$) then the gain and burn efficiency varies linearly with ρR_k : $G \sim \rho R_k/T_k^{3/2}$, $g \sim \rho R_k/\sqrt{T_k}$. With allowance for depletion effect, the expression of burn efficiency takes the form [151]: $g = \rho R_k/(\rho R_k+A\sqrt{T_k})$. In the case of a thin target, only the temperature of TN flash ($T_k \sim$ 15-20 keV) is relevant. That is why the temperature dependence of g is not considered as a rule, and the corresponding expression is extended to the range of $\rho R_k > 1\ g/cm^2$. It should be noted that TN fusion rate and α particle energy losses in deuterium - tritium plasma are combined in such a way that considerable depletion is accompanied by considerable self-heating. Both depletion and self - heating affect the burn efficiency in a similar way. Because the constants in the expressions for g are treated as fitting parameters it is no wonder that the well-known evaluation of burn efficiency [151] obtained without allowance for target self

- heating: $g = \rho R_k/(\rho R_k + 7)$ is in a good agreement with eq.(3.5.3a) (see figure 3.5.6). The temperature dependencies of the gain presented in figure 3.5.5, however, show that the effective TN flash represents a self-regulated process, which once initiated, is independent of the initial temperature. Therefore an approach based on allowing for the self-heating effect and leading to Eqs.(3.5.3), (3.5.3a) is more suitable in the range under consideration. As a matter of fact the fuel depletion affects the energy gain considerably and should be taken into account simultaneously with the self-heating effect only in the range of large density $\rho R_k >> 1$ g/cm^2.

The same reasons remain in force and rise in importance in the case of ecologically cleaner fuel with a small number of neutrons among the products of its burning (for instance a mixture of deuterium with He^3). In this case the fraction of TN energy responsible for target self-heating increases considerably, and as a result the relative contribution of fuel depletion in the ρR_k-dependence of the gain is reduced.

Fast Ignition Criteria

There is no heating by α particles, and as a consequence no burn wave in rather thin targets with $\rho R_k < 1$ g/cm^2. For thick targets with $\rho R_k > 1$ g/cm^2 the TN combustion wave initiated by a spark makes it possible to produce an effective TN flash at an initial temperature substantially below the critical temperature of volume ignition. Fixing the value of T_k in this range, we consider a set of model isochoric targets with different igniter parameters T_s and ρR_s at given ρR_k. For every target the energy gain G is obtained by means of TERA code simulation of TN burning. The total number of targets considered exceeded 500.

Consider the typical results of the calculation for targets with $\rho R_k = 2$ g/cm^2. The obtained values G as a function of igniter dimension ρR_s at different T_k and T_s are presented in **Figure 3.5.5.** It may be seen that for every value of igniter temperature T_s there are some critical values of igniter dimension, corresponding to the sharp increase in the gain within relatively narrow range of ρR_s. A similar behavior of the gain-dimension dependencies is obtained at other ρR_k , and the range of critical ρR_s narrows with increase of ρR_k. The same results may be presented in terms of gain-temperature dependency for fixed ρR_s. In this case the range of critical temperature T_s corresponding to the sharp increase in the gain is less than 0.5 keV.

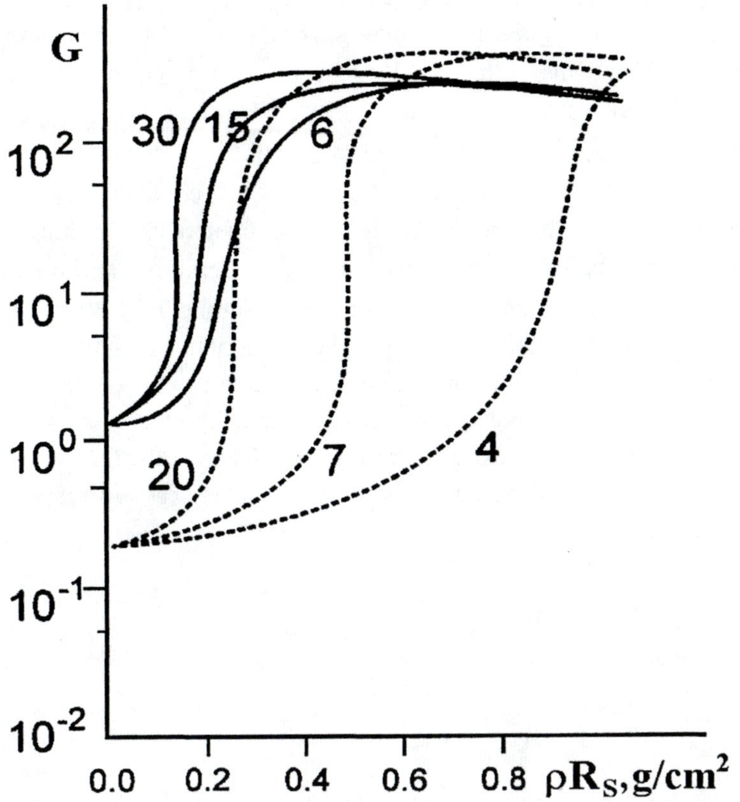

Fig. 3.5.5. The energy gain G as a function of the igniter density ρR_s at different igniter temperature T_s (indicated in the figure) for target with $\rho R_k = 2$ g/cm^2. Solid and dashed lines correspond to $T_k = 3$ and 2 keV respectively.

From the physical point of view this increase is associated with the fact that the time of TN burn extension over the whole target, under conditions near to the critical is of the order of the target lifetime. Up to the critical igniter parameters the low energy gain corresponds to the "smoldering" of a homogeneous target with temperature T_k and is independent on igniter presence. There is no spark ignition. In the critical case, TN flash results in effective burning with $G \sim 10^2$-10^3. If a TN flash takes place, the energy release is practically independent of the history of ignition. The calculated values of energy gain in the limit with overcritical ρR_s, T_s tend to ones given by Eq.(3.5.3) with the obvious correction on the additional igniter energy:

$$G = G_H / (1 + \Delta_E) ,$$

where G_H is the energy gain of homogeneous target with ρR_k, T_k; Δ_E is a dimensionless ratio of energy coupled to igniter due to an extra laser pulse E_s to the total energy of corresponding homogeneous target with the same ρR_k, T_k.

Thus only the values of critical igniter parameters are relevant. As for the parameters of the main fuel we intend to obtain the condition under which the gain is maximized. As shown by Eq.(3.5.3), it is necessary to reduce the target temperature as much as possible. On the other hand, as was argued above, plasma is supposed to be in non-degenerate state whereas in dense DT plasma with $\rho \sim 10^2$ g/cm^3 the Fermi energy runs into hundred of eV ($E_F \cong 14\ \rho^{2/3}$ eV, where ρ is measured in g/cm^3). Therefore, the initial temperatures T_0 are chosen in the range $T_k \sim 0.5 - 1$ keV. We carried out numerical calculations of critical igniter parameters ρR_s, T_s for targets with area density ρR_k equals to 3, 4 and 6 g/cm^2.

Our calculations show that the critical igniter parameters are practically independent of the target area density ρR_k in the range under consideration. The obtained values of ρR_s, T_s for targets with different T_k are presented in **Figure 3.5.6.** It is obvious that the critical igniter temperature increases with the reduction of igniter area density in agreement with the curves in figure 3.5.8. More interesting is the behavior of the corresponding additional thermal energy E_s coupled to igniter because it is connected immediately with energy of the extra laser pulse. In other words, for every target there is a minimum value of additional absorbed extra pulse energy corresponding to some optimum igniter.

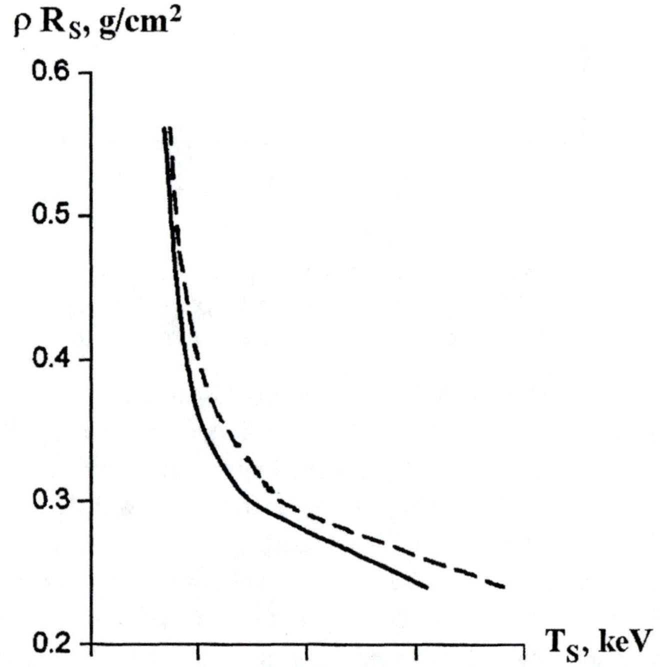

Fig.3.5.6. The critical igniter parameters for targets with the main fuel temperature $T_k = 1$ keV (solid line) and $T_k = 0.6$ keV (dashed line).

The presence of optimum igniter may be consistently explained taking into account the temperature behavior of TN fusion reaction rate. Until the critical temperature reaches the values corresponding to the maximum $<\sigma v>$ at $T \sim 15 - 20$ keV a considerable reduction in critical igniter dimension is compensated by a small variation of critical temperature due to the sharp increase of $<\sigma v>$ with temperature. As a result the critical value of E_s is reduced in this range. The further reduction of igniter dimension and increase of igniter temperature gives no additional advantage in energy because TN fusion reaction rate remains approximately constant and the electron heat conductivity spreads the absorbed energy over the optimum igniter size corresponding to the temperature of maximum <σv>. For illustration, the critical igniter parameters given in figure 3.5.6 are presented in **Figure 3.5.7** in terms of Δ_E, T_s ($\rho R_k = 3$ g/cm^2).

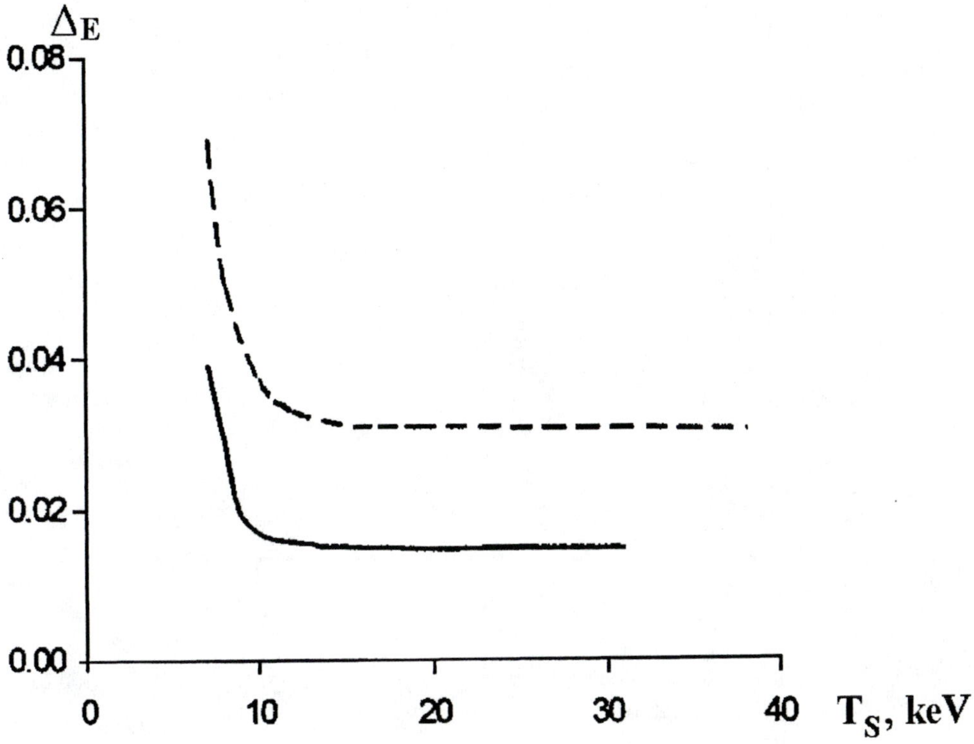

Fig. 3.5.7 The relative value of energy $\Delta_E = E_s / E_0$ corresponding to the critical parameters of igniter as a function of igniter temperature. Solid and dashed lines correspond to the main fuel temperature $T_k = 1$ keV and $T_k = 0.6$ keV respectively.

As may be seen Δ_E tends to the constant in the range $T_s > 12$ keV, according to the above-stated arguments. The minimum values of igniter energy E_s run about a few percent and under of the total target internal energy E_0. We calculate the values of minimum absorbed energy under fast ignition for targets with different temperature T_k. As was mentioned above, these values depend on the density ρ whereas the values of $E_s^* = \rho^2 E_s$ are approximately scale-invariant. To verify this statement, the calculations were carried out at ρ equal to 10 and 100 g/cm^3. At tenfold change in density the corresponding change in critical igniter energy E_s^* is less than five percent. The calculated values of minimum additional absorbed extra pulse energy E_s corresponding to optimum igniter are presented in **Figure 3.5.8** in the form of energy - temperature dependencies for targets with $\rho = 316$ g/cm^3. With account for scale invariance, the presented curves allow us to evaluate the minimum extra laser pulse energy at arbitrary condition of basic driver implosion and may be used examine the fast ignition scheme of ICF.

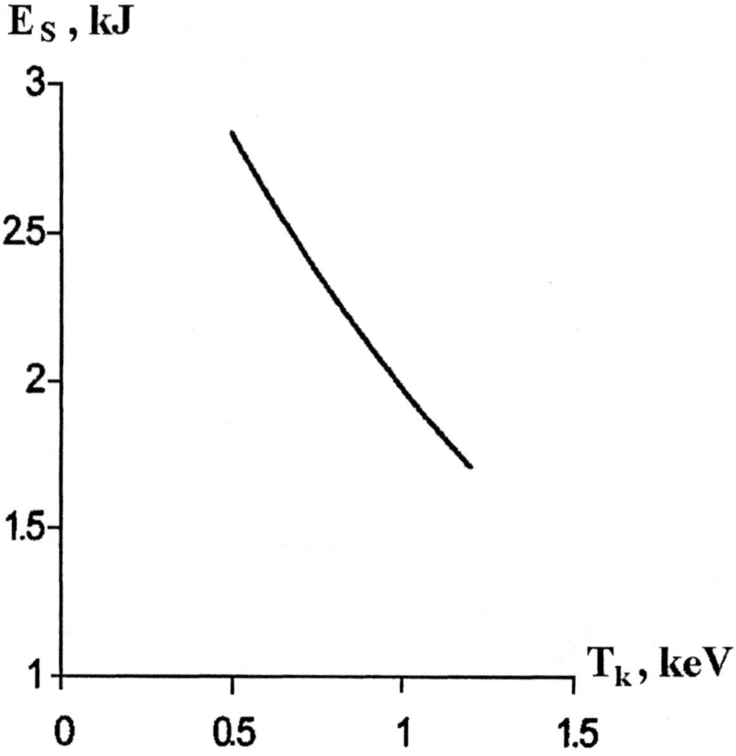

Fig. 3.5.8. The minimum value of the energy coupled to the igniter E_s as a function of temperature of isochoric target with $\rho = 316$ g/cm^3.

3.5.3 "Fast Igniter" ICF Scheme Using Laser Triggered Fast Ions

We consider now the possibility of organizing FI scheme with help of a fast ion beam created by laser pulse foil interaction as in section 3.4. The general approach to fast ignition involves a powerful external source and aims to define the igniting parameters for a beam and core. A first estimate of the pulse parameters for FI was presented in [24] and we followed it in 3.5.1.

As was pointed out in [158], this conception has dealt with nearly isobaric process and underestimated the energy required for the igniting which is more relevant to nearly isohoric one and is somewhat larger than firstly proposed.

In the model [153] the energy and temperature required to create an ignition spark in DT plasma at density ρ have been written by the next formulas:

$$E_s \approx 10\left(\frac{\rho}{100g/cm^3}\right)^{-2} kJ\,,\ T_s\left[keV\right] \approx 10 \text{ KeV, for } (\rho R)_s = 0.3 \text{ g/cm}^2$$

The optimum conditions for spark area according to [158,161]:

$$1.2 \text{ g/cm}^2 \geq \rho R \geq 0.3 \text{ g/cm}^2,\ E_s \approx 10 \text{ kJ},\ W_s \approx 10^{15} \text{ W},\ R_s \approx 10\ \mu\text{m},\ t_s \approx 10 \text{ ps}$$

In accordance with [158,159] the minimum intensity of the pulse of fast particles, for the ignition is

$$I_s \approx 10^{20} \text{ W/cm}^2 \qquad (3.5.4)$$

Let's consider the next plasma parameters in compressed target pre-ignition stage

$$\rho R = 0.3 \text{ g/cm}^2,\ T_e = T_i = 5 \text{ keV} \qquad (3.5.5)$$

Ion beam interacts with plasma core at these parameters to ignite hot spot. We consider the ion beam parameters according to the model developed in Sec. 3.4. To exam if the required ion energy is consistent with the ignition windows for $(\rho R)_s$ we accept I_s as given in (3.5.4) and estimate the typical ion density from quasi-neutrality condition $n_i = 10^2 \lambda_\mu^{-2} Z^{-1}\sqrt{1+0.7I_{18}}$ choosing the hot electron density equal approximately critical value.

For ion energy the scaling $\mathcal{E}_i = Z\lambda I_{18}^{1/2}\left[MeV\right]$, can be inferred from the recent experiment on high-energy ion generation and developed model in Section 3.4. Then at $Z = 1$, $\lambda = 1\mu$m and $I_{18} = 100$ ion energy is $\mathcal{E}_i = 10$MeV.

Ion bunch duration and spot size determined from laser parameters:

$t_i = t_L$, spot size $S_i = S_L$

Using these assumptions for the estimation of ion energy flux one can get

$$I_i \approx 1.7 \cdot 10^{-2} I_{18}^{5/4} \sqrt{Z/A\lambda} \ \text{ W/cm}^2,$$

where we suppose that ions are non relativistic.

Our key issue includes an estimation of ion penetration depth into the dense compressed DT core with density and temperature.

For the ions with energy higher than hundred KeV the penetration depth are due to their collisions with electrons, i.e. fast ions heat electrons of a core and loose their energy in accordance with the equation for ion free path in dense plasma with temperature T_e [150]:

$$l_i = \sqrt{\mathcal{E}_i}\,\frac{T_e^{3/2}}{Z^3 n_i}\left(\frac{m_e}{m_i}\right)^{1/2},\quad T_e \approx T_i \text{ – in compressed part of target}$$

Concerning the penetration depth of the heating particles this should be comparable to the size of the hot spot $R_s = (\rho R)_k / \rho_k$ at the optimal $(\rho R)_k = 0.3$ g/cm^2. Then we obtain the following equality:

$$\rho_k l_i = 0.1(AT_e^3 Z^4 \Lambda^{-2}\ \mathcal{E}_i)^{1/2} = 0.3 \text{ g/cm}^2,$$

where $\Lambda \approx 6$ Coulomb logarithm, and fore temperature we should take pre-ignition value: $T_e = T_i = 5$ keV, then for $Z=A=1$ we obtain $\mathcal{E}_i \cong 7$ MeV.

The conversion efficiency of ion intensity to laser one:

$$\frac{I_i}{I_L} = 310^{-2} I_{L18}^{1/4}\sqrt{Z\lambda / A}\,.$$

This equation predicts that a conversion efficiency of laser energy into energetic ions scales with an intensity as $I^{1/4}$ and for protons at intensity $I_s \approx 10^{20}$ W/cm^2 gives 10% for the conversion efficiency. From the condition $I_i = I_s$ we define the threshold intensity for 1 μm laser which is found to be practically the same for protons, deuteron and beryllium beams with typical ion energies ≈ 7 MeV, 8 MeV and 30 MeV correspondingly. In this case for ion energy $E_i \approx 1$ KJ we get laser energy $E_L \approx 10$KJ

3.6 Laser Nucleonic

It is well known already that high intensity laser pulse can produce nuclear reactions when interacting with different targets. To date, many applications call for using a source of γ -radiation, and the vast scope of applications range from non destructive gamma flaw detection to medical examinations [162]. It should be noted that, presently, the intensity of ordinary radioactive sources cannot be higher than 10-100 GBk, whereas much higher intensity is necessary for many applications. Commonly, synchrotron radiation is used as a high-intensity γ -source, its activity being about PBk. However, significant success notwithstanding, the use of the synchrotron radiation as a γ -source has some essential drawbacks. First, it is impossible to obtain high intensity γ -radiation in a very narrow frequency range, because the synchrotron radiation spectrum is continuous. Hence narrow-frequency-band experiments with a high-power γ -source appear to be impossible, because the power of such a γ - radiation source drops as the frequency range narrows and it is difficult to cut a narrow MeV spectral range with any filters. The second reason is that synchrotron radiation source is a expensive for some applications.

In this Section, we consider two possibilities for laser excited monochromatic γ - source. First [163]: using nuclear isomers as a high-intensity frequency-conversion γ - source. The nuclei are prepared in the isomeric state. Then the isomeric ground-state nuclei are pumped to an excited isomeric state, wherefrom an active γ -transition arises by the X -ray radiation from laser plasma produced by the action of high-intensity laser pulse on a solid-state target. The second method [164] is laser-plasma acceleration of fast particles to generate nuclear reactions with high γ yield on interaction of fast electrons or ions with different targets.

Basic Concept of a Laser Triggering Monochromatic Nuclear γ - Source

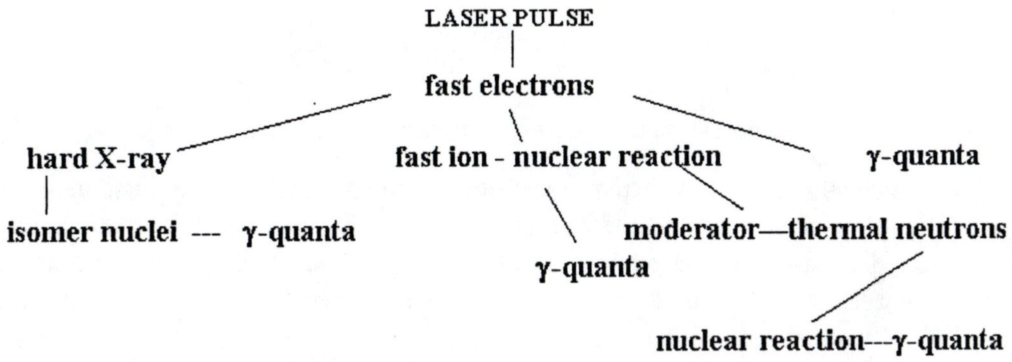

Figure 3.6.1. Laser - γ-radiation conversion schema.

To produce MeV range energy photon emission we should use nuclear excitation instead of atomic excitation because even for high charge Z ions the energy of quanta estimated as RyZ^2 can not exceed 100 keV.

As direct excitation of nuclear by laser field has very low efficiency we will consider some indirect processes whereby laser energy is transformed into electron energy effectively enough at first and then through another channels to cause nuclear excitation. On the **Figure 3.6.1** we suggest for future analysis the different methods of γ – photon production involving high power laser radiation.

3.6.1 Laser Induced γ - Fluorescence of Isomeric Nuclei

If we have not very much laser intensity we can use the isomer nuclear method [165]. Let's first consider the main idea of a high-intensity γ - source using isomer nuclei (see **Fig. 3.6.2**).

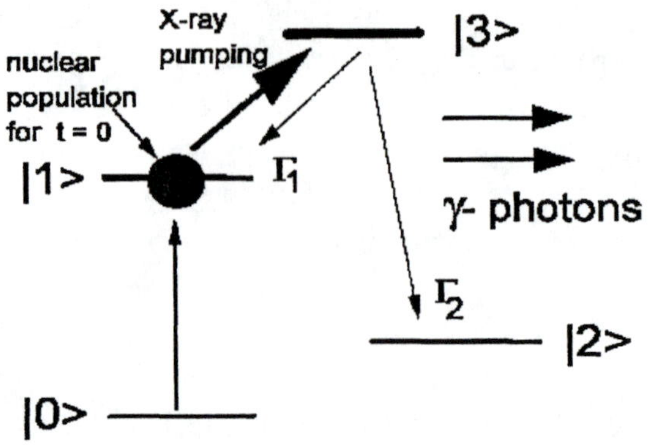

Figure 3.6.2. Three level schema for isomer nuclei pumping X-ray.

We assume that the |3>-|2> active γ -transition is pumped in two steps. First, atoms in the |1> nuclear isomeric state should be prepared. We can estimate the γ - radiation pulse power suppose that all isomeric nuclei in the target being pumped. We assume that the confinement region of the target is ~ 10^{-3} cm^{-3}, whereas the number of isomeric nuclei is N_0 10^{12}. Then the γ-radiation power, on condition that all isomeric nuclei in the target are pumped, is 10^{12} γ-photons during plasma lifetime τ_i $1/\Gamma_0$ – lifetime of the |3> upper excited state. Nuclear level width Γ_0 can be estimated, very roughly, from matrix element of nuclear magnetic momentum: $\Gamma_0 = \Gamma_1 + \Gamma_2 \approx \frac{4\omega_{13}{}^3}{3\hbar c^3}\mu_{13}$, where $\omega_{13} = \omega_X$ –frequency of X-ray pumping. Also the natural width of the |3> excited state, Γ_0, can be obtained from a one-partial Waiskopth's approximation as [166]: Γ_0 = T(M1) = $1.76*10^{13} E_\gamma{}^3 B(M1)$, where $B(M1)=(7.5)^2/(10\pi)$ has to be taken in units $(eh/(4\pi Mc))^2$, while the γ-photon energy - E_γ in MeV, if we consider transition |3> - |2> as M1 [167]. Then the natural width value is Γ_0 10^9 s^{-1} for λ_γ $6*10^{-9}$ cm, $E_\gamma = \hbar\omega_\gamma$ =20 keV. Then the total energy of the γ -radiation will be about $2*10^{-3}$ J.

At the same time, the width of the γ -radiation spectrum, which is emitted by the |3> - |2> transition, is governed by the Doppler effect caused the recoil during absorption of the X-ray pump. So Doppler shift of γ - line from recoil effect is the same for all nuclei $\Delta\omega_{\gamma D} = \hbar k_\gamma^2 / Am_p$ and Doppler width is Γ_D = $\hbar k_x k_\gamma / Am_p$ because nuclei obtain velocity $\hbar k_x / Am_p$ at absorption of pump X-ray quantum Where $k_{\gamma,x}$ is the wave vector of the γ, X -radiation, m_p -is the proton mass, and A is the atomic number. For the above values, the width of the γ -radiation spectrum is $\Gamma_D = 10^{13}$ Hz. Line width is increased at isomer ionization from addition nuclear heating because all Z* ionized electrons with energy $\sim \hbar\omega_x$ give velocity to nuclei $\sim Z^* \sqrt{m\hbar\omega_x / Am_p}$ and it increases line width up to $(Z^* \sqrt{mc / \hbar k_x}) \hbar k_\gamma k_x / Am_p$.

Finally ionized electrons heat nuclear by collisions, but in our case time of e-i collisions much more compare to γ - quantum generation time and line width does not increase up to $k_\gamma \sqrt{T_e / Am_p}$.

X-Ray Pumping of the Isomeric Nuclei

In this section, we discuss the conversion efficiency of emission from hot laser plasma of X-ray photons to produce γ -photons in isomeric nuclei confined in a target. Presently, the X-ray pumping with such high intensity (when the number of resonant X-ray photons is at least equal to the number of the isomeric nuclei in the target) can be obtained only when a laser radiation pulse with a high power acts on a solid-state target. We will consider laser plasma thus produced as a source of blackbody radiation.

Let us now estimate the number of the isomeric nuclei which are activated in a target when pumped by X-radiation of a laser plasma. The X-ray photon flux from a unit area on the surface of a hot plasma spot is equal to

$$\mathrm{d}N_x/\mathrm{d}t\, \mathrm{d}\omega_x\, \mathrm{d}S) = (\omega_x^2 / \pi^2 c^2)/[\exp(\hbar\, \omega_x/T_e) - 1] \tag{3.6.1}$$

where T is the temperature of the laser plasma, and ω_x and N_x are the frequency and the number of X-ray photons, respectively. Assuming the plasma spot area to be S and the thickness of a cloud of the isomeric nuclei along the pump propagation direction to be d, it is possible to estimate the overall cross section for X - ray pump resonant absorption by an ensemble of nuclei with concentration n_i in a target by

$$\sigma_f = \Delta\Omega n_i\, S\, d\, \sigma_x, \tag{3.6.2}$$

where $\Delta\Omega$ 1 – geometrical factor, $\sigma_x = 2\pi c^2 / \omega_x^2$ is the X-ray resonant absorption cross section (we have assumed that the laser plasma, which is produced under the action of a sub-picosecond high-power laser pulse on a solid-state target, is near the region where the activated atoms are located). Then, during plasma lifetime τ_i, N_s isomeric nuclei in the volume of a target will rise to the |3> upper active level

$$N_s = (\mathrm{d}N_x/\mathrm{d}t\, \mathrm{d}\omega_x\, \mathrm{d}S)\, \sigma_f \Gamma_0 \tau_i \quad \Delta\Omega(2/)\, N_0 \Gamma_0 \tau_i / [\exp(\hbar\, \omega_x/T_e) - 1] \tag{3.6.3}$$

As can be seen from (3.6.3), only a small fraction of nuclei found in the target can be activated. For example, for τ_i 10^{-11} sec, $\Gamma_0 = 10^9$ sec^{-1}, that is, the number of activated nuclei is about 10^{-4} of the initial number of nuclei N_0.

It should be emphasized that (3.6.3) defines the number of activated nuclei in a target without any losses. To determine the real concentration of the isomeric nuclei in the target, N_γ, it is necessary to take into account the X-ray photon loss via two main processes: photo-effect in a continuous X-ray pump spectrum with cross section $\sigma_{ph.eff}(\omega_x) = 15.4\pi\, \alpha^4 Z^5 e^2/(m_e c^2)^2 (2\pi m_e c^2/\hbar\omega_x)^{7/2}$, at $\hbar\omega > J_z$ – ionisation potential of atom;

$$\sigma_{ph.eff}(\omega_x) \quad \pi R_{Bohr}, \text{ at } \hbar\omega < J_z; \tag{3.6.4}$$

$$\sigma_{ph.eff}(\omega_x) \quad \pi \lambda_x^2, \qquad \text{at resonance}$$

and the absorption of X-ray photons at other nuclear and electronic levels.

Note that the atomic electron shells feature resonance absorption cross sections much larger than σ_X. Therefore, it is necessary to cut off the low-frequency portion of the pump spectrum. This can be done using a filter which cuts off pump frequencies lower than ω^*. As a result, the actual concentration of activated isomeric nuclei can be found from the equation

$$N_\gamma = N_S/\{1+\Sigma_i (\gamma_i/\Gamma_0)(exp(E_x) - 1)/(exp(E_i) - 1) +$$
$$+ \quad {}_{\omega^*}\sigma_{ph.eff}(\omega_0)(\omega_0/c)^2)(exp(E_x) - 1)/(exp(E_0) - 1)\, d\omega_0/2\pi\,\Gamma_0\} \tag{3.6.5}$$

Here $E_j = \hbar\omega_j/2\pi T$ ($j = 0, x, i$), where both nuclear and electronic levels, which have frequencies lying in the pump spectral interval, are included in the sum over i from 1. Thus, each of such resonance has absorption cross-section $\sigma_i = 2\pi c^2/\omega_i^2$ and width γ_i. If we assume the energy of pump photons to be $\hbar\omega_j > T$, $\hbar\omega^*$, then $\omega_i > \omega_x$ for all i ($\omega^* < \omega_x < \omega_i$), the terms in the sum in the denominator of fraction (3.6.5) will be exponentially small.

Next we estimate loss of pump photons due to ionisation. To this end, it is necessary to calculate the integral in the denominator of fraction (3.6.5). As a result, the condition for small losses of activated atoms due to ionization can be written as [168]:

$$K_I = 7.4\, Z^5\alpha^6 (2\pi m_e c^2/\hbar\omega^*)^{3/2}(T/\hbar\gamma_0))(exp(E_x) - 1)exp(-\hbar\omega^*/T_e) \tag{3.6.6}$$

where $\alpha = 1/137$. By substituting $\hbar\Gamma_0 = 10^{-6}$ eV ($\Gamma_0 = 10^9$ c^{-1}) $\hbar\omega^* = 511$ eV, $\hbar\omega_x - \hbar\omega^* < T_e$ and $T_e = 400$ eV in (3.6.6), we obtain an estimate for the charge of the isomeric nuclei: Z < 10. Therefore, to obtain the nuclei with the charge much larger than Z=10, it is necessary to increase X-ray pumping by increasing laser pulse intensity. In another case we will have the decreasing of the number of active nuclei in K_I times.

Pumping by K_α Line Radiation

As we already mentioned in Sect.3.3, it is possible to increase X-ray yield at oblique incidence of P-polarized laser radiation by resonant absorption. Anyway the spectral brightness of such X-ray pump in nuclear transition is not so high. From this reason we analyse the possibility of plasma line X-ray emission at energy approximately equal to nuclear transition energy. For this case the best candidate is K_α - line because the energy of such X-ray quanta can be very high. In Sect.3.3 we estimated the intensity of K_α - line emission from a flow of fast electrons interacted with over dense plasma at resonant absorption of laser radiation. For non-relativistic laser intensity we calculated the concentration of fast electrons and K_α line intensity is determined by the next approximation:

$$I_{k_\alpha} \approx 4\cdot 10^8 \left(\frac{\Delta E_{k_\alpha}}{1keV}\right)\left(\frac{J_z}{1keV}\right)^{-1/2}\left(\frac{Z_{nu}}{10}\right)^5\left(\frac{E_L}{0.1J}\right)\left(\frac{d}{20\mu m}\right)^{-2}\left[\frac{W}{cm^2}\right]$$

The comparison of this formula with intensity of black-body radiation in the same band of quantum energies gives the following result:

$$I_T \approx 2.5\cdot 10^{11}\left(\frac{\Gamma_{k_\alpha}\Delta E^3_{k_\alpha}}{\exp(\Delta E_{k_\alpha}/T_e)-1}\right)\left(\frac{d}{20\mu m}\right)^{-2}\left[\frac{W}{cm^2}\right],$$

where $\Gamma_{K\alpha}$ - a line width, estimated as $\Delta E_{K\alpha}$ (v_i/c).

For T=1 keV, E_L=1J, $\Delta E_{K\alpha}$ =3.4 keV, Z_{nu}=16, J_z =50, A=32 the ratio $I_{K\alpha}/I_T$ $2.4\cdot 10^2$. Thus we have magnification of pumping intensity of two orders of magnitude.

Why do we choose the element with Z_{nu}=16, A=32 (This is Sulphur - S). We can estimate an energy of the basic state for –hydrogen like spectrum $\Delta E_{K\alpha}$ = 3,482 keV. It is closest to 3.4 keV in the nuclei Rb^{86}, besides the following statements are valid: the atom of sulphur in the target is partially ionised, therefore high atoms levels are free. There is $E_{k_\alpha} = (3.482\ \text{keV})(1-1/n^2)$ at transition between these levels and ground state. By selecting the number n of high level it is possible to hit in a nuclear level. We can also remark, that Doppler width of X-ray pumping is about one electron volt at $\omega_x < \omega_l$ and it can help us to be very close near nuclear level. Another examples of such agreement is shown on this table [165]:

Coincidences between the Energies of Nuclear Transitions and the Energies of Characteristic Emission Lines

Nucleus	ΔE_n, keV	Element	X-ray line	ΔE, keV	Linewidth, eV
$^{159}Tb_{65}$	57.995	W_{73}	$K_{\alpha 2}$	57.982	43.2
$^{165}Ho_{67}$	94.699	U_{93}	$K_{\alpha 2}$	94.665	107.0
$^{169}Tm_{69}$	8.401	W_{73}	$L_{\alpha 1}$	8.398	5.0

It should be specially noted that such a γ -source has not substantial difficulties being faced in its experimental implementation. In experiments on observation of such a high activity of nuclei in the target, the ^{58}Co, ^{84}Rb, ^{93}Mo or ^{152}Eu nuclei could be used, because they feature the necessary scheme of nuclear transitions [167].

For example, it is preferable to take ^{84}Rb nuclei for the first experiment, because such atoms are very well known in respect of both laser pumping and trapping into Magnetic Trap [169]. As a result, a concentration of 10^{12} cm^{-3} can be readily obtained in MT. This number of cooled nuclei allows 10^6 resonant γ -photons to be emitted. At the same time, the noise of the natural γ -activity can be eliminated, by using time-selective signal detection. For the above number of nuclei, the noise of the natural γ -activity of isomer nuclei with a half-life of one hour typically corresponds to 10^7 decays per second. However, during 10^{-6} sec, the number of naturally emitted γ -photons is 20, whereas the number of γ -photons produced by X-ray pumping of an active γ -transition will exceed 10^6. Therefore, signal-to-noise ratio of 10^6 can be readily observed in such a time-selective experiment.

We also analyse the possibility of nuclear excitation with help of solid target. In this case we use X-ray pump from laser plasma to illuminate solid film mixture of KBr and Rb^{84} (decay time 1 our). There is 10^9 total number of Rb isomer nuclei in this target at the range of X-ray transparency. Geometrical loses of X-ray emission is near 0.1 and 200 keV decay γ –photons from excited isomer nuclei can be registered by a detector with photomultiplier.

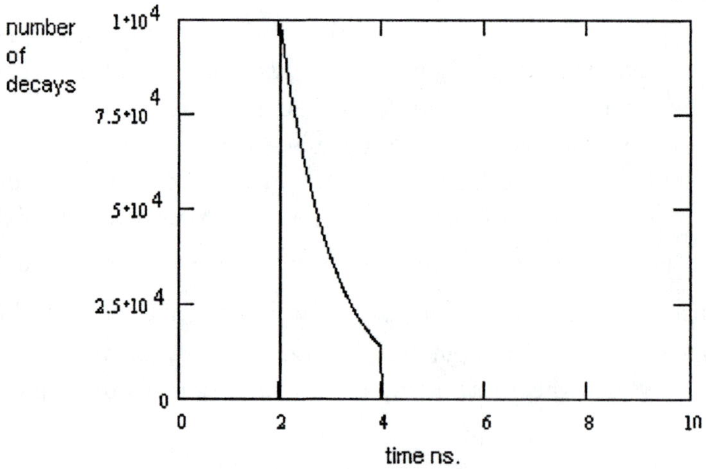

Figure 3.6.3. Time dependence of γ-emission detected signal for isomer Rb target pumping by X-ray picosecond laser plasma radiation.

In the **Figure 3.6.3** there is expected signal of γ – detector, which was switched on for 2 ns after laser shot. Time of life of excited level is 1 ns. We can see that approximately 10^4 γ –photons should be produced in this experiment and this is enough to be measured.

1. Amplification of γ – radiation in active media

$$G = (\lambda_\gamma^2 / 4\pi)(\Gamma_0 / \Gamma_D) n_{nu} L > 1$$

2. Losses of X-ray pump in laser media
The main losses are from photo-ionization in our case

$$\sigma_{ph.eff.}(\omega_x) \quad \pi \lambda_x^2(\Gamma_0/\Gamma_D)$$

3. Optimal thickness of laser media for X-ray pump

$$d \quad 1/(n_{nu} \lambda_x^2)$$

4. Diffraction losses of γ - radiation

$$L < k_\gamma d^2$$

As $\Gamma_0 \approx \dfrac{4\omega_x^3}{3\hbar c^3}\mu_{13}$, where $\Delta\omega \quad \Gamma_D = \hbar k_x k_\gamma / A m_p$ we have two possibilities for active media and pump parameters:

1. $Z < 30$, $n_{nu} \quad 10^{17}$ cm^{-3}, d $\quad 10^{-3}$ cm, $L < 100$ cm, $\mathcal{E}_\gamma \quad$ 10 keV, $\mathcal{E}_x \quad$ 1 keV; it can be gas Co in "Magnetic Trap";
2. Z $\quad$ 100, $n_{nu} \quad 10^{23}$ cm^{-3}, $d \quad 10^{-5}$ cm, $L < 10$ cm, $\mathcal{E}_\gamma \quad$ 1 MeV, $\mathcal{E}_x \quad$ 100 KeV; it can be solid Ag^{110}.

3.6.2 Laser Triggering Nuclear Reaction γ - Source

According to Fig. 3.6.1 we now consider fast electron generation by laser pulse, its conversion into fast ions and then nuclear reactions for γ - photon production with help of these ions.

We already estimated the number of fast electrons produced by a laser pulse. It has been shown in Sect.3.1 that the absorption coefficient, in the range 10^{18}-10^{20} W/cm^2, $\eta \propto L$, so we will use the scaling of $\eta(L,I)$ as: $\eta(L,I) \quad (0.1+0.01L)\, I_{18}/(30+I_{18})^{0.7}$. At pre-pulse intensity 10^{12} W/cm^2 we have L=10 at t_{pl}=10 ps. In Sect. 3.1, we considered the physical mechanism of formation of fast electron jets due to the ponderomotive light pressure. We proposed that $N_{ef} \quad K_e(I)\mathcal{E}_L/\mathcal{E}_e$ electrons are accelerated during the laser pulse where ε_L is the laser pulse energy, $K_e(I)$ is the transformation coefficient of laser energy into fast electron energy, and $\mathcal{E}_e$ is the energy of fast electron. It has been shown before that for laser intensities more then 10^{20} W/cm^2, $K_e(I) \approx \eta$. The energy of fast electron is specified by laser intensity inside the skin-layer $\mathcal{E}_e \approx mc^2[(1+(2-\eta)\, I_{18})]^{1/2}$. From these formulas we can estimate N_{ef} for the following plasma and laser pulse parameters: $I_{18}=10^2$, $\lambda=10^{-4}$ cm, $t_L = 100$ fs, $S = 10^{-6}$ cm^2, so for the chosen parameters N_{ef} 10^{10} electrons.

Hard γ-Quanta Production by Fast Electrons

We consider the hard component of the bremsstrahlung radiation of high-speed electrons formed under the action of a laser pulse on a target, the quantum energy being higher than the rest energy of an electron. Photons with such energy take part in the further production of electron-positron pairs and in nuclear reactions. We estimate the total energy of γ-quanta produced E_γ and laser energy-γ-quantum one conversion factor K_γ. The bremsstrahlung radiation energy is defined by the known Bete-Gaitler's equation [170]

$$E_\gamma = \frac{4e^6 Z^2 n_{nu}}{m^2 c^4 \hbar} \int_0^{\tau_{ef}} dt \int_{V_{ef}} dV \int_{mc^2}^{\infty} d\mathcal{E} f(\mathcal{E},t) \mathcal{E}\left[\ln\left(2\mathcal{E}/mc^2\right) - 1/3\right], \qquad (3.6.7)$$

where the space integral of the distribution function $f(\mathcal{E}, t)$ gives the total number of the high-speed electrons at the given instant of time t, and τ_{ef} is the effective lifetime of high-speed electrons, which comprises the laser pulse length and the time of free path of the electrons produced $(n_{nu} c \sigma^{f}_{tot})^{-1}$. For the γ-quantum energies considered, $\hbar\omega_\gamma$, the screening and density effects in the bremsstrahlung radiation are insignificant. By virtue of the weak logarithmic dependence of third multiplier in integral (3.6.7) on electron energy $\mathcal{E}_e$, this integral is estimated as

$$E_\gamma = N_f \mathcal{E}_e \frac{4e^6 Z^2 n_{nu} \tau_{ef}}{m^2 c^4 \hbar} \left(\ln\left(2\mathcal{E}_e/mc^2\right) - 1/3\right) = K_\gamma(I)\mathcal{E}_L \qquad (3.6.8)$$

For electron energies of several mega-electron-volts, the time of free travel is substantially larger than the laser pulse duration. To calculate τ_{ef}, we use Bete-Bloch's equation [168]

$$\tau_{ef} = \frac{3\mathcal{E}_e}{4mc^3 Z n_{nu} \sigma_T \ln\left(\mathcal{E}_e^3 / 2J_z^2 mc^2\right)} \qquad (3.6.9)$$

where σ_T is the Thompson scattering cross-section and J_z is the mean ionisation potential of an atom. Then we obtain the following simple expression for the conversion factor:

where σ_T is the Thompson scattering cross-section and J_z is the mean ionisation potential of an atom. Then we obtain the following simple expression for the conversion factor:

$$K_\gamma(I) = K_e(I)\frac{9Z^2\varepsilon_e\left(\ln\left(\varepsilon_e/mc^2\right)+0.36\right)}{8\pi\hbar cmc^2\ln\left(\varepsilon_e^3/2J_z^2mc^2\right)} \qquad (3.6.10)$$

The laser radiation intensity enters in Eq. (3.6.10) through the characteristic electron energy ε_e and $K_e(I)$. Dependence $K_\gamma = K_\gamma(I,L)$ from laser intensity for L=0,10 at $Z = 10$ $J_z = 50$ KeV are shown on **Fig. 3.6.4.**

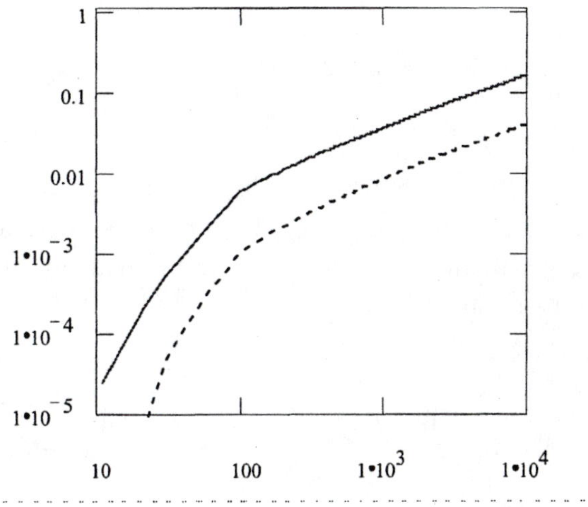

Figure 3.6.4. Dependence $K_\gamma = K_\gamma(I,L)$ from laser intensity I for L=0,10 at Z = 10, here K_γ at L=0 - blue line, at L=10 - red line.

With these parameters and the radiation intensity I_{18}=100, K_γ reaches 0.01 and the total number of γ-quanta appears to be $N_\gamma \sim 10^8$. As for ultra-relativistic electrons the main energy losses channel is radiation losses and almost all absorbed laser energy transformed into fast electrons we have some saturation in $K_\gamma(I)$ up to absorption coefficient at ultrahigh intensity.

In the case of a target with small Z, ions are accelerated under the action of ponderomotive pressure and electrostatic force. Moreover, at given energy the nuclear reaction cross-section drops as Z increases; therefore, this channel works for light-atom elements. If we suppose that return current has a time to neutralize positive ion charge arriving from pondermotive pressure, then (see Sect.3.4), ion energy takes the form:

$$\varepsilon_{ih} \approx Z(n_e/n_c)mc^2\,(2-\eta)I_{18}[1+(2-\eta)\,I_{18})]^{-1/2}.$$

Production of γ- Photons via Laser Fast Ion-Ion Collisions

As the result of laser ion acceleration there is some distribution in ion energy which depends on laser time and space profile. Nuclear reaction cross-sections have complicated behavior and depend on particle energetic distribution parameters. We model these cases by the next formula for the number of nuclear reactions:

$$N_{nur} = n_a \int_0^{\infty} d\mathcal{E} \frac{dN_i}{d\mathcal{E}} \int_0^{\mathcal{E}} d\mathcal{E}\sigma(\mathcal{E}) \left| \frac{d\mathcal{E}}{dx} \right|^{-1} \tag{3.6.11}$$

here from Bete formula[168]:

$$\frac{d\mathcal{E}}{dx} = -\frac{m_a m_i}{m_a + m_i} \frac{2\pi e^4 Z_i^2 Z_a}{m\mathcal{E}} n_a \ln(4m\mathcal{E}\frac{m_a + m_i}{J_Z \; m_a m_i}) \text{, where } \quad n_a, m_a, Z_a -$$

concentration, mass and charge of target atoms, J_Z – their ionisation potential.

If fast ions leave skin depth area before laser pulse is finished we have stationary ion distribution in energy. In this case from ion kinetic equation we obtain the following equation for fast ion number [171]:

$$F\frac{\partial N_i}{\partial \mathcal{E}} = \frac{\partial N_i}{\partial x} \approx n_{i0} S \text{, where } F \approx -Zmc^2 \frac{\partial}{\partial x}\sqrt{1+I_{18}} \approx \frac{1}{2l_s} \frac{\mathcal{E}(\mathcal{E}+2Zmc^2)}{\mathcal{E}+Zmc^2} \text{ -}$$

force acting on ions.

From these equations we obtain the next ion distribution:

$$\frac{\partial N_i}{\partial \mathcal{E}} \approx n_{i0} S l_s \frac{(\mathcal{E}+Zmc^2)}{\mathcal{E}(\mathcal{E}+2Zmc^2)} \theta(\mathcal{E}-\mathcal{E}_m) \tag{3.6.12}$$

Here $\Theta\left(\mathcal{E}-\mathcal{E}_m\right)$ – step function and $\mathcal{E}_m$ – maximum of ion energy from (see Sect.3.4)

$$\mathcal{E}_m \approx Z\, mc^2\, [(1+\eta I_{max})^{1/2} - 1].$$

By way of example, we consider the reaction p + t = γ + ^{4}He, whose maximum cross-section is $\propto$ 2 mbarn for a proton of 8-MeV energy and photon energy is near 5 MeV [167]. Such a reaction can proceed in a tritium target on a substrate. The number of photons produced in proton tritium nuclear collisions can be estimated from the equation (3.6.9):

$N_\gamma^{(t+p)} = N_p\, \sigma^{(t+p)}\, n^t_{nu}\, l^i_{ef}$,

where N_p is the total number of high-speed protons with energy $\mathcal{E}$ and l^i_{ef} is their free path length

$$l^i_{ef} = \mathcal{E} / 2\pi Z^t n^t_{nu} \ln\left(4\mathcal{E} J_z\right)$$

For numerical estimates, we can use the following parameters: $n_{nu} = 6x10^{22}$ cm^{-3}, ionization potential $J_Z = 10$ eV. The thickness of the T- ice target is near 1 mm then $l^i_{ef} =$ 1 mm and does not depend from energy. The number of fast protons is $N_p \approx N_{he}$ - number of fast electrons, from quasi-neutrality. The dependencies of photon number $N_\gamma^{(t+p)}$ from laser intensity I are shown on **Fig. 3.6.5** for L=10.

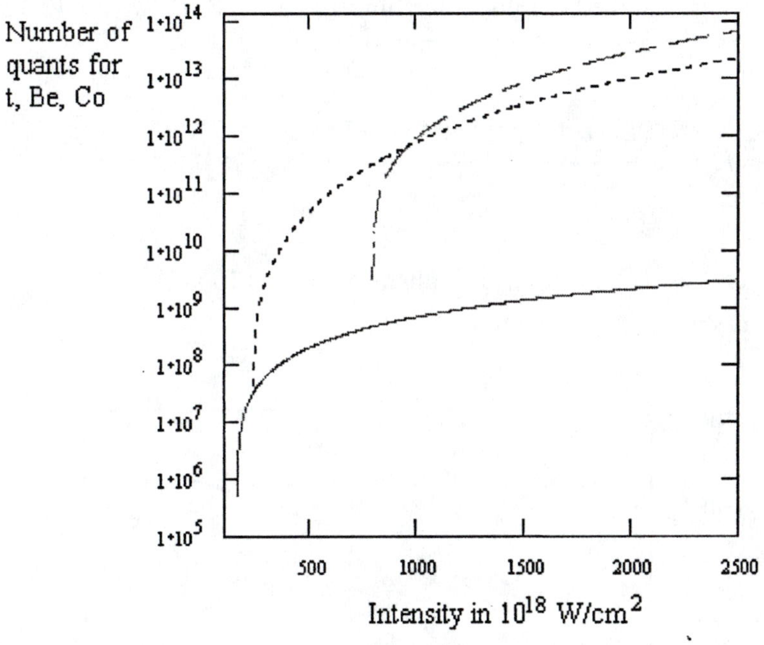

Figure 3.6.5. N_γ-quanta yield versus laser intensity, here $N_\gamma^{(t+p)}$ at L=10 - red line, $N_\gamma^{(Be+p)}$ - blue line, $N_\gamma^{(Co+p)}$ - green line.

On Fig. 3.6.5 there are the number of γ-quanta for Be^{10} ($E_\gamma = 0.2$ MeV) and Co^{60} ($E_\gamma =$ 2 MeV) targets with thickness greater than l_{ef} in the following nuclear reactions: $p+Be^{10} \rightarrow B^{11}+\gamma$; $p+Co^{60} \rightarrow Ni^{59}+\gamma+2n$. In these targets the γ-quanta yield is more compare to T-ice target from reaction cross-section and more thickness, but the largest neutron yield has a higher laser intensity threshold because the high Z nuclei have higher Coulomb barrier.

Finally we estimate the energy of γ-quanta produced in spectral range for γ-photon registration ΔE_γ and laser energy - γ-quantum one conversion factor K_γ. The bremsstrahlung radiation energy is defined by following formula:

$$\Delta E_\gamma = \frac{4e^6 Z^2 n_{nu}}{m^2 c^4 \hbar} \int_0^{\tau_{ef}} dt \int_{V_{ef}} dV \int_{mc^2}^{\infty} d\mathcal{E} f(\mathcal{E},t)\mathcal{E}\left[\ln\left(2\mathcal{E}/mc^2\right)-1/3\right], \qquad (3.6.13)$$

where the space integral of the distribution function $f(\mathcal{E}, t)$ gives the total number of high-speed electrons at a given time t, and τ_{ef} is the effective lifetime of fast electrons, which comprises the laser pulse length and the time of free path of the electrons produced $(n_{nu}c\sigma^{e}_{tot})^{-1}$, $(\mathcal{E}_2 - \mathcal{E}_1)$ – spectral range for γ-photon registration.

For electron energies of several mega-electron-volts, the time of free travel is substantially larger than the laser pulse duration. To calculate τ_{ef}, we use (3.6.9), then we obtain the following expression for the conversion factor:

$$K_{\Delta\gamma}(I) = K_e(I)\frac{9Ze^2\varepsilon_e\left(\ln\left(\varepsilon_e/mc^2\right)+0.36\right)}{8\pi hcmc^2\ln\left(\varepsilon_e^3/2J_z^2mc^2\right)}\left(\varepsilon_2-\varepsilon_1\right)/mc^2$$

Dependence $K_\gamma = K_\gamma(I,L)$ from laser intensity for L=0,10 at Z = 10, J_Z = 50 keV, $(\varepsilon_2 - \varepsilon_1)/mc^2 \approx 10^{-6}$ is shown on **Fig. 3.6.6.**

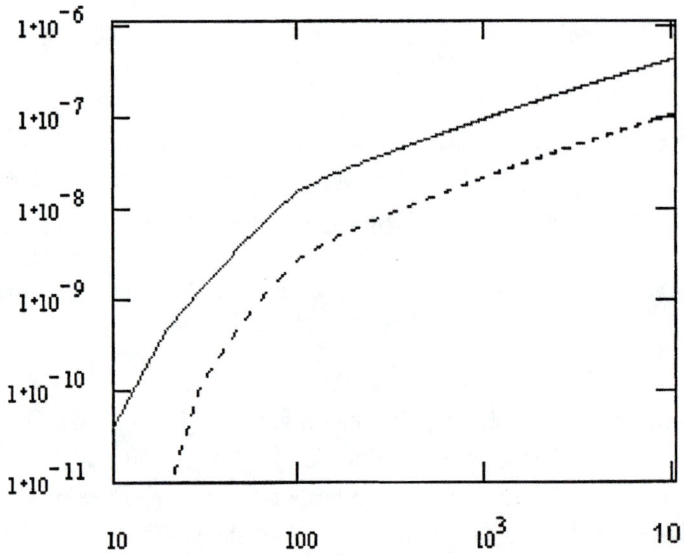

Figure 3.6.6. Dependence of conversion efficiecy $K_{\Delta\gamma}$ from laser intensity I for different plasma scale length, here $K_{\Delta\gamma}$ at L=0 – dashed line, at L = 10 solid line.

With these parameters and radiation intensity I_{18}=100, $K_\gamma \approx 10^{-8}$ and the total number of γ-quanta appears to be $N_\gamma \sim 10^2$. From these calculations we can conclude that number of the MeV photons in the line spectral range is much less than for nuclear transition emission.

3.6.3 High-Power Laser Plasma Source of Nuclear Reaction

Electron-Positron Pair Production by a Laser Pulse

Now we turn to estimating the number of electron-positron pairs produced via reactions e + Z = 2e + e $^+$+ Z and e + Z = γ + e + Z = 2e + e $^+$+ Z. The electron-positron pairs can also be produced via other reaction channels, for example, γ + γ = e + e $^+$, e + e = e $^+$+ 3e, and γ + e=e $^+$+ 2e, however, these effects are insignificant if intensity I_{18} 10^3. The positron production via electron-ion collisions has been considered in a number of papers [137,172,173]. Therefore, we present here only the main results for this reaction, neglecting the consideration of targets of finite thickness, cascade processes, and similar complicating factors. The cross section for the pair production by an electron in a nuclear field is well known [174]

$$\sigma_{ei} = \frac{7}{18\pi^2}\sigma_T \alpha^2 Z^2 n^3 \left(0.66 + \mathcal{E}_e / 3.5\right), \qquad (3.6.14)$$

where α = 1/137 is the fine structure constant and $\mathcal{E}_e$ is the electron energy in mega-electron-volts. Electron free path length l_{ef} is related to the above-defined time τ_{ef} via the equation $l_{ef}=c\tau_{ef}$. Thus, for a thick target and ignoring cascade processes, the number of e $^+$ e$^-$ pairs in the *ei* -channel can be estimated from the equation

$$N_{ei}^P = N_i \sigma_{ei} n_{nu} l_{ef}\,. \qquad (3.6.15)$$

Now we proceed to the consideration of the second pair production channel γ + Z = e + e $^+$+ Z. From formula Bete-Haitler [168] we obtain the photon energy distribution $dN_\gamma/d\varepsilon_\gamma$ and the number of pairs produced via this channel equals:

$$N_{\gamma i}^P = \int_{2mc^2}^{\varepsilon_e} \frac{dN_\gamma}{d\varepsilon_\gamma} \frac{\sigma_{\gamma i}^P\left(\mathcal{E}_\gamma\right)}{\sigma_{tot}^\gamma\left(\mathcal{E}_\gamma\right)}, \qquad (3.6.16)$$

where $\sigma_\gamma^P\left(\mathcal{E}_\gamma\right)_i$ is the cross section for the pair production by a photon with energy $\mathcal{E}_\gamma$ found in a nuclear field with charge Z [168]

$$\sigma_\gamma^p = (7/6\pi) Z^2 \alpha \sigma_T \left(\ln\left(2\mathcal{E}_\gamma / mc^2\right) - 109/42 \right), \tag{3.6.17}$$

and σ_{tot}^γ is the total cross-section for inelastic γ -quantum scattering, which, in addition to the pair production, includes Compton's cross-section σ_k and the photo-ionisation cross-section for atoms in the target. Estimates of photo-ionisation cross-sections for photons with energy of several mega-electron-volts [168] yield the values by an order of magnitude smaller than Compton's cross-section

$$\sigma_k\left(\mathcal{E}_\gamma\right) = \left(3Z\sigma_T mc^2 / 8\mathcal{E}_\gamma\right)\left(\ln\left(2\mathcal{E}_\gamma / mc^2\right) + 1/2\right). \tag{3.6.18}$$

Let us comparatively assess the above two pair production channels. On the order of magnitude, the relation $N_{ei}^p / N_{\gamma i}^p$ makes up

$$N_{ei}^p / N_{\gamma i}^p = \frac{K_e(I)\sigma_{tot}^\gamma \sigma_{ei}^p}{K_\gamma(I)\sigma_{\gamma i}^p \sigma_{tot}^e}, \tag{3.6.19}$$

where $\sigma_{tot}^e = 1/l_{ef} n_{nu}$ is the total cross section for the high-speed electron losses. The electron-to- γ -quantum conversion factor K_γ was calculated by Eq. (3.6.10), K_γ being found to vary from 10^{-4} to 0.1 for intensities I_{18} lying in the10-10^3 range. The partial cross section for the γ -quantum conversion to an electron-positron pair, which enters in Eq. (3.6.19), reads

$$\sigma_\gamma^p / \sigma_{tot}^\gamma = \left(1 + mc^2 / Z\alpha\mathcal{E}_e\right)^{-1}. \tag{3.6.20}$$

For an Au atom (Z = 79) and radiation intensity $I_{18} = 10^2$, the partial cross section is close to unity, whereas the partial cross section for the production of high-speed electron pairs at the same parameters appears to be

$$\sigma_{ei}^p \sigma_{tot}^e = Z\alpha^2 \mathcal{E}_e / 2\pi^2 mc^2 \approx 1.3\ 10^{-3} \tag{3.6.21}$$

Substituting the partial cross sections and the γ -quantum conversion factor in Eq. (3.6.19), we see that the electron-positron pair production channel, attributed to the γ -quantum conversion to an electron-positron pair in a nuclear field, prevails in materials with the large Z. In an Au target, this channel yields the number of pairs by an order of magnitude larger than do commonly considered electron-ion collisions ($N_{ei}^p / N_{\gamma i}^p \sim 0.1$).

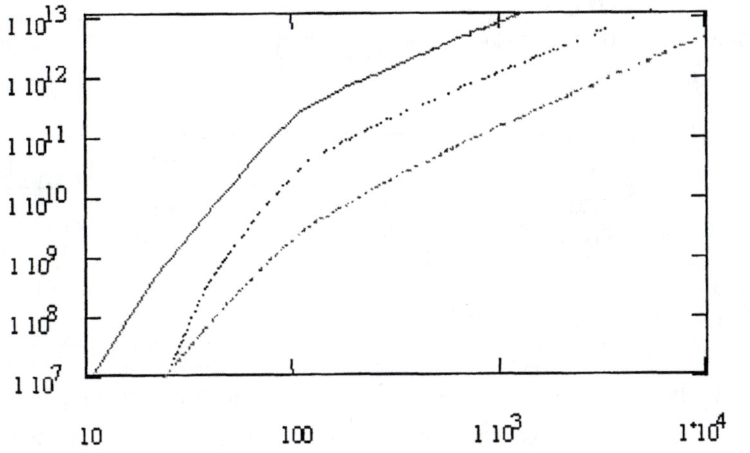

Figure 3.6.7. Dependence of electron-positron pair number yield N^p from laser intensity for Au target, here N^p at *L*=10 – solid red line, N^p at *L*=0 - doted blue line and N^p_{ei} at *L*=10 – dashed green line.

Dependences N^p_{ei} and total number of pair $N^p = N^p_{ei} + N^p_{\gamma i}$ from laser intensity are shown on **Fig. 3.6.7** for laser pulse duration 100 fs and *L*=0,10. This figure shows that for heavy materials the number of electron positron pair due to γ-quanta scattering is more then a number pair from electron-ion collisions considered in [137].

Neutron Production in Laser Plasma

When the energy of high-speed electrons exceeds several mega-electron-volts, that is, when the intensity is higher than 10^{19} W/cm^2, nuclear reactions producing neutrons are possible in laser plasma formed. The neutrons can be produced via three independent channels: the direct high-speed electron-nucleus interaction (electro-fission of nuclei), the γ - quantum-nucleus interaction (photofission), and the high-speed ion-nucleus interaction. Let us consider each of these channels in detail.

A. Nuclear photo-fission

The cross section for the photo-neutron production from a nucleus is defined by Breit-Wigner's equation [174]

$$\sigma(n,\gamma) = \max\sigma(n,\gamma)\Gamma_{nu} / \left[(\varepsilon - \varepsilon_0)^2\right]^{1/2}, \tag{3.6.22}$$

where Γ_{nu} is the width of a nuclear level with energy ε_0 and ε is the full width of a γ - quantum absorption line, which includes Compton's scattering and ionization

$$\Gamma = \Gamma_{nu} + n_e(hc/2\pi)\sigma_k + n_e(hc/2\pi)\sigma_{ion}$$

σ_k being Compton's scattering cross section.

Typical magnitudes of the parameters entering in Eq. (3.6.22) are listed in table below for various nuclei.

Nucleus	$\mathcal{E}_0$	Γ_{nu}	$\max\sigma$	$\sigma_k(\mathcal{E}_0)$
Be^9	1.71 MeV	50 keV	1.15 mb	1790 mb
$_{71}Lu^{175}$	16 MeV	8.4 MeV	255 mb	370 mb

The cross sections are given here in millibarns (1b = 10^{-25} cm^2).

Data from the above table is evidence that photo-neutrons can be observed only in media with the high Z, Compton's losses (σ_k) being of the same order of magnitude as $\max\sigma$ and width Γ_{nu} being large enough (on the order of energy $\mathcal{E}$ itself). Such a material can be exemplified by Lu^{175} in the vicinity of the giant resonance (16 MeV). Using Eq. (3.6.7) to estimate the number of γ-quanta emitted within width Γ_{nu}, we derive the equation for the total number of photo-neutrons

$$N_n^{(\gamma)} = \max\sigma(n,\gamma)(\Gamma_{nu}/\mathcal{E}_0)(hc/2\pi\Gamma)n_{nu}N_\gamma\,, \qquad (3.6.23)$$

where N_γ is the total number of γ-quanta with energy higher than mc^2.

Thus the γ-quantum-to-neutron conversion factor is defined by the equation

$$K_{\gamma n} = (\Gamma_{nu}/\mathcal{E}_0)\max\sigma(n,\gamma)/\left[\max\sigma(n,\gamma)+\sigma_k\right]$$

For $^{71}Lu_{175}$, $K_{\gamma n}$ reaches 0.15, and for light nuclei, such as Be^9, $K_{\gamma n}$ is as small as 2 10^{-5}.

B. High-Speed Electron-Induced Nuclear Electro-Fission

The heavy-nuclear electro-fission process has been considered in Ref. [137]. In this case, the necessary threshold energy amounts to 10 MeV, which corresponds to a laser intensity of 10^{21} W/cm^2. The electro-fission cross section for heavy nuclei makes up as small a fraction of the photo-fission cross section as 0.01. Therefore, the number of neutrons produced by photo-fission amounts to

$$N_n^{(efis)} = 0.01\, N_n^{(phfis)}\, K_e/K_\gamma,$$

as compared with the number of photo-neutrons. For an intensity of 10^{21} W/cm^2, the K_e/K_γ less then 10^2 and the electro-fission process contributes less to the neutron production than does the photo-fission.

C. Neutron Production via Ion-Ion Collisions

In the case of a target with the small Z, ions are vigorously accelerated under the action of the ponderomotive pressure force. Moreover, at the given energy the nuclear reaction cross-section drops as Z increases; therefore, this channel works for light-atom elements. As we know, the characteristic speed of an ion takes the form $u \approx 0.5\ c\ (n_c m_e I_{18}/ n_e m_p)^{1/2}$. When laser intensity $I_{18} = 10^3$, the ion energy amounts to 500A keV. If A 10, such energy is sufficient for nuclear fission reactions to proceed via ion-ion collisions. By way of example, we consider reaction t + ^{7}Li = n + ^{9}Be, whose cross-section is $\sigma \approx 1$ mbarn for a 1MeV energy of a tritium nucleus. Such a reaction can proceed in a tritium target on a lithium substrate. The number of neutrons produced in tritium-lithium nuclear collisions can be estimated from the equation

$$N_n^{(t+Li)} = N_t\ \sigma^{(t+Li)}\ n_{nu}^{\ Li}\ l^i_{ef}\ , \tag{3.6.24}$$

where N_t is the total number of fast ions with energy $\mathcal{E}_{ih}$ and l^i_{eff} is their free path length $l^i_{ef} = \mathcal{E}_{ih}^{\ 2}/2\pi Z^{Li} n_{nu}^{\ Li}\ \mathcal{E}_{ih}^{\ 4} \ln(4\mathcal{E}_{ih}/J)$.

For numerical estimates, we can use the following parameters: n_{nu} = 6 10^{22} sm^{-3}, ionization potential J_z = 20 eV, and $\mathcal{E}_{ih}$ = 1 MeV. Then l^i_{ef} = 1 cm and $N_n^{(t+Li)} = 0.01\ N_t$. If, from the quasi-neutrality condition, we assume the number of high-speed ions to be equal to the number of high-speed electrons divided by the nuclear charge, then $N_n^{(t+Li)}$ = 0.01 N_e .

The dependencies of neutron number $N_n^{(t+Li)}$ from laser intensity I_{18} are shown on **Fig. 3.6.8** for L=0,10. In [173] was obtain ~ 10^2 neutrons when the laser pulse at intensity < 10^{19} W/cm^2 interacted with D_2 solid target.

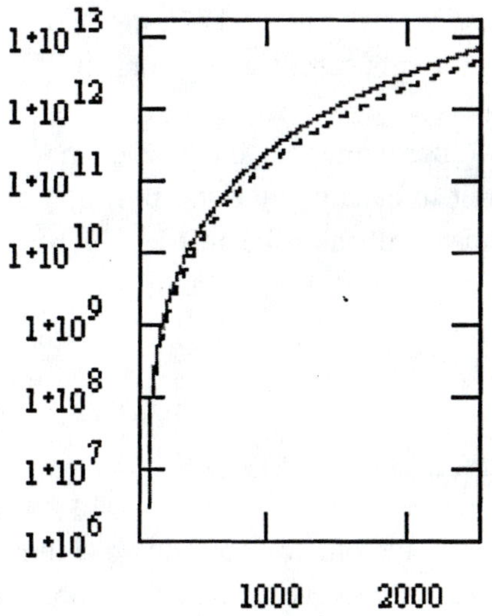

Figure 3.6.8. Dependencies of neutron number yield from laser intensity, here $N_n^{(t+Li)}$ at L=10 – solid line, $N_n^{(t+Li)}$ at L=0 - dashed line.

Estimations from (3.6.24), with the cross-sections for this material, confirm this result. Thus, a high-power laser pulse can be used as a γ photon-emitting source for investigating physics of solids, for biology, and also for stimulating nuclear reactions.

3.7 Non-Linear Optical Phenomena and Damage of Vacuum in the Field of Super-Strong Laser Radiation

One of the most interesting effects predicted by quantum electrodynamics is photon-photon scattering, and the cause of it is vacuum polarization. This phenomena was successfully studied since the 1930s [175]. Now scientific interest in these processes is experiencing a new peak. And the reason is as follows: development of high-power lasers with femtosecond pulses of petawatt power made actual the investigation of influence of vacuum polarization on parameters of radiation.

The central effect of vacuum polarization is the electron-positron pairs production in the strong light fields. Although modern lasers are unable yet to produce them and the maximum of this phenomenon is laying in the X-ray region, some other effects are already observable. Indeed, if we consider, already discussed in Part 2, laser parameters, such as energy of 1 MJ , pulse duration of 10 femtoseconds and focus area of 10^{-8} cm^2 , than we receive the radiation intensity of 10^{28} W/cm^2 · This means that the value of electric field intensity in the focal area is only several times smaller than the critical one (we call the value of electric field critical when it starts to produce electron-positron pairs).

In the recent papers the processes arising from the interaction of intensive polarized light beams were considered. Among them "four-wave" interaction [32,176], double refraction of vacuum and polarization turning [177], etc. However most of these works were done in the frames of Heisenberg-Euler approach [175], that considers the local limit of vacuum polarization [170].

In this section we reject approximation and use quantum electrodynamics in the quasi-classical (non-local) limit to calculate the vacuum polarization. This allows us to find and analyze some non-linear optical effects as higher harmonic generation of laser radiation in vacuum [178] or physical threshold of vacuum transparency for super-intensive laser pulse [31].

3.7.1 Electromagnetic Processes

As a model of the present situation, we consider quantum electrodynamics in the field of a classical electromagnetic wave. For description of pure electromagnetic processes the formalism of photon Green functions can be used. The most convenient representation for the generating functional *G(J)* of these functions is one of functional integral:

$$G(J)=\int e^{iS(\bar{\psi},\psi,A)+JA}D\bar{\psi}D\psi DA \tag{3.7.1}$$

Here the standard quantum electrodynamics (QED) action $S(\bar{\psi},\psi,A)$ [170] has the form:

$$S(\bar{\psi},\psi,A)=\bar{\psi}\left(i\hat{\partial}+e\hat{A}+eA_{cl}-m\right)\psi-\frac{1}{4}F_{\mu\nu}F^{\mu\nu} \tag{3.7.2}$$

where $F^{\mu\nu}\equiv\frac{\partial A^{\nu}}{\partial x_{\mu}}-\frac{\partial A^{\mu}}{\partial x_{\nu}}$, A_{cl}^{μ} is vector potential of classical external field, A^{μ} is the quantized electromagnetic field, and $\bar{\psi},\psi$ are the fermionic spinor fields of electrons and positrons. The vector potential A_{cl}^{μ} models laser radiation. Here we used units where are $h=c=1$. The term $\frac{1}{4}F_{\mu\nu}F^{\mu\nu}$ is the ordinary free electromagnetic part of the action and the other represents the interaction of photonic field with electron-positron field. After derivation $G(J)$, with respect to J, n times and then letting $J=0$ we receive n-point photon Green function.

Using the representation (3.7.1) the fermionic fields contribution can be taken into account exactly. In other words an integral over fermionic fields can be taken. As a result we receive $G(J)$ in the form of functional integral over field A :

$$G(J)=\int e^{iS_{eff}(A)+JA}DA \tag{3.7.3}$$

where $S_{eff}=\bar{S}(A)-\frac{1}{4}F_{\mu\nu}F^{\mu\nu}$, $\bar{S}(A)$ is the correction to the electromagnetic part of the action, that arises because of interaction of photonic field with fluctuations of electron-positron vacuum. It can be written in the following form:

$$i\bar{S}(A)=Tr\ln(i\hat{\partial}-m+e(\hat{A}+\hat{A}_{cl}))-Tr\ln(i\hat{\partial}-m) \tag{3.7.4}$$

The functional $\bar{S}(A)$ is non-local and for sufficiently small field A_{μ} the right hand side of (3.7.4) can be presented by the convergent power series

$$\bar{S}(A)=\bar{S}(0)+i\sum_{n=1}^{\infty}\int\frac{(ie)^{n}}{n}\prod\nolimits_{\mu_1\ldots\mu_n}\left(x_1\ldots x_n\middle|A_{cl}\right)A^{\mu_1}(x_1)\ldots A^{\mu_n}(x_n)\,dx_1\ldots dx_n \tag{3.7.5}$$

Here

$$\prod_{\mu_1\ldots\mu_n}(x_1\ldots x_n|A_{cl}) = Tr\left(\gamma_{\mu_1}D(x_1,x_2)\gamma_{\mu_2}D(x_2,x_3)\ldots\gamma_{\mu_n}D(x_n,x_1)\right) \tag{3.7.6}$$

and $D(x,y) \equiv D(x,y|A_{cl})$ denotes the Green function for the Dirac equation in the field A_{cl}:

$$\left(i\hat{\partial} - m + e\hat{A}_{cl}\right)D(x,y) = -i\delta(x-y) \tag{3.7.7}$$

The total electromagnetic field A_{tot} consists of the laser radiation field and vacuum polarization contribution: $A^{\mu}_{tot} = A^{\mu}_{cl} + A^{\mu}_{vac}$. In the quasi-classical approximation the equations on the field A_{vac} can be obtained by applying the Hamilton principle to the total action: $\dfrac{\delta S_{eff}(A)}{\delta A^{\mu}} = 0$. The field A_{cl} in formula (3.7.4) is taken into account exactly, assuming it is a strong field. If $\overline{S}$ is to be expanded in a row with respect to the sum of the fields $\tilde{A}^{\mu} = A^{\mu} + A^{\mu}_{cl}$, then due to the Furry theorem [170] we obtain a row with only even powers:

$$\overline{S}(A) = i\sum_{n=1}^{\infty}\int\frac{(-ie)^{2n}}{2n}\prod_{\mu_1\ldots\mu_n}(x_1\ldots x_n)\tilde{A}^{\mu_1}(x_1)\ldots\tilde{A}^{\mu_{2n}}(x_{2n})dx_1\ldots dx_{2n} \tag{3.7.8}$$

where $\prod_{\mu_1\ldots\mu_n}$ can be received from definition (3.7.5) and equation (3.7.7) when $A_{cl} = 0$. For the case of slowly changing field A (i.e if for the field frequency we have ω/m<<1) the electric field amplitude, corresponding to vector-potential A, can be substituted by constant value. Thus the well-known Heisenberg-Euler approximation can be obtained.

A simple derivation of these expressions can be found, for example, in [178]. The first correction to the main approximation in (3.7.8) come from the fourth-order term $S^{(4)}$ of the total action $\overline{S}$:

$$S^{(4)} = \frac{ie^4}{4}\int\prod_{\mu_1\mu_2\mu_3\mu_4}(x_1,x_2,x_3,x_4)\tilde{A}^{\mu_1}(x_1)\tilde{A}^{\mu_2}(x_2)\tilde{A}^{\mu_3}(x_3)\tilde{A}^{\mu_4}(x_4)dx_1dx_2dx_3dx_4 \tag{3.7.9}$$

In the local limit ω/m<<1 (or in other words when the field is slowly varying with time) the formula (3.7.9) has the following form [170]:

$$S^{(4)} = \frac{1}{180}\frac{\alpha^2}{m^4}\int\left[-5\left(\tilde{F}_{\mu\nu}\tilde{F}^{\mu\nu}\right)^2 + 14\left(\tilde{F}_{\mu\nu}\tilde{F}^{\nu\lambda}\tilde{F}_{\lambda\sigma}\tilde{F}^{\sigma\mu}\right)\right]dx \tag{3.7.10}$$

where α is the fine structure constant and

$$\tilde{F}_{\mu\nu} = \partial_\mu \tilde{A}_y - \partial_y \tilde{A}_\mu$$

If we vary the formula (3.7.10) with respect to the field, we will obtain the equations for A^μ_{vac}. These equations were used for calculating non-linear effects in the papers [29,170].

In order to receive the explicit equations for $A^\mu \equiv A^\mu_{vac}$ in quasi-classical approximation we rewrite the formula (3.7.7) as:

$$D_{\mu\nu}A^\nu + \frac{\delta \overline{S}}{\delta A} = 0 \tag{3.7.11}$$

where

$$D_{\mu\nu} = \partial^2 g_{\mu\nu} - \partial_\mu\partial_\nu, \quad \partial_\mu = \frac{\partial}{dx^\mu}.$$

Equation $D_{\mu\nu}A^\nu = 0$ represents the ordinary Maxwell equations in our notations and the second term in (3.7.11) represents the correction that appeared because of the vacuum polarization.

In the case of weak field A^μ, equation (3.7.11) can be replaced by the linear approximation:

$$K_{\mu\nu}A^\nu = e\Pi_\mu. \tag{3.7.12}$$

Here

$$K_{\mu\nu} \equiv D_{\mu\nu} - ie^2\Pi_{\mu\nu}.$$

and $\Pi_{\mu\nu}, \Pi_\mu$ are defined by (3.7.5).

Finally for A^μ we have an integro-differential equation:

$$D_{\mu\nu}A^\nu(x) - \int \Pi_{\mu\nu}(x,y)A^\nu(y)\,dy = -ie\Pi_\mu(x). \tag{3.7.13}$$

As a model for the laser radiation field A_{cl} a simple plane wave can be considered:

$$A_{cl}^{\mu} = e^{\mu} e^{ikx}$$

Here $^{\mu}$=0 is the polarization vector: ^{0}= 0, $\vec{e}\vec{\kappa} = 0, \kappa = (\kappa_0, \vec{\kappa})$ is the wave 4-vector of laser radiation, $\kappa^2 = \kappa_0^2 - \vec{\kappa}^2 = 0$, $\kappa_0 = \omega / c$, ω is the laser frequency, c is the velocity of light. For such a field A_{cl}, $\Pi_{\mu} = 0$ and the explicit form of $\Pi_{\mu\nu}$ [178] was obtained from definition (3.7.5) using $D(x,y|A_{cl})$ constructed in [179]. Therefore now we have an explicit equation for A^{μ}:

$$D_{\mu\nu} A^{\nu}(x) - \int \Pi_{\mu\nu}(x,y) A^{\nu}(y) dy = 0 \tag{3.7.14}$$

The solution of equation (3.7.14) has the following form [178]:

$$A^{\nu}(x) = e^{iqx} a^{\nu}\left(e^{ikx}\right). \tag{3.7.15}$$

Here $a^{\nu}(0) \neq 0$ and q^{ν} are the arbitrary parameters of the solution and satisfy the following condition:

$$q^{\nu} a_{\nu}(0) = 0.$$

The field $A^{\nu}(x)$ is transverse, i.e. satisfies the Lorenz gauge condition: $\partial_{\nu} A^{\nu}(x) = 0$. If these parameters are specified, all coefficients of the functions

$$a^{\nu}(\xi) = \sum_{i=0}^{\infty} a_i^{\nu} \xi^i$$

into power series can be found.

For $\rho = \frac{e^2 (ee)^2}{2m^2} \frac{kq}{m^2} \approx 1$, the field has the following form:

$$A_{\nu}(x) \approx e^{iqx} \left(a_{\nu}(0) + \frac{\rho e^{ikx}}{2} \tau_{\mu\nu} a^{\mu}(0) \right),$$

where the value of vector a^{ν} has the same order of magnitude as vector value $a_{\nu}(0)$. It is clear that the interaction of a plane wave $a(0)e^{iqx}$ with the laser radiation field results in appearance of a new wave with wave vector $q+2k$ and amplitude in ρ times smaller. The ratio of the amplitudes $|a_1|/|a_0|\rho$ can represent the measure of vacuum non- transparency for a plane wave with the wave vector q in the presence of the laser radiation A_{cl}.

The received solution shows that interaction of a strong laser wave with a weak one leads to generation of harmonics of the strong wave.

3.7.2 Vacuum Damage

The phenomena of vacuum non-stability in strong laser field can be quantitatively evaluated on the base of calculation of $G(0)$. It is an amplitude of transition from vacuum-state to vacuum-state. Therefore Re$\ln G(0)$ determines the pair production probability, i.e. damage of vacuum in the laser radiation field. If $W(x,t)$ is a probability of a pair production in place with coordinates x and in time moment t, than $W(x,t)dxdt$=2Re$\ln G(0)$. For the case of slowly changing field A_{cl} , i.e for $\omega/m <<1$, Re$G(0)$ can be properly evaluated as -Im S_{eff} that calculated for the case of a constant field with vector-potential A . Since for the constant field S can be calculated explicitly, the exact formula for probability of pair production can be obtained:

$$W \approx \alpha E^2 \sum_{n=1}^{\infty} \frac{1}{n^2} \exp\left(-\frac{n\pi c^3 m}{|eE|\hbar} \right) \tag{3.7.16}$$

where E is electric field intensity, α is the fine structure constant, e is the electron charge. The derivation of this formula is presented in [170]. The expression (3.7.16) can be used for rough evaluation of vacuum damage probability. Evaluating the quantity of electrons in formula (3.7.16) one can obtain that for intensity of 3 10^{28} W/cm^2 the laser beam would scatter on produced electron-positron cloud.

The investigation of the phenomena of non-stability of vacuum in a strong laser field can be based on analyzing the electron Green functions $D(x, y|A_{cl})$, defined in (3.7.6). For the case of a plane wave, i.e. when $A_{cl} = A^{\mu}(kx)$, this Green function can be calculated explicitly [178]. For purpose it is more convenient to make a Fourier transformation of the Green function with respect to (x – y):

$$D(x, p|A_{cl}(k)) = i\int_0^{\infty} ds(\hat{p}-m)e^{is(p^2-m^2)} \exp\left[ie^2 \int_0^s \left(2p_\mu + \sigma_{\mu\nu} ik^\nu \frac{\partial}{\partial(kx)} \right) A_{cl}^{\mu}(kx-2kp\xi)d\xi + ie^2 \int_0^s A_{cl}^{\mu^2}(kx-2kp\xi)d\xi \right]. \tag{3.7.17}$$

We used the standard designation $\sigma_{\mu\nu}$ for the expression $(1/2)(\gamma_\mu\gamma_\nu - \gamma_\nu\gamma_\mu)$, where γ_μ are the Dirac matrices. Presence of the external field A_{cl}^{μ} violates the translation invariance, and therefore the right-hand side of the formula (3.7.16) depends on space coordinate x.

If we use for the classical external field the form $A_\mu = \mathrm{e}_\mu \sin(kx)$, the main term of this expression is:

$$D(x, p|A_{cl}) \approx i\int_0^\infty ds\,(\hat{p}-m)\,e^{is\left(p^2-\left(m^2-(1/2)e^2\mathrm{e}^2\right)\right)}. \tag{3.7.18}$$

Green function of the Dirac equations in the "free" field case (i.e. without external field) has a form:

$$\hat{D}(p) = i\int_0^\infty ds\,(\hat{p}-m)\,e^{is\left(p^2-m^2\right)}.$$

The calculations in QED require the Feynman diagram technique, where the green function $\hat{D}$ (3.7.18) represent the fermionic $\hat{D}$ propagator. For the case of presence of the strong external field A_{cl} the same technique can be applied. But instead the fermionic propagator $\hat{D}$ we have to use $D(x, y|A_{cl})$. Since the main approximation of $D(x, y|A_{cl})$ has the form (3.7.17), we can conclude that all the formulas received in ordinary QED for A_{cl} =0 can be used with the minimal modification in our case. From expressions (3.7.17) and (3.7.18) it is clear, that in our case m^2 has to be substituted by $m_{eff} = m^2 - (1/2)e^2$ ı ĕ

As an example we can consider the question about electron-positron pairs generation in a constant electromagnetic field E in presence of the external laser field A_{cl}. Repeating the calculations for the probability of pair production [170], we have to obtain the approximation for $\tilde{S}^{(4)}$. If in this expression the propagators $D(x, y|A_{cl})$ are replaced with (3.7.17), then we obtain the same formula for action as in [170] but with m_{eff} in place of m. Afterwards we can continue calculation along the way shown in [170]. The probability of the electron-positron pair production in the constant field looks like (3.7.16) where m is changed for m_{eff}. This probability grows when we increase the laser field intensity (i.e. degrease m_{eff}).

Replacing m by m_{eff} in the well-known formulas for the differential cross-section of e^+ - e^- pair production by photon in the Coulomb field [180] and the total cross-section of e^+ - e^- pair production by two photons [170,180] we obtain:

$$d\sigma = \frac{Z^2\alpha^2}{2\pi^2}\frac{|\vec{p}_+||\vec{p}_-|}{\omega^3}\frac{de_+do_+do_-}{\vec{q}^2}\frac{4}{m_{eff}^4}\left[-4\left(\vec{k}\left(\frac{e_+\vec{p}_-}{k_1}+\frac{e_-\vec{p}_+}{k_2}\right)\right)^2+\right.$$

$$\left.+\vec{q}^2\left(\vec{k}\left(\frac{\vec{p}_-}{k_1}-\frac{\vec{p}_+}{k_2}\right)\right)^2+\frac{2\omega^2}{k_1k_2}\left(\vec{k}\left(\vec{p}_-+\vec{p}_+\right)\right)^2\right] \quad (3.7.19)$$

$$\sigma = \frac{\pi r_0^2 m_{eff}^2}{\omega_0^2}\left(2+\frac{2m_{eff}^2}{\omega_0^2}-\frac{m_{eff}^4}{\omega_0^4}\right)\ln\left(\frac{\omega_0}{m_{eff}}\left(1+\sqrt{1-\frac{m_{eff}^2}{\omega_0^2}}\right)-\left(\sqrt{1-\frac{m_{eff}^2}{\omega_0^2}}\left(1+\frac{m_{eff}^2}{\omega_0^2}\right)\right)\right) \quad (3.7.20)$$

Here r_0 is the Compton electron radius and

$$\omega_0^2 = -1/4\left(k+k'\right)^2,$$

were k, k' are the photon momentum. In the expression for the total cross-section (3.7.20) the threshold of reaction $\omega_0 > m_{eff}$ diminishes with increasing external field. In the formula (3.7.19) $m_{eff}^2 k_1 = -2p_-k$, $m_{eff}^2 k_2 = -2p_+k$, $\vec{p}_\pm$ are the electron and positron momentum, that lay in the solid angles $o_\pm$; $\varepsilon_\pm$ are the corresponding energies, k is photon momentum, Z is the nuclear charge, ω is the photon frequency, $\vec{q} = \vec{k} - \vec{p}_- - \vec{p}_+$. The differential cross-section increases with increasing laser radiation intensity. We see that requirements for damage of vacuum are weakened in the presence of laser radiation. This can have an importance in methodology of experimental studies of QED vacuum.

REFERENCES FOR PART 3

1. Delone N.B., Fedorov M.V. *Sov. Phys. Usp.,* v.32, p.500, 1989.
2. Landen O., Campbell E., Perry M. *Opt. Commun.*, v.63, p.253, 1987.
3. Agostini P., Fabre F., Mainfray G., Petite G. *Phys. Rev. Lett.,* v.42, p.1127, 1979.
4. Ferray M., L`Huillier A., Li X.F.et al. *J. Phys.,* B 21, p.31, 1988.
5. Kulander H.C., Shore B.W. *Phys. Rev. Lett.,* v.62, p.524, 1989.
6. Eberly J.H., Su Q., Javanainen J. *Phys. Rev. Lett.,* v.62, p.881, 1989.
7. Kim A.V., Lirin S.V., Sergeev A.M. et al. *Phys. Rev.,* A 42, p.2493, 1990.
8. Andreev A.A., Semenov V.E. *Opt. Spectrosc.*, v.78, p.533, 1995.
9. Suckever S., Skinner C. *Science*, v.247, p.1553, 1990.
10. Matthews D., Hagelstein P., Rosen M. *Phys. Rev. Lett.*, v.54, v.110, 1985.
11. Corcum P., Burnett N. *Phys. Rev. A*, p.56, 1988.
12. Milchberg H., Freeman R., Davey S. et al., *Phys.Rev.Lett.*, v.61, p.2364, 1981.

13. Fedosejev R., Ottman R., Sigel R. et al. *Phys. Rev. Lett.*, v.64, p.1250, 1990.
14. Murnane M.M., Kapteyn H.C., Falcone R.W *Phys. Fluids*, B 3, p. 2409, 1991.
15. Murnane M.M, Kapteyn H., Rosen M.D., et al., *Science*, v.251, p.531, 1991.
16. Meyerhofer D., Chen H., Delettrez J. et al. *Phys. Fluids,* B5, p.2584, 1993.
17. Pettit G.H., Sauerbrey R. *Appl. Phys.*, A 56, p.51, 1993.
18. Fews A.P., Norreys P.A., Beg F.N. et al. *Phys. Rev. Lett.*, v.73, p.1801, 1994
19. Andreev A.A., Limpouch J., Semakhin A.N. *Pros. SPIE,* v.1980 75 1992.
20. Tajima T., Dawson J.M. *Phys. Rev. Lett.*, v.43, p.267, 1979.
21. Joshi C., Mori W.B, Katsouleas T. et al. *Nature (London)*, v.311, p.525, 1984.
22. Gorbunov L.M. *Priroda*, v.5, p.15, 1989.
23. Boyer K., Luk T.S., Rhodes C.K. *Phys. Rev. Lett.*, v.60, p.557, 1988.
24. Tabak M., Hammer J., Glinsky M.E. et al., *Phys. Plasmas*, v.1, p.1624, 1994.
25. Pukhov A., Meyer-ter-Vehn J., *Phys. Rev. Lett.*, v.79, p.2686, 1997.
26. Wilks S.C., Kruer W.L., Tabak M.et al., *Phys. Rev. Lett.*, v.69, p.1833, 1992.
27. Askary`yan G.A. *JETP Lett.*, v.48, p.193, 1988.
28. Rivlin L.A., *Book of abstracts 12 LIRPP*, Osaka, Japan, p.332, 1995.
29. Moore C.I., Knauer J., D.D.Meyerhofer J.P. *Phys. Rev. Lett.*, v.74, p.2439, 1995.
30. Ritus V.I. *Tr. Fiz. Inst. Akad. Nauk SSSR*, v.111, p.3, 1979.
31. Andreev A.A. et al. *Proc. SPIE*, v.3735, p.170, 1998.
32. Rozanov N.N. *ZhETPh*, v.76, p.991, 1993.
33. Rozmus W., Tikhonchuk V.T. *Phys. Rev.* A 42, p.7401, 1990.
34. Pert G. *Phys. Rev.*, E 51, p.4778, 1995.
35. Andreev A.A., Bayanov V.I., Vankov A.B. et al. *Proc. SPIE,* v.2770, p.82, 1995.
36. Ginzburg V.L. *Propagation of electromagnetic wave in plasma*, New York: Gordon &Breach 1961.
37. Kruer W.L. *The Physics of Laser Plasma Interaction*, New York, 1988.
38. Estabrook K., Kruer W.L. *Phys. Rev. Letts*, v.40, p.42, 1978.
39. Andreev A.A., Semakhin A.N. *Proc. SPIE*, v.2097, p.326, 1993.
40. Brunel F. *Phys. Rev. Lett.*, v.59, p.52, 1987.
41. Andreev A.A., Gamaly E.G., Novikov V.N. et al. *ZhETF,* v.101, p.1808, 1992.
42. Gibbon P. and Foerster E. *Plasma Phys. Control. Fusion*, v.38, p.769, 1996.
43. Lichters R., Meyer-ter-Vehn J., Puchov A. *Phys. Plasmas*, v.3, №.9, p.3425 1996.
44. Gibbon P. *Phys. Rev. Lett.*, v.76 , p.50 1996.
45. Andreev A.A., Platonov K.Yu., Tanaka K. *Proc. JAERI-Conf 98-004 37,* 1998.
46. Andreev A.A., Platonov K.Yu., Gauthier J.C. *Phys. Rev.*, v.A58, p.2424, 1998.
47. Andreev A.A., Zapysov A.I., Charukhchev A.V. et al., *Izv. AN, ser. Phys.*, v.63, p.1239, 1999.
48. Komarov V.M., et al. *Tech. Program IX Conf. Laser Optics*, SPb, p.61, 1998.
49. Estabrook et al. *Phys. Rev. Lett.*, v.50, p.2082, 1983.
50. Stamper J. et al. *Phys. Rev. Lett*, v.26, p.1012, 1971.
51. Sudan R. *Phys. Rev. Lett.*, v.70, p.3075, 1993.
52. Ruhl H., Mulser P. *Phys. Lett.* A 205, p.388, 1995.
53. Bulanov V., Naumova N. M., Pegoraro F. *Phys. Plasmas*, v.1, p.745 1994.

54. Askay`yan G.A., Bulanow S.V., Pegoraro F. et al., *Fizika plazmy*, v.21, p.884, 1995.
55. Andreev A.A., Novikov V.N., Platonov K.Yu. *Proc. SPIE*, v.2770, p.153, 1995.
56. Ruhl H. *Phys. Plasmas*, v.3, p.3129, 1996.
57. Andreev A. A., Limpouch J. et al. *Proc.of SPIE*, v.2790, p.150, 1996.
58. Andreev A.A. et al. *Proc. SPIE*, v.2790 p.82, 1996.
59. Andreev A. A., Limpouch J. et al. *Proc.of SPIE*, v.2790, p.150, 1996.
60. Andreev A.A., Limpouch J. *J.Plasma Physics*, v.62, p.179, 1999.
61. Lee R.W. et al. *Quant. Spectrosc. Radiat. Transfer*, v.32 p.91, 1984.
62. Ammosov M.V., Delone N.B., Krainov V.P. *ZhETF*, v.91, p.2008, 1986.
63. Lee Y.T., More R.M. *Phys. Fluids*, v.27, p.1273, 1984
64. Andreev N.E., Auer G. et al. *Phys. Fluids*, v.24, p.1492, 1981.
65. Andreev A.A., Limpouch J., Semakhin A.N. *Izv. AN, Ser. Fiz.*, v.58, p.167, 1994.
66. Estabrook K., Valeo E.J., Kruer W. *Phys. Fluids*, v.18, p.1151, 1975.
67. Rae S.C., Burnett K. *Phys. Rev.*, A 44, p.3835, 1991.
68. Dragila R., Gamaly E.G. *Phys. Rev.*, A 44, p.6828, 1991.
69. Price D.P., More R.M.,Walling R.S. et al. *Phys. Rev. Lett.*, v.75, p.252, 1995.
70. Andreev N.E., Silin V.P., Stenchikov G.L. *ZhETF*, v.78, p.1396, 1980.
71. Davis J., Dark R., Guiliani J. *Laser and Particle Beams*, v.13, p.3, 1995.
72. Kieffer J.-C., Matte J.-P., Belair S. et al. *IEEE J.Quant.Electr.* QE-25, №12, p.2640, 1989.
73. Gibbon P., Bell A.R. *Phys.Rev.Lett.*, v.68, p.1535 1992.
74. Aleksandrov A.F., Rukhadze A.A., Bogdankevich L.S. *Course of Plasma electrodynamics*, Moscow, 1988.
75. Van'kov A., Kozlov A., Chizhov S. et al., *Proc. of SPIE*, v.2095, p.87, 1994.
76. Yang B., Kruer W.L., Langdon A.B. et al., *Phys. Plasmas* v.3, p.2709, 1996.
77. Andreev. A.A., Gauthier, J.C. Platonov, K., *Proc. ECLIM*, Madrid, PM43, 1996.
78. Andreev, N.E. et al. *Proc. SPIE*, v.2790, p.27, 1996.
79. Brian Yang, Kruer W.L., More R.M. et al. *Phys. Plasmas*, v.2, p.3146, 1995.
80. Beitman M., Erdei J. *Hypergeometric functions*, Moscow: Nauka 1963.
81. Meierovich B., Pitaevskii L.P. Liberman M.A. *ZhETF*, vol.62, p.1737, 1972.
82. Kondratenko A.N. *Penetration of an Electromagnetic Field into Plasma*, Moscow: Nauka 1979.
83. Bourdier A., *Phys Fluids*, v.26, p.1804, 1983.
84. Karpman V.I. *Nonlinear waves in disperse mediums*, M.: Science, 1973.
85. Langdon A., Lasinski B. in: *Methods of Comp. Physics*, v.16, p.327, New York: Academic Press, 1976.
86. Deletrez J., Bonnaud G., Audebert P. et al. *Bull. APS*, v.10, p.1987, 1993.
87. Gibbon P., Andreev A.A., Lefebvre E. et al. *Physics of Plasma*, v.6, p.947, 1999.
88. Gorbunov L.M. *Usp.Fiz. Nauk*, v.109, p.631, 1973.
89. Tikhonchuk V.T. *Kvant. Elektron.*, v.1, p.151, 1991.
90. Baton S.D., Rousseau C., Mounaix P. et al. *Phys. Rev.*, E 49, p.3602, 1994.
91. Hinkel D.E., Yvillia E.A., Berger R.L. *Phys. Plasmas*, v.2, p.3447, 1995.
92. Baldis H.A., Villeneuve D.M., La Fontaine B. et al. *Phys. Fluids*, v.5, p.3319, 1993.

93. Andreev N.E., Chegotov M., Tikhonchuk V.T. *Preprint 169*, M., FIAN, 1988.
94. Andreev N.E., Silin V.P., Tikhonchuk V.T. *Fiz. Plasmy*, v.14, p.851, 1988.
95. Zozulja A.A., Silin V.P., Tikhonchuk V.T. *ZhETF*, v.12, p.96, 1984.
96. Andreev A.A., Andreev N.E., Sutjagin A.N. et al., *Fiz. Plazmy*, v.15, p.944, 1989.
97. Andreev A.A., Kurnin I.V. *JOSA*, v.13, p.2, 1996.
98. Kalashnikov M.P. et al, *Phys. Rev. Lett.*, v.73, p.260, 1994.
99. Andreev A.A., Platonov K.Yu. *Plasma Phys. Rep.*, v.24, p.26 1998.
100. Zeldovich Ya.B., Raiser Yu.P., *Physics of shock waves and high-temperature hydrodynamic phenomena*, M.: Science, 1966.
101. Wilks S.C. *Phys.Fluids*, B 5, p.2603, 1993.
102. Miyakoshi T., Tanaka K.A., Kodama R. et al. *AHPLA'99, HPL02, 3886-68*, p.80, Osaka Univ., Japan, November 1999.
103. Andreev A.A., Bayanov V.I., Vankov A.B.et al., *AIP Conference Proc. 426 Int. Conf. "Superstrong Field in Plasma"*, p. 61, Varenna, Italy, August, 1997.
104. Plaja L., Roso L., *Phys. Rev.*, E. v.56, p.7142, 1997.
105. Andreev A. A, Platonov K.Yu., Tanaka K.A. *JAERI-Conf 98-004*, p.37, 1998.
106. Andreev A. A. *Dr. of Sci. Theses*, Vavilov State Optical Inst., L., 1990.
107. Erokhin N.E., Moiseev S.S., Mukhin V.V., *Nuclear Fusion*, v.14, p.333, 1974.
108. D. von der Linde, Schuler H., *Appl. Phys. Lett.*, v.66, p.807, 1995.
109. Cairns R.A. *J. Plasma Phys.*, v.22, p.149, 1979.
110. Altenbernd D., Teubner U., Gibbon P., et al. *J. Phys.*, B 30, p.3969, 1997.
111. Wülker C., Theobald W., Schäfer F.P. et al. *Phys. Rev.*, E 50, p.4920, 1994.
112. Bergsma J.P., Coladonato M.H., Edelsten P.M., et al. *J. Chem. Phys.*, v.84, p.6151, 1986.
113. Solem J.C., *JOSA*, B 3, p.1551, 1986.
114. Andreev A.A., Semakhin A.N., Akulinichev V.V. *Probl. Nauchn. Priborostr.*, v.3, p.35, 1993.
115. Limpouch J., Andreev A.A, Semakhin A.N. *Proc.SPIE*, v.1980, p. 75, 1993.
116. Dzhidzhoev M.S., Gordienko V., Kolchin V., et al. *JOSA*, B 13, p.143, 1996.
117. Mihalas D., *Stellar Atmospheres*, Freeman, San Francisco, 1978.
118. Apruzese J.P., Davis J., Duston D., et al. *Phys. Rev.*, A 29, p.246, 1984.
119. Peyrusse O., *Phys. Fluids*, B 4, p.2007, 1992.
120. Teubner U., Theobald W., Wülker C. *J.Phys.*, B 29, p.4333, 1996.
121. Wilhein T., Altenbernd D., Teubner U. et al. *JOSA*, B 15, p.1235, 1998.
122. Vainshtein L., Sobelman I., Yukov E.: *Atom excitation and broadening of spectral lines*, (in Russian), Nauka, Moskow, 1979.
123. Teubner U., Wagner U., Gibbon P. Andreev A.A. et al. *OSA Proc. on "Applications of High Field and Short Wavelength Sources VIII"*, Potsdam, 1999.
124. Broughton J.N., Fedosejevs R., *J. Appl. Phys.*, v.74, p.3712, 1993.
125. Andreev A.A. et al. *Appl. Phys.*, v. B 70, p.505, 2000.
126. Nakano H., Nishikawa T., Ahn H. et al., *Appl.Phys.*, B 63, p.107, 1996.
127. Nakano H., Lu P., Nishikawa T, Uesugi N. *Inst. Phys. Conf.*, №159, p.535; *X-ray Laser Conf.* Kyoto, Japan, September, 1998.
128. Andreev A.A. et al. *Quantum Electronics*, v.27, p.76, 1997.

129. Andreev A. A. et.al. *Quantum Electronics*, v. 26, p.884, 1996.
130. Andreev A.A. et al. *Proceedings SPIE*, v.3934, p.52, 2000.
131. Teubner U., Kühnle G., Schäfer F.P. *Appl. Phys. Lett.*, v.59, p.2672, 1991.
132. Landau L.D. and Lifshits E.M. *Theory of Field*, Moscow: Nauka, 1974
133. Key M.H., Cable M.D., et al. *Phys. Plasmas*, v.5, p.1966, 1998.
134. Sentoky Y., Mima K., et al., *Phys. Plasmas*, v.5, p.4366, 1998.
135. Kmetec J.D. *Phys. Rev. Lett.*, v.68, p.3431, 1992.
136. Andreev A.A. et al. *Proceeding of SPIE*, v.2270, p.157, 1995.
137. Wilks S.C., Kruer W.L. *J. Quant. Elect. IEEE*, November, p.238, 1997.
138. Lefebvre E., Bonnaud G., *Phys. Rev.*, E 55, p.1011, 1997.
139. Gurevich A.B, Mescherkin A.P. *ZhETF*, v.80, p.1810, 1981.
140. Wickens L.M., Alien J.E., Rumsby P.T. *Phys. Rev. Lett.*, v.41, p.243, 1978.
141. Kishimoto Y. et al. *Phys. Fluids*, v. 26, p.2308, 1983.
142. Komarov V.A. et al. in: *Book of Abstracts, LIRPP,* Monterrey, CA, 1997.
143. Zhidkov A. et al. *Phys. Rev.*, E 61, p.2224, 2000.
144. Pearlman J.S., Morse R.L. *Phys. Rev. Lett.*, v.40, p.1652, 1978.
145. Borodin V.G., Gilev O.N., Zapysov A.L. *Pis'ma w ZhETF*, v.71, p.354, 2000.
146. Cowan T.E. at al. *Tech. Digest Lasers'97*, New Orleans, December, 1997.
147. Andreev A.A., Mak A.A., Yashin V.E. *Kvant. Elektronika*, v.24, p.99, 1997.
148. Bell A.R., Davies J.R. *Phys. Rev.*, E 58, p.2471, 1998.
149. Andreev A.A., Litvinenko I.A., Platonov K.Yu. *JETP*, v.116, p.1, 1999.
150. Afanas'ev Yu.V., et al. *Tr. FIAN*, v.134, p.11, 1982.
151. Duderstadt J.J., Moses G.A. *Inertial Confinement Fusion*, J.Wiley, NY, 1982.
152. Andreev A.A., Mak A.A., Solovyev N.A. *An introduction to hot laser plasma physics*, (V.233 in Horizons in World Physics) Nova Sci. Publ., Inc., 163 p., 2000.
153. Kidder R., *Nucl. Fusion*, v. 16, p. 405, 1976.
154. Andreev A.A., Charuhchev A., Yashin V. *Usp. Fiz. Nauk,* v. 169, p. 72, 1999.
155. Antonsen T., Mora P. *Phys. Rev. Lett.*, v. 69, p. 2204, 1992.
156. Gamaly E.G. *Preprint*, Canberra: Australian National University, 1993.
157. Gamaly E.G. *Phys. Fluids*, B 5, p. 944, 1993.
158. Atzeny S., *Physics of Plasmas*, v. 6, p. 3316, 1999.
159. Andreev A., Il'in D., Levkovskii A. et al., *ZhETF*, v. 92, p. 69, 2001.
160. Cauglan G., Fowler W., Harris M., Zimmerman B., *Atomic Data and Nucl. Data Tabl.*, v. 32, p. 197, 1985.
161. Piriz A., Sanchez M. *Phys. Plasma*, v. 5, p. 4373, 1998.
162. Baldwin G.C., Solem J.C. *Reviews of Modern Physics*, v. 69, p. 1085, 1997.
163. Andreev A., Platonov K., Rozhdestvenskii Yu., *JETP Lett.*, v. 68, p. 704, 1998.
164. Andreev A.A., Platonov K.Yu. *Laser and Particle Beams*, v. 18, 2000.
165. Letohov V.S. *Kvantovaya Elektronika*, p. 125, 1974.
166. Bor O., Mottelson B. *Atomic nuclear structure*, v. 1, 1976.
167. *Nuclear Data Sheets*, 1989 - 1998.
168. Landau L.D., Lifshits E.M. *Quantum mechanics*, Moscow: Nauka, 1974.
169. Ienbner V. et al. *Proc. SPIE*, v. 2700, p. 170, 1995.
170. Ahiezer A. I., Berestetskiy V. B. *Quantum Electrodynamics*, Nauka: M., 1969.

171. Bychenkov V.Yu, Tikchonchuk V.T., Tolokonnikov S.V. *JETP*, № 6, 1999.
172. Gryaznykh D.A. et al. *Pis`ma w ZhETF*, v. 167, p. 342, 1998.
173. Pretzler G. et al. *Technical Digest of conference Lasers`97*, p. 22, 1997.
174. Berestetskii V.B., Lifshits E.M., Pitaevskii L.P., *Quantum Electrodynamics,* Moscow: Nauka, 1980.
175. Heisenberg W., Euler H. *Zs. fur Phys.*, v. 98, p. 714, 1936.
176. Rosanov N. N. *ZhETF*, v. 113, p. 513, 1998.
177. Aleksandrov E. B., Anselm A. A., Moskalyov A. *ZhETF*, v. 89, p. 1181, 1985.
178. Andreev A. A, Pis'mak Yu. M. *Preprint of St. Petersburg State University*, SPbU-IP-96-43, 1996.
179. Barbashov B.M. *ZhETF*, v. 48, p. 607, 1965.
180. Itzykson C., Zuber J.-B. *Quantum Field Theory*, New York: McGraw-Hill Book Company, 1980.

INDEX

F

G

H

I

J

K

L